D0889946

From Lab to Market

Commercialization of Public Sector Technology

From Lab to Market

Commercialization of Public Sector Technology

Edited by

Suleiman K. Kassicieh and
H. Raymond Radosevich

Anderson School of Management
University of New Mexico
Albuquerque, New Mexico

Plenum Press • New York and London

Library of Congress Cataloging-in-Publication Data

Technology Commercialization Conference (1993: Sante Fe, N.M.)
From lab to market : commercialization of public sector technology / edited by Suleiman K. Kassicieh and H. Raymond Radosevich.
p. cm.
"Proceedings of the Technology Commercialization Conference, held March 1993, in Sante Fe, New Mexico."
Includes bibliographical references and index.
ISBN 0-306-44717-7
1. Technology transfer--Economic aspects--United States--Congresses. 2. Technology and state--United States--Congresses. 3. Government business enterprises--United States--Congresses. I. Kassicieh, Suleiman K. II. Radosevich,H. Raymond. III. Title. IV. Title: Commercialization of public sector technology.
HC110.T4T396 1993
338.973'06--dc20 94-14316
CIP

Proceedings of the Technology Commercialization Conference, held March 1993, in Santa Fe, New Mexico

ISBN 0-306-44717-7

A Division of Plenum Publishing Corporation
233 Spring Street, New York, N.Y. 10013

Printed in the United States of America

PREFACE

The topic of this book, the commercialization of public-sector technology, continues to grow in importance in the United States and similar societies. The issues involved are relevant to many roles including those of policy makers, managers, patent attorneys, licensing agents, and technical staff members of public technology sources. Institutions increasingly involved in the process include federal and other governmental laboratories and their related agencies, public universities and their state governments, public and private transfer agents and, of course, all the private recipients of public technology.

Scarcely a day goes by without a significant event related to technology transfer and commercialization. The popular business press is regularly carrying articles addressing the issues, explaining new initiatives and describing events of notable success or failure.[1] As an example of current important events, the Technology Reinvestment Project (TRP) is formulating its initiatives to transfer public technology and promote technology-based public/private partnerships as a collaboration between the Advanced Research Projects Agency (ARPA), the National Institute of Standards and Technology (NIST), the National Science Foundation (NSF) the National Aeronautics and Space Administration (NASA), and the Department of Energy, Defense Programs (DOE/DP).

Many analysts of Japan's economic growth have attributed a significant portion of it to the cooperation between public institutions and private sectors, especially micro-electronics, telecommunications, robotics and advanced manufacturing. Thus, most popular prescriptions for U.S. economic growth based on the Japanese phenomenon recommend that public sources of technology learn to cooperate in much more effective fashion with the key industries that can drive the U.S. economy. In some, government agencies are themselves major markets, as their captive laboratories try to develop advanced technologies to solve their problems. For example, estimates of the expenditures to be made by the Department of Energy and the Department of Defense for environmental remediation of their facilities have reached combined levels over one-half trillion dollars projected for the next several decades. DOD, and especially DOE, laboratories are currently spending huge sums to develop advanced technologies in waste management and remediation. Being both technology sources and potentially lucrative markets, these agencies have unprecedented opportunities to work with the industrial firms which will eventually become the contractors for the remediation efforts. Technical skills acquired in serving these agencies should make these American firms the world's leaders in international waste management markets, if effective technology transfer and commercialization can be accomplished.

Not all technology transfer events have been harmonious and productive, nor should they be expected to be so. For example, accusations of apparent conflicts of interest, unfair access to licensing opportunities, and ineptitude accompany efforts to improve the process. At least one government official has been fired because of accusations of overzealous support of entrepreneurial spin-off activities from one government laboratory.[2] However, the success of experimental efforts to improve and increase the incidence of technology commercialization is so important to the economic health of our nation that we must not be deterred by occasional failures, but rather learn from the experiences.

To the extent that the process of commercializing public-sector technology is increased and improved significantly, a number of pressing problems are addressed. The international competitiveness of U.S. industry can be enhanced with public-sector technology according to the intent of the National Competitiveness Technology Transfer Act of 1989. With forecasts of shrinking budgets for federal laboratories (especially those that are defense related), incremental funding from intellectual property licenses and other commercialization mechanisms will be eagerly sought by these technology sources. Rewards for cooperation will also be available in the form of new markets and a better bottom line for the private participants, as the processes of public-sector technology commercialization become more effective.

Most of the papers in this book were presented at a conference held in Santa Fe, New Mexico, during March, 1993. Papers were invited by the editors of this volume from leading practitioners, academics and policy makers. The exchange of views during the conference was then reflected in modifications to the original papers. Thus, these papers reflect the state-of-the-art in theory and practice as perceived by a diverse group of experts.

Although the primary focus is federal technology transfer and commercialization, the principles espoused are just as relevant to universities and other public technology sources. Indeed, several papers present examples from these other sources.

Each section of the book is designed to:

a. raise important issues and factors affecting the success or failure of moving technologies from the research stage to commercialization across disparate organizational boundaries,

b. explain current systems and roles involved in technology commercialization and indicate the changes necessary in the systems to enhance the success of endeavors from the inception of the research idea to its implementation in a production setting

c. suggest models by which problems can be overcome to achieve better ways of solving the complex set of issues that have been identified and

d. provide examples of successful and failed attempts at commercialization with the underlying factors ultimately providing prescriptions for successful implementation of technology transfer and commercialization in a variety of institutional settings.

The book is designed to be used by the professional who is interested in knowing more about the technology commercialization process and how to increase success rates. It also may be used as a readings book for a senior-level or graduate-level course in technology commercialization or technology transfer. In addition, it may serve the researcher who is looking for a current compendium to establish the state-of-the-art of technology transfer and commercialization roles, mechanisms, processes and efforts.

The introductory section, "Public-Sector Technology Commercialization: Problems, Practices and Prescriptions," includes articles explaining the major issues facing the United States in moving technology from the research environment of public-sector laboratories through the commercialization process in the private sector. This section deals with current factors that make decision-making in technology transfer an issue of public policy with substantial socioeconomic consequences for the country. This section also includes a description of the programs and mechanisms used at different laboratories and their effects on technology commercialization.

The second section of this book, "Participant Roles in Public-Sector Technology Commercialization," examines the roles of various types of institutions and individuals in the process of technology commercialization. Each article describes the roles of key players and the current status and practices of the role performers, and also offers suggestions for changing those roles. The roles described include researchers, intermediaries and agents, entrepreneurs, consortia, and other strategic alliances between technology sources and users.

The third section, "Public Sector Technology Commercialization: Mechanisms and Processes," examines the techniques, systems and mechanisms that can be used to improve the effectiveness of the technology transfer and commercialization process. Improving the process involves experimentation with a variety of techniques and mechanisms with some understanding of the context in which each is likely to be successful. Advances are needed in systems that increase the efficacy of both technology-push and technology-pull mechanisms. The articles in this section describe techniques for matching technical solutions with market needs as well as for enhancing the value of technology in different organizational settings.

The last section, "Prescriptive Paradigms in Public-Sector Technology Commercialization," provides some case examples and conclusions as to the kinds of change that will be necessary to reach the goals of multiple parties involved in the process. The suggestions pertain to laboratory spin-offs, consortia, extant businesses and high-growth ventures in existing organizations.

As always, an effort such as the development of this book depends upon the cooperation and support of many institutions and individuals. The authors wish to express their appreciation to the authors of the papers and the participants who were so actively involved in the panels and discussions during the conference. We also would like to thank the sponsors of the conference for their generous support. The sponsors included:

- The Anderson Schools of Management, University of New Mexico
- Sandia National Laboratories
- Los Alamos National Laboratory
- Phillips Laboratory
- The RGK Foundation and the Institute for Constructive Capitalism, The University of Texas–Austin
- The Economic Development Department, State of New Mexico, and
- The Offices of U.S. Senators Jeff Bingaman and Pete Domenici

In addition to these institutions, several individuals deserve special thanks for their involvement in the conference and the preparation of this book. Especially notable was the work of Ms. Peggy Merrell from the Anderson Schools who performed invaluable editing. Ms. Roberta Murray performed exemplary word processing efforts on numerous drafts. The primary administrator of the conference was Ms. Doris Rhodes from the Anderson Schools. Without her help, the conference would have been much less successful. Members of the conference committee who represented the federal laboratories were extremely helpful in the program design and in the recruitment of participants. These individuals were Mr. Mark Allen (Sandia National Laboratories), Ms. Sue Fenimore (Los Alamos National Laboratory), and Mr. Pat Rodriquez (Phillips Laboratory).

REFERENCES

1. As we write this preface, the current issue of *Business Week* contains an article which thoughtfully summarizes several of the issues in and barriers the process of public-sector technology commercialization

(Cares, J., Hof, R. D., & Atchison, S. D., [1993, June 7]. Firefight over the weapons labs. *Business Week* 104–106).

2. Barber, J., (1990, June). DOE official's role in license examined. *McGraw-Hill's Tech Transfer Report* 4.

CONTENTS

IV. Prescriptive Paradigms in Public-Sector Technology Commercialization

I

THE IMPORTANCE OF PUBLIC-SECTOR TECHNOLOGY COMMERCIALIZATION

INTRODUCTION

Although the majority of the approximately $70 billion annual research and development budget of the United States federal government is performed by private contractors, the government spends more than $16 billion annually on R&D within its own laboratories. This federal spending is equal to the combined R&D total of all U.S. industry. Almost 200,000 scientists and engineers are employees of the federal R&D system. With this magnitude of capabilities devoted to improving science and technology, the public sector represents a sizable resource for partnerships with industry. Increasingly, federal policy is charging these partnerships with the mission of increasing technological innovation in the United States and thus improving the global competitiveness of our private sector.

The decline of U.S. industrial competitiveness in the global marketplace has been well documented [1]. Even popular literature now recognizes the seriousness of the "economic war" being waged on global fronts. Books like *Rising Sun* by Michael Crichton are causing people in general to become aware of and concerned with the ineffectiveness of past U.S. policies and practices in the support of industry engaged in the international marketplace. The success of the Japanese in this area has become obvious in everyday American life.

In the short term, it is unlikely that some suspected causes of American noncompetitiveness can be dramatically cured. Federal deficit reduction, defense conversion, improvements in education, increased personal savings, etc., will probably take many years, if ever accomplished on a competitive scale. U.S. labor is still paid much higher rates than labor in many competitor nations. Can we improve the productivity and quality of our human, physical and capital resources so that our products and services become more valuable in the marketplace? Some policy makers believe that the storehouse of federal technology, as well as its potential for future developments, can play a significant role in improving our productivity in the short and medium term if we can effectively transfer this technology and use it in the private sector.

Legislative/policy intent has been in place for more than a decade although the effectiveness of implementation varies considerably from one agency or laboratory to another. Changes in public law and policy and the resulting variety of practices in federal agencies and their laboratories are well documented in the article by Robert Carr in this first section (*Doing technology transfer in federal laboratories*, p. 61). The most recent legislative efforts demonstrate a shift in policy which recognizes the importance of relationships between the public and private sectors. The perceived success of alliances between these sectors in Japan appears to have influenced U.S. policy makers to adopt the Japanese system as a model from which the U.S. should learn some important lessons.

Recently, many federal laboratories have expanded their technology transfer operations in response to federal initiatives which include incremental funding, facilitative policies and legislation as well as additional personnel. For example, Sandia National Laboratories, DOE's largest laboratory, has increased its small technology transfer office staff of three years ago to over thirty professionals today.

THE ISSUES

The article by Radosevich and Kassicieh (*The King Solomon role in public-sector technology commercialization*, p. 9) in this first section describes the primary issues requiring resolution before publicly-sponsored technology can be easily transferred to and commercialized by the private sector. Many of these issues are currently being considered as government agencies experiment with new forms of partnerships such as research consortia. Newer mechanisms, such as cooperative research and development agreements (CRADAs), are being formed at a truly impressive pace, given the unproven nature of potential benefits. Clearly, industrial firms are intrigued with the potential of using existing public technology and forming partnerships to leverage their internal resources.

While we should be encouraged by the number of partnerships being formed, we should also note that many of these endeavors are breaking new ground. We must guard against the possibility of discouragement and recrimination as the inevitable (and, let us hope, isolated) failure occurs. In some laboratories, the scale of alliance formation is awesome, given the unproven process of technology transfer and commercialization through alliances. Alliances are being formed between laboratories and individual firms, both large and small. Individual laboratories, as well as consortia of laboratories, are working with consortia of firms, many of these arrangements being newly formed.

One prime concern of policy makers is the emphasis on basic research as the basis for the preponderance of laboratory/industry partnerships. From the perspective of the scientists at the laboratories, this is ideal. Fears that these partnerships might drive the laboratories toward "mundane" applied efforts seem unjustified based on this early experience. Laboratory management can find comfort in the possibility that the private sector will supplant waning federal budgetary support. Industry's willingness to support basic research partnerships must be a welcome sign to the laboratories. However, for those policy-level administrators who are seeking short- and medium-term benefits to the economy and international competitiveness, basic research is too remote in its impact.

As these early efforts to increase government-industry cooperation are being implemented, many key issues are being identified and addressed. For example, Radosevich and Kassicieh describe the heterogeneity of practices and policies across government agencies and even program offices within the same agency. As firms discover the most effective processes for working with one office or federal laboratory, this learning is less transferable than is desirable when dealing with another government agency or contractor. However, the newer programs such as AMTEX (described by Peterson in *Current practices, coming changes*, p. 97), involve an entire industry (in this case, the textile industry) and associated research institutions in an integrated fashion. The AMTEX program also involves multiple DOE program offices. Radosevich and Kassicieh, in addition, address the issue of increasing the relevance of federal R&D results by involving industry in programming government research. AMTEX manages industry participation in the R&D planning process through the cooperation of industrial members, federal laboratories and DOE program offices. Thus, many of the issues raised by Radosevich and Kassicieh are not unrecognized; they are, however, not yet resolved with any certainty. Some of the novel programs that are being designed as experiments in federal technology transfer must fail, as is true of all experimentation.

Startup firms based on technology developed by the federal government are not the typical method of transfer from the national laboratory system. An extensive study by the Federal Laboratory Consortium found only three cases, out of 55 studied, using the mechanism of new-venture startup [2]. Although this study is almost a decade old and may not be representative of actual current entrepreneurial commercialization efforts by federal laboratory employees, the actual incidence is clearly insufficient to affect the economy to the degree desired. Since several studies have demonstrated the greater effectiveness at job and wealth creation of startup and small businesses, compared to large firms in the United States, it is clearly important for government agencies and laboratories to understand and use these mechanisms [3]. Carr's article suggests that more use should be made of the entrepreneurial method of public technology commercialization, especially by those laboratories concerned with making a positive local impact.

All the federal laboratories are concerned about their responsibilities to the local communities, especially those which have resulted in significant environmental infractions over the years. Kay Adams of the Industrial Partnership Center (IPC), Los Alamos National Laboratory (LANL), in her article (*Accelerating technology development for economic competitiveness*, p. 89), lists job creation through technological entrepreneurship as one of three challenges being addressed by the IPC. Spinoff activities are likely to have a considerable local impact, at least in the short run. In the case of Los Alamos, isolation from markets, suppliers and investors has caused many local laboratory spinoff companies to leave the area after being successfully launched. LANL assistance to the local incubator has somewhat ameliorated the situation, but the community and the laboratory face a continuing challenge in creating a dynamic business environment in a remote small community. This situation is not unlike that of a number of federal laboratories for which isolation was a criterion of their original establishment.

Adams also recognizes the importance of laboratory-industry alliances to meet the other two challenges of the IPC. Industry must become more involved in the technology aspects of solutions to major national problems and related opportunities such as environmental clean-up and the use of alternatively fueled vehicles. Another challenge which can be addressed by alliances is the design and implementation of market-driven, well-defined programs that are cost-shared by laboratories and private firms and which may result in products, services and processes with export potential. Adams' article provides recommendations for actions which can facilitate the formation of these alliances. The article by James Botkin in section III of this book (*The commercialization of public sector technology: How to form, manage and evaluate effective strategic alliances*, p. 225), discusses the processes most likely to result in a productive collaborative relationship between a laboratory and a private firm.

Of all the issues most worrisome to the managements of government-owned laboratories, those receiving the most attention are the possibilities for conflicts of interest and lack of fairness in access to federal technology. The article by Albert Sopp in this section (*The impact of federal technology transfer on the commercialization process conflict of interest*, p. 115), examines the issue of conflict of interest and explains why its careful resolution is so critical to eventual success in transferring public technologies. As Sopp explains, conflict of interest can apply to a wide variety of activities and effects, from dilution of researchers' attention to the laboratories' primary missions to competition between public and private concerns for intellectual property. Even the appearance of conflicts of interest can be damning in the highly-politicized and emotional realm of wealth creation and assignment.

CURRENT PRACTICES

One of the primary purposes of this book is to establish a benchmark of current practices in this rapidly evolving field. Because so many experiments are under way on new types of policies, programs and mechanisms, it is important occasionally to document progress to date. Carr's first article, *Doing technology transfer in federal laboratories*, presents the results of a comprehensive survey of practices. The article by Beverly Berger, Washington representative to the Federal Laboratory Consortium, (*Technology transfer in a time of transition*, p. 29), both summarizes current practices and suggests issues that must be addressed to improve technology transfer from the current federal laboratory system.

Carr's article portrays current practices in laboratory technology transfer as being focused on the large established firm as the most likely commercialization institution. Radosevich and Kassicieh argue that the small, entrepreneurial firm may be the most effective mechanism to meet goals of job and wealth generation. In section II, the role of entrepreneurs (or intrapreneurs) is discussed in several articles.

Changing current practices in many federal laboratories seems unnecessarily slow and awkward unless one understands the complexity of the overall organizational structure. In its efforts to implement the 1989 National Competitiveness Technology Transfer Act, which was aimed primarily at DOE, that agency needed almost a year to establish positions on several dozen critical issues and to translate these positions into policies and orders. A gargantuan task force was necessary to incorporate representation from various program offices field offices and contractors as well as headquarters functions. To effect the new policies, the maintenance and operating contracts had to be renegotiated with those contractors who manage DOE's laboratories. (Most of DOE's laboratories are government-owned, contractor-operated GOCOs.) It took nearly a year to complete the negotiations with all of the contractors.

To get an accurate sense of the dynamics of federal technology transfer and commercialization, one need only survey an issue or two of the newsletters which report the important related events. Good examples of these are the newsletter of the Technology Transfer Society, *T Squared* and *McGraw-Hill's Federal Technology Report* which incorporates the *Tech Transfer Report.* Berger's article describes recent activities and important events defining the dynamics of the federal technology transfer processes. These sources also demonstrate the plethora of current practices across different agencies and even among different contractors within the same agency.

THE NEED FOR NEW MODELS AND MECHANISMS

At present, technology in the public sector laboratories is perceived as a factor imbued with intrinsic value for the innovation process. To facilitate the "push" mechanisms most commonly used in the past, applications had to be envisioned primarily by the technology source. The intrinsic value of technology was enhanced by the creative process of envisioning some socially useful purpose, hopefully manifested in market-driven products and services. More recent methods of technology transfer, such as CRADAs, have been designed to form partnerships and promote the use of "pull" mechanisms.

Before either "push" or "pull" mechanisms can be used effectively, potential partners must become aware of each other's resources and needs. While most laboratories have been expanding their outreach activities recently, even higher rates of "handshaking" will be needed to achieve the desired volume of cooperation. Knowledge of the existence of intellectual property or unique facilities and equipment available for shared use by private partners is essential to the development of alliances, licenses or technical assistance relationships. Organizations such as the Federal Laboratory Consortium (FLC) and, more recently, the National Technology Transfer Center (NTTC) have been formally chartered to serve as clearinghouses. The article by Rivers (*The role of the researcher*, p. 103), describes the purpose, role and activities or the NTTC. There is also movement within a number of professional, scientific and trade associations and societies to become involved in the clearinghouse functions.

There seems to be an inherent assumption, unspoken and perhaps unrealized by many technology sources, that a good partnership will dramatically increase the value of the "raw" federal technology by contributing matching money, people, physical assets and technology. And yet, industry by itself has always had difficulty moving its own internally-developed technology to the marketplace. Almost every large firm has experimented with new mechanisms and processes to improve the incidence and rate of success of internal technological innovation. In the last decade, the codeword for this experimentation has been "intrapreneurship." In the previous two decades, it was "venture management" or "innovation management."

Radosevich and Kassicieh argue that the technology sources must assume more responsibility in creating greater commercial value potential in public-sector technology. As more definitive industrial policy is developed by the Clinton administration, critical industries and areas of technology may be defined. Technology sources may be pressured to move toward more applied and developmental work in these critical areas. This may raise new issues about the relative roles of laboratories and universities in the advancement of science. Creating more value within the technology source will undoubtedly require additional skills and knowledge in the scientific and technical staff. Improving the rate and quality of technical disclosures may compel laboratories and universities to provide new incentives as well as training in intellectual property issues. Indeed, the laboratories and government agencies have, in some instances, been negligent about pursuing intellectual property rights, especially foreign rights. Considerably greater resources must be devoted to this process in many agencies and laboratories if commercial value is to be increased within the province of the technology source.

This internal experimentation for the improvement of technological innovation in industry continues, with most firms developing idiosyncratic mechanisms, organizational processes and management concepts. Yet many public-sector technology sources feel their technology will be dramatically increased in value and effectively commercialized if only they can form a partnership to develop the technology further. To demonstrate the fallacy of this precept, one must only recognize that the vast majority of CRADAs being signed between national laboratories and industrial firms are for the mutual performance of basic research.

Industrial firms, within the boundaries of their own organizations have enough difficulty finding concepts of commercial applications for internal applied research. The process is well recognized as complex, risky and ill-structured. Each successful event seems to be sufficiently unique that generally-applicable mechanisms are rare. And yet we expect, in spite of the added complexity, that a strategic alliance between a public laboratory and one or more industrial firms will result in effective technology commercialization.

We are clearly expecting too much from these alliances in the short term. We should recognize the fact that U.S. industrial leaders, because of the more stringent U.S. anti-trust environment, were prohibited from forming the alliances that Europeans and Japanese industrialists have been making for decades. Our leaders, therefore, may not have the same skills and networks to apply to these new alliances that the industrial leaders of our competitor nations have. We need to experiment and learn to avoid being discouraged if not all alliances prove to be highly satisfactory to all parties. Section III discusses the ways in which alliances are being developed and suggests some additional considerations for the improvement of future partnerships.

Clearly, there will be no one best way to commercialize public-sector technology. It should also be apparent to us all that a CRADA, or any other new mechanism, will not by itself increase the value of raw technology. We must be as creative in developing new experimental mechanisms as we are at developing new technology. Unfortunately, while we continue to spend billions of dollars annually on creating new science and technology, proportionately very little is being spent to devise, test and evaluate new mechanisms and processes.

In addition to the development and testing of new models and mechanisms, common practices at most technology sources can be improved by sharing lessons learned from previous experiences—within the source organizations and at other similar ones. Roger Lewis, (*The impact of federal technology transfer on the commercialization process*, p. 109) provides a sampling of the lessons learned from the DOE experience of implementing the 1986 Federal Technology Transfer Act (FTTA) and the 1989 National Competitiveness Technology Transfer Act (NCTTA). Although many of these lessons are specific to the mechanism of CRADAs, many new policy issues have evolved while DOE policies, regulations and orders are being implemented.

In his article, Carr not only describes the current state of practice but also cites the "best practices" that he found in his survey of federal laboratory technology transfer. His selection of best practices was subjective, in that few formal evaluations have been done except on such macro-metrics as number of CRADAs approved or dollar volume of CRADAs being executed. In another article in the last section of this book, Carr argues for the assignment of resources to more formal and comprehensive measurement and evaluation systems to facilitate the identification and understanding of best practices. The article by Berger in this first section describes the issue of measuring progress in the federal technology transfer initiatives.

Some of the implementation of innovative models and mechanisms results from the formation of new organizations. Some of these new organizations are intended to become permanent entities with significant resources such as the National Technology Transfer Center described by its executive director, Lee Rivers (*The role of the researcher*, p. 103). Both the NTTC and the FLC have responsibilities that span the boundaries of all federal agencies involved in research and development. Other organizations are new program entities, such as FORUM or AMTEX, described by Norman Peterson (*Current practices, coming changes*, p. 97). Whether these new organizations will serve as effective models depends on the extent of their successes, the existence of an evaluation system and the development of communication processes that make them more visible.

Later sections of this book discuss various concepts of value in technology and suggest new forms of experimentation to be undertaken. The second section also examines the respective roles, and potential modifications to these roles, that various participants in technology commercialization can develop.

NOTES

1. See for example, Drucker, P. F. (1989). *The New Realities.* New York: Harper & Row; Chakrabarti, A. K. (1991). Competition in high technology: Analysis of patents of U.S., Japan, U. K., France, West Germany, and Canada. *IEEE Transactions on Engineering Management, 38* (1), 78–84; and Dertouzos, M. L., Lester, R. K. & Solow, R. M. (1989). *Made in America: Regaining the productive edge.* The report of the MIT Commission on Industrial Productivity. Cambridge, Mass. : MIT Press.
2. Federal Laboratory-Industry Interaction Working Group of the Federal Laboratory Consortium (1985). *Interagency study of federal laboratory technology transfer organization and operation* DOE/METC-85/6019, 21.
3. Rothwell, R. & Zegveld, W. (1982). *Innovation and the small and medium-sized firms.* London: Frances Pinter. See also: Flender, J. O. and Morse R. S. (1977). The role of new technical enterprises in the U.S. economy. In the Report of the Select Committee on Small Business, U.S. Senate *Small business and the quality of American life.* Washington: U.S. Government Printing Office, 72–90, and Birch, D. L. (1987). *Job creation in America: How our small companies put the most people to work.* New York: Free Press.

THE KING SOLOMON ROLE IN PUBLIC-SECTOR TECHNOLOGY COMMERCIALIZATION*

Raymond Radosevich and Suleiman Kassicieh

R. O. Anderson Schools of Management
University of New Mexico
Albuquerque, New Mexico 87131

INTRODUCTION

Since 1980, U.S. legislation has been regularly enacted to facilitate the transfer and commercialization of federally-sponsored technology. For example, the 1989 National Competitiveness Technology Transfer Act created an official technology transfer mission for the federal laboratories previously excluded—the Department of Energy (DOE) Defense Program laboratories. Since then, DOE and the contractors operating its laboratories have been modifying the management and operating contracts, defining policy, and beginning implementation of the technology commercialization activities. The objective of all this activity is to improve significantly the global competitiveness of U.S. industry.

The impact of public-sector technology on U.S. competitiveness obviously depends on the development of alliances between federal laboratories and contractors and U.S. industry. To achieve a high probability of fruitful industry-laboratory interaction, it is imperative that industry managers understand the challenges implied by the current system and the intended changes. On the other hand, laboratory managers must recognize these strategic challenges in order to guide the process of change and learn from their early interactions with industry.

The federal government is defining a new role for itself in promoting the commercialization of public-sector technology. It has long been the primary sponsor of research and development in this country, but it now sees itself as needing to transfer, and assist in the commercialization of, technology which it has sponsored or performed. The annual real funding of R & D by the private sector has decreased during recent economic downturns.[1] The innovations needed to rejuvenate traditional industries and launch new ones are in jeopardy, unless the basic research sponsored by the federal government is more closely

---

* An abridged version of this paper has been published in the *California Management Review* as "Strategic Challenges to Competitiveness Through Public-Sector Technology," copyright 1993 by the Regents of the University of California. Portions reprinted from the *California Management Review* Vol. 35, No. 4. By permission of the Regents.

coupled to private R&D efforts. The degree of success achieved by the government and its intended partner, industry, in this collaborative effort could be a significant factor in American competitiveness. But are we going about it properly?

THE SITUATION

The degree, cost and effectiveness of government interference in the free-enterprise process is constantly debated. Many policy makers believe that substantial potential commercial value resides in government-owned or government-sponsored R&D. If this assumption is valid, the process of assigning proprietary rights to various private, public and quasi-public organizations may affect the U.S. economic system in significant and currently unanticipated ways.

In effect, a small number of decision makers represented by government employees and contractors are actively seeking and entering into cooperative research and development agreements, licensing proprietary technology, and using various other mechanisms to transfer technology. What issues are critical to their ability to make wise decisions? What are appropriate policies, orders and regulations to guide their actions? Should public funds be spent to advance proprietary positions toward specific applications that are of interest to individual private firms? What specific criteria should they use to choose between specific contenders for proprietary technologies? Indeed, if federal technology transfer is truly to improve the nation's economic competitiveness, is there sufficient thought and planning behind this experiment to direct and evaluate the efforts?

For some time, most government agencies were reluctant to enter arrangements to acquire rights to their technology on an exclusive basis, because of the potential impact on relative competitive positions of private companies. Such an impact would be viewed by many as government interference. Should the government award an exclusive position in a federally-owned technology to the industry leader in hopes that this leader might effect a more substantial position in the global marketplace? Or should the technology be provided to a second-position firm (or even further down the pecking order) in order to stimulate domestic competition?

In the last few years, many agencies have realized that creating significant value through strong intellectual property rights, and assigning those rights on some basis of exclusivity, is necessary to provide industry the incentives to use the technology and incur the significant costs of commercialization. The change has been difficult for some agencies where technology-transfer policies have historically been based on free and wide dissemination of all scientific and technical knowledge. NASA, for example, created a huge public-information search and retrieval system based on a desire to disperse technology as widely as possible. Changing the culture in these agencies has been difficult. Some progress has been made because of the recognition that U.S.-generated technology frequently goes abroad whenever it is widely and inexpensively disseminated domestically.

THE STRATEGIC CHALLENGES

There are a number of critical issues and associated strategic challenges associated with establishing public/private partnerships to commercialize federal technology. Some of these issues relate to the research and development process itself. Others involve the identification and promotion of specific commercial opportunities resulting from federal research programs. Some challenges are derived from the methods used to motivate and reward the government researchers who must be involved in the technology transfer if it is to be successful. Other challenges result from the selection of recipients or agents for the technology transfer, to the exclusion of others. And many of these challenges must be resolved in the context of poorly-defined models of technology-driven, international economic competition.[2]

Presented below are thirteen strategic challenges which must be met by the government, its industrial partners, and any intermediaries who intend to participate in and improve the federal technology commercialization process. Associated with each challenge are responses to be considered by the management of the involved organizations. Table 1 summarizes these challenges and proposed responses. Although there are other challenges also requiring resolution, those listed below are intended to provide a sense of the difficulty and urgency of action required to improve the process.

Strategic Challenge Number One: How Do We Foster Industry Involvement in Federal Research Programming?

A number of observers of the federal technology transfer process have suggested that industry should be involved before a technology is actually developed. "Technology push" transfer mechanisms are often considered less effective than "technology pull" mechanisms. Thus, it seems only logical that the government (except in specific missions such as defense) should listen to industry at the earliest stages of R&D programming to encourage industrial participation downstream. Paul Magnusson, in a commentary in *Business Week*, suggested, "What's needed is an approach more like Japan's. Tokyo invites industry to help design research for maximum market impact. Companies share in the discoveries, then compete in product development."[3]

In some instances, the U.S. government has done just that. DOE's Clean Coal Program has had significant industry input during many stages of the research. In most other research programming instances, however, the programming is done to further the government's missions or policy makers' perceptions of national needs. "Dual-use" applications or spinoffs are sought in cooperation with industry after the fact. What mechanisms could the government use at the earliest planning stages to ensure adequate and appropriate representation of industry?

The difficulties of government R&D programming intended to provide direct market impact were vividly demonstrated during the late 1970s and early 1980s when alternative forms of energy were being sought to solve the national energy "crisis." The government effectively stimulated the demand for alternative energy systems with significant federal and state tax incentives. At the same time, DOE was reprogramming significant portions of its nuclear weapons laboratories and increasing the budgets of its energy laboratories in order to develop and transfer technology in solar, wind, geothermal, clean coal, coal gasification, ocean waves and other energy fields. Early systems were adopted, developed and promoted by the private sector in an entrepreneurial groundswell that few nations other than the U.S. could generate. Many of the systems were unproven and uneconomical to the end user, even with tax incentives. Gradually, the tedious process of technology transfer and diffusion began to make progress. Some systems were developed and introduced which incorporated substantial technological advances and could be justified economically.

Unfortunately, at about this time the availability and lower prices of traditional energy sources caused Congress, state legislatures, and the administration to declare that the energy crisis had passed; rescinding tax credits killed the demand for alternative energy systems just as the technology was becoming relevant. Most of the new private-sector ventures failed, and the nuclear weapons laboratories returned to their primary missions. Once again, the private sector learned that responding to government incentives is very risky. Those who were drawn into innovation by the excitement of federally-sponsored technology and tax incentives

Table 1. Challenges and Proposed Responses for Commercialization of Public-Sector Technology

Challenges	Responses
1. Private involvement in public research programming	a) Industry identification of critical technologies. b) Industry participation in agency and laboratory research programming. c) More cooperative research and development and funds-in agreements.
2. Control of technology recipient nationality	a) Rights awarded on basis of U.S. economic impact.
3. Support for small entrepreneurial firms	a) Small firm preference in technology awards. b) Broker small firm/large firm relationship. c) Simplify procedures for small firm interactions.
4. Creation of commercial value in public-sector technology	a) Train researchers in intellectual property issues. b) Focus on technology applications and demonstration projects. c) Establish a strong intellectual property position in each public technology. d) Award some form of exclusivity.
5. Provision for technology maturation	a) Match public/private investment. b) Use intermediaries for technology advancement.
6. Creation of consistent T^2 policies across agencies and laboratories	a) Devise uniform policies through supra-agency. b) Use services of FCL and NTTC.
7. Avoidance of conflict of interest in technology commercialization	a) Use objective agents and intermediaries. b) Provide better policy guidance and training. c) Improve opportunity outreach.
8. Tolerance of failure in public-sector technology commercialization	a) Design and evaluate experiments in technology commercialization. b) Provide incentives to innovate.
9. Identification of desired socio-economic impacts	a) Define T^2 goals and metrics. b) Awards based on accomplishment of goals. c) Develop better models.
10. Improvement of interaction mechanisms	a) Experiment with and evaluate new mechanisms.
11. Fairness of opportunity	a) Expand outreach communication. b) Centralized information base.
12. Preservation of national security	a) Restrictions applied at applications level. b) Expand definition of "security."
13. Facilitation of organizational learning to improve technology commercialization	a) Demand commercialization planning. b) Provide for third-party evaluation.

frequently lost their investments. Furthermore, the many millions of taxpayer dollars that went into government-sponsored research in alternative energy have lain fallow.

When the primary government role in technology was confined to funding and performance of basic, nondirected research, industry was left with the responsibility for interpreting scientific advances and identifying their potential for productive applications. Now, however, the government is seeking to expand its role to the point of developing applications or collaborating with industry to provide new materials, products or processes. How can industry's perceptions of future demands for new products, services and systems be effectively incorporated into government programming of research? Many of those industrial perceptions are proprietary, especially if they achieve a significant level of specificity. This suggests that feedback from industry or trade associations or from special multi-firm groups, whether organized by themselves or the government, will be able to provide only very general information on industry needs or market opportunities.

Can the U.S. government cooperate with industry leaders to support sweeping advances in the science and technology of critical industries such as computing, telecommunications, or manufacturing equipment? If the government listens at the policy level, will it listen at the implementation level of specific research programs? What mechanisms might be tried to facilitate the required communication before research programming and budgeting is accomplished in the federal system?

Many industry leaders have a healthy skepticism about cooperative efforts with the government. Thus, they are unlikely to become the initiators of innovative methods of collaboration. Indeed, a study of CEO priorities reported that research and development (whether internal or external to the firm) is a surprisingly low priority of senior management.[4]

Most observers would agree with Donald Hornig's assessment,

> In the context of scientific and technological innovation to develop industry and promote the general welfare, the present question is whether it [the government] can or should attempt to promote and stimulate designated industries and play a role such as that of the Ministry of International Trade and Industry in Japan. One must conclude that we have not yet found a way to do so.[5]

In spite of the complexity of the challenge of fostering industry involvement in federal research programming, there are several options to be considered.

Strategic Response Number One: Federal Technology Policy and Consensus on Critical Fields

Should agencies with significant research budgets set up policy-level advisory boards as well as technical advisors at the working level? First, there must be a well-defined, coherent national technology policy. The effect of such a policy should be to set a climate for technology commercialization, to survey existing policies worldwide and note their effects, to coordinate the many institutions involved (both public and private), and to perform a gap-filling process with directly-performed research by the government as well as contracted and cooperative research.[2] Secondly, there must be near-consensus regarding a set of critical technology fields like the 94 critical fields identified by the Council on Competitiveness.[6] Industry leaders should be surveyed to ascertain the priorities reflected in private-sector plans.

At the institutional level, laboratories can invite industrial input at both the senior management level and at the programmatic level. Advisory committees are being used more frequently, but their advice is not yet manifested in research programming. The top-down research programming in budgeting processes (based on agency objectives and perceptions of societal needs) found in most agencies can easily incorporate input from advisory groups because of its centralized structure. More difficult is the diffused bottom-up programming which responds to investigators' proposals at the laboratories. These proposals should be required to describe the relationship of the proposed work to the associated activities and interests of industry. Such a requirement may drive the R&D mix toward more applied work. This may be beneficial in the short term, as the effects of increased government/industry collaboration will be manifested more rapidly.

Individual firms, dealing with specific government laboratories, can provide proprietary and detailed information about their own research efforts and needs, once they have been assured of the laboratory's ability to preserve confidentiality. Here, government policy seems to be moving in the proper direction, with some agencies providing five-year data rights on joint work and assurances of complete confidentiality of the firm's proprietary information. As detailed information becomes available on a firm-by-firm basis at the laboratory level, this information must be aggregated across as many firms as possible to be most useful for policy making.

Many federal and contractor-operated laboratories are successfully pursuing cooperative research and development agreements (CRADAs) with industry. Over time, these efforts will create a more general understanding of mutual decision processes for research programming and specific priorities within technological fields. However, at this time the majority of CRADAs have resulted in *basic* joint research projects. The effect on product and process innovations in industry will not be known for some years.

Strategic Challenge Number Two: How Do We Control the Nationality of the Technology Recipient?

Because of the U.S. government's natural interest in using its technology to increase U.S. international economic competitiveness there has been reluctance to assign rights or establish collaborative work with *foreign-owned* enterprise. This presents several dilemmas. First, how does one determine the extent of foreign ownership in large, complex firms? If the firm is owned principally by U.S. citizens today, who will own it tomorrow? If ownership changes hands does the government recall its intellectual property rights? Who should then monitor the ownership of each firm having access to proprietary federal technology? Technology transfer staffs of federal agencies, laboratories and contractors in most instances do not have sufficient resources to "make the deals" on a scale large enough to affect national competitiveness. Monitoring the consummated relationships is far beyond current resource levels. Secondly, much of the government-sponsored proprietary intellectual property is basic or early applied research, far from commercial applications. Only those firms with a very long-term perspective in their managerial decision making are likely to desire partnerships with government laboratories if commercial products or processes are more than two or three years away.

There is some evidence that foreign firms, most notably Japanese, are willing to take a longer-term perspective in emerging technologies than are U.S. managers. This long-term perspective may be essential for continued product innovation.[7] As Robert Reich has so ably pointed out, some foreign-owned firms are making more significant positive impacts on the U.S. economy than are average U.S.-owned firms.[8] Analysts like Reich argue that the technology should go to the firm that demonstrates the potential to make the greatest economic and social impact within the U.S., regardless of ownership. However this would initially require a change in policy for most federal agencies.

Finally, federal policies governing access to indigenously-developed technology can be interpreted negatively by foreign governments, and thus result in host-country policies discriminating against U.S. firms operating abroad. Intervention has increased in the target-market countries of many U.S. firms, and as competition increases in these markets, the bargaining power of U.S.-owned MNCs may well decrease.[9]

Strategic Response Number Two: Award of Intellectual Rights Based on Impact on US economy

With recognition of these concerns of government policy makers, progress towards this challenge can be made by awarding intellectual property rights on the basis of competitive bids which detail the impacts on the U.S. as well as local economies. Rather than relying primarily on ownership, rights should be allocated by focusing on the proposals which create the most jobs, pledge the highest investment in new plant and equipment (or use of idle capacity), commit to workforce training, and promise to protect the environment and address local concerns.

Finally, Joiner has proposed an increase in partnering to improve the competitive position of U.S. firms in those circumstances in which non-American firms have greater skills in deploying and marketing new technologies.[10] Policies which restrict foreign access to U.S. technology must be carefully constructed to avoid the deterrence of this proposed partnering.

Strategic Challenge Number Three: Whom Should We Choose as the Technology Recipient: the Entrepreneurial Firm or the Industry Leader?

Several well-publicized studies have concluded that small and medium-sized entrepreneurial firms innovate and commercialize technology faster and more cheaply than large firms.[11,12] However, the government has generally been less effective in transferring technology to small businesses.

There are a number of reasons why the government and its contractors find it more difficult to cooperate with small and embryonic firms. Unless they have a larger-than-average capital base, small firms typically don't have the resources to pay national laboratories for collaborative work. Scientific and technical work in most national laboratories is not performed cheaply, especially if federal policy requires full-cost reimbursement including depreciation and amortization of very expensive equipment and facilities. Indeed, the "inexpensive, work-in-the-garage" syndrome is the comparative advantage of the small firm. Thus, subcontracting technical work to the expensive national laboratories would defeat the purpose of the special considerations (to small firms) that are required by some federal policies and laws. Can agencies establish meaningful policies and incentives for laboratories and their contractors to work with small firms, if it is in the national interests to do so? The National Competitiveness Technology Transfer Act of 1989 requires federal agencies to provide special considerations for small businesses, but very few agencies have met the challenge. There is a high fixed cost associated with establishing mutual interests, defining joint work statements, generating and getting approval for cooperative research and development or funds-in agreements and organizing to implement agreements. Thus, small contracts are simply not worth the effort to large national laboratories. Furthermore, government policy makers have exhibited a paucity of understanding and empathy for the circumstances of small and embryonic firms and the decision-making styles of entrepreneurs.

Strategic Response Number Three: Advocates in Technology Transfer Organizations for Small Business

Technology transfer personnel at the laboratories should include people with small business experience, to ensure understanding small firms' needs and situations. This might be effected by including small-business advocates in the technology transfer functions, not unlike the small-business advocates frequently found in the purchasing functions of large organizations. In any event, the technology source organization, when awarding rights to a small business, should be prepared to provide ample and continuing assistance to ensure the successful transfer and commercialization. Examples of successful assistance methods are provided by Radosevich.[13] For example, opportunities exist to designate "user facilities" at federal laboratories. This gives private industry access to unique and expensive laboratory facilities and equipment. Such arrangements are particularly valuable to local small businesses. This access would help minimize the cost of joint work to the recipient of proprietary technology.

Local economic development infrastructure can occasionally facilitate interaction between laboratories and small firms to encourage the evolution of local entrepreneurial enterprises. Local business assistance organizations have occasionally been able to intercede with the federal bureaucracy and, as intermediaries, simplify the process of interaction.

Laboratory technology transfer personnel can also serve a "brokerage" function by facilitating an alliance between a small and a large firm each of which has interest in a lab's technology. Such a partnership, if properly constructed, can reap the benefits of swift entrepreneurial action in the small firm as well as the technical resources and market presence of the large firm.[14]

Strategic Challenge Number Four: How Can the Government Create Commercial Value in Its Technology?

Because the government is trying, through technology transfer, to get the best possible return on technology developed with taxpayer funds, many policy makers believe the technology should be prepared and presented in its highest-value form. However, the best way to accomplish this is often obscure.

Does the U.S. government have the obligation or right to restructure the basis of competition in a given industry if it has a truly important proprietary technology? Should it award intellectual property rights on the basis of exclusive field-of-use to those companies already in the market? Some technology transfer staff members argue that this does not disturb private markets if the restricted fields of use are carefully defined. Exclusivity is necessary to create real value and provide incentives for adoption, adaption, and commercialization. The logical process of examining how best to award exclusivity with minimal disruption to the private sector have not been well developed.

Strategic Response Number Four: Establish and Manage a Strong International Intellectual Property Position

Creating value in technology and managing the technology asset are complicated managerial functions for which few guidelines are available in either theory or practice. Experimentation with and evaluation of novel methods are necessary to improve practice.[15] Public-sector laboratory managers must begin to think of technology as a valuable asset and must create the capability within the laboratory to establish strong international positions in intellectual property. For example, one of the nation's largest national laboratories, with over 8500 employees, has one FTE patent attorney and had fewer than 100 patent disclosures last year. The laboratory's limited budget for prosecuting patent applications resulted in fewer than one dozen. Clearly, in this and similar situations, the intellectual property office needs substantially greater resources, especially if foreign patent rights are to be secured. Obviously, such rights are critical to the pursuit of international competitiveness.

Typical laboratory scientists, engineers and technicians know little about creating proprietary positions or other facets of value in their technologies. There is a strong incentive for them to assist in the creation of value in their inventions, since the 1986 Technology Transfer Act ensures their participation in the royalties secured by licensing the technology. Obviously, even short-term training in the process of creating value in technology would be useful for most laboratory employees.

A stronger focus on applications and demonstration projects by the laboratories would not only create more value in their technologies but also improve the interests of prospective industrial partners. To accomplish this strategy, the laboratories may need to increase the proportion of applied research and development as compared to basic research. Some policy makers feel this is inevitable if closer alliances with industry are desired. Some of these policy makers argue that universities should be the primary institutions conducting basic research.

Strategic Challenge Number Five: How Can We Provide for Technology "Maturation?"

At present, the taxpayer is paying for the maturation of some technologies developed by the government or its contractors, especially in fields for which technology transfer is the primary objective. For example, DOE laboratories are still spending millions of dollars on alternative energy research which is largely incompatible with the nation's energy policy. Agency decision makers are correctly assuming that many technologies in the federal laboratories are as yet inadequately developed to be of obvious benefit to industry. Federal laboratory efforts to extend technology beyond the current charter of applied research may be necessary in many instances. However, the selection of technologies to be advanced and the consequent creation of commercial value in these technologies should involve consequential private-sector participation in both decision making and funding support.

Why can the public agencies and their laboratories not develop suitable relationships with private risk capitalists, who constantly complain that there is an insufficient flow of "good" deals to absorb the immense amount of risk capital available in the U. S? Again the fault lies partially with the federal agencies, who do not understand the investment practices of risk capitalists. And the financial institutions managing risk capital need to understand better the potential of federal technology partnerships and the issues to be resolved before collaboration becomes pragmatic.

Orientation toward investments such as leveraged buyouts has minimized private-sector interest in early seed capital deals for technology-driven new ventures. Clearly, some private-sector institutions and their managers must change, as well as the federal systems.

How can experienced private-sector management contribute to a federal R&D project in order to advance the technology to a point of commercial interest? Most experienced venture capitalists pride themselves on their ability to package and repackage deals that offer sufficient investment excitement but don't meet all their criteria. Their networks of successful entrepreneurs, technical resources for technology assessments, and potential industrial partners for takeouts and strategic alliances, can contribute great value to the process of spinning off federal technology to commercial applications. However, their skepticism toward any relationship with the government has kept most venture capital managers from exploring this potential. In addition, many private-sector managers lack an understanding of the issues which must be resolved to culminate a successful relationship with a federal laboratory.

Strategic Response Number Five: Public and Private Partnerships to Achieve Technology "Maturation"

Public/private partnerships for technology maturation are starting to increase because of the emphasis being placed upon CRADAs. This process can be improved even further by the establishment of sound federal industrial and technology policies. Such policies are likely to identify, prioritize and fund opportunities, like the Human Genome Project, to which billions of public dollars have been committed. Those projects of sufficient commercial interest to draw private investment should be matched by public investment in R&D. And at the micro-level within the laboratories, charters should be revised to continue technology maturation beyond the proof of concept. Rewards must be built into laboratory incentive systems so that inventions with commercial potential are recognized as well as the performance of good scientific work.

In the U.S., a number of potential intermediaries exist between the public laboratories and industry. Some of these possess substantial technical competence and the ability to advance and apply public-sector technology. Many of these intermediaries have extensive experience in working with one or more specific industries and hence have knowledge of industry needs and interests. Greater use of these intermediaries may be a viable alternative to spending public funds to advance technologies within the federal laboratories. Early evidence suggests that industrial research consortia may serve as effective intermediaries, once internal issues are resolved in their operations.[16]

Strategic Challenge Number Six: Can We Create Consistency in Technology Transfer Practices Across Government Agencies?

Industry would like to have consistent policy and practice in the various federal agencies so that it could efficiently interact with multiple agencies without having to learn the idiosyncracies of each. Unfortunately, that is unlikely to happen. Consider, for example, the Department of Energy, which has one of the largest captive R & D budgets in the federal system. DOE efforts to coordinate the policies and practices of the disparate DOE program offices have made progress but certainly can not be represented as uniform. To compound the problem further, the vast majority of DOE laboratories are managed by contractors, each of which has incorporated into its technology transfer activities the guidance of its corporate parent's policies and procedures. Thus, if a company interacts with Martin Marrietta Energy Systems, which operates the Oak Ridge National Laboratory for DOE, it will find a very different system than that employed by Sandia National Laboratories (operated by AT&T) or Los Alamos National Laboratory (managed by the University of California). The problem is even greater if one tries to discover the best way to interact with NASA or its contractors, the National Institute of Standards and Technology, the National Institutes of Health, the Environmental Protection Agency, the Department of Defense or its contractors, or the Department of Agriculture, all of which invoke different policies and procedures for technology transfer.

Various mechanisms exist to help the potential industrial partner seeking an entry point and guidance into the federal labyrinth. Congress first assigned the charter for coordination of the federal system to the Federal Laboratory Consortium (FLC), which operates a clearinghouse providing a network to many of the laboratories in the federal system. More recently, Congress authorized another entity to perform a similar integrative function, the National Technology Transfer Center (NTTC). The NTTC is in a nascent stage but the FLC has been operating effectively for a number of years. Nevertheless, after its initial introduction (by whatever method), a firm will soon learn that detailed negotiations leading to a meaningful relationship must occur in substantially distinct systems for each agency, and often for each laboratory. This is clearly an inefficient process, and one seemingly designed to discourage government/industry collaboration. Unfortunately, given the power of the individual fiefdoms within the federal bureaucracy, a consistent and uniform system is unlikely to evolve in the foreseeable future. Firms interested in securing federal technology, or developing collaborative relations to advance technologies of mutual interest, must acquire the knowledge and skills to work with each agency and contractor.

Strategic Response Number Six: Assign a Lead Agency to Define and Implement Uniform Practices

One resolution of this challenge which is discussed from time to time is the creation of a supra-agency within the government, perhaps within the Department of Commerce, to define a uniform set of practices. If the NTTC becomes in reality more than an NASA technology transfer coordinator, it could join the FLC in ensuring greater uniformity. In the interim, advice on effective procedures for interaction with specific laboratories can be secured through the FLC or NTTC networks.

Another option is to secure the services of one of the private intermediaries who are trying to keep up with the changing practices of each agency. Many of these private organizations are especially effective in dealing with specific agencies, such as the Department of Defense.

Strategic Challenge Number Seven: How Do We Avoid Conflicts of Interest that Can Deter the Technology Transfer and Commercialization Process?

Mention "Dingell's committee," and panic reigns in the public bureaucracy. As Chairman of the House Energy Subcommittee on Oversight and Investigations, Rep. John Dingell has scrutinized technology transfer arrangements, such as the n-Chip licensing agreement from DOE's Lawrence Livermore National Laboratory (LLNL).[17] Because three former employees from LLNL received an exclusive license for a laser pantography technology which enables ultra-dense packing of computer chips, a controversy arose when an existing firm expressed interest in the technology.

Some policy makers perceive a paradox involving personal conflicts of interest if individual scientists employed with taxpayers funds can invent and then claim ownership of intellectual property. The perceived paradox occurs because many experts feel that the best chance of commercializing technology is through entrepreneurial spin-offs involving the inventor. Furthermore, this mechanism is consistent with the small business priorities established by Congress.

Since some government R&D contractors are required by law to share royalties with the inventors of licensed technology, there is a potential for confusion between conflicts of interest and proper incentives for motivating researchers. Some policy makers feel that monetary rewards might encourage government laboratory employees to withhold disclosures or critical information about commercializable technology in order to gain personally later. The same potential for conflict of interest has existed for years with the policies of many universities about the rights of professors with respect to intellectual property generated by federally-sponsored grants.

There is also an opportunity for organizational conflicts of interest as government contractors develop federally-sponsored technology. Various arrangements exist for splitting royalties received between 1) the contractors, for use as discretionary funds, and 2) the U.S. Treasury. Some private contract research laboratories and universities are becoming concerned that the government is not adequately protecting them from competition by the national laboratories. Basically, the laboratories are expected to police themselves to make sure that they are not competing with the private sector. Perhaps the greatest concern should be that policy makers will become so concerned about perceived conflicts of interest that rigid procedures are installed and initiatives supportive of technology commercialization are inhibited.

Strategic Response Number Seven: Use of External Agents and Intermediaries to Avoid Conflicts of Interest

Conflict of interest is one area in which uniform policies across all agencies would be beneficial. Conflict should not be defined in such a manner that technical staff members of government laboratories or contractors can not play a role in the commercialization of their inventions. They should be allowed to benefit personally as long as their roles are recognized, disclosed, publicized and approved. Claims of conflict can be mitigated by sufficient information dissemination by the technology source, so that all potential recipients have a reasonable chance to learn about commercial opportunities. The source may also serve a valuable function as a "broker", putting together industrial partners with ex-employee inventors who wish to be involved.

The use of external agents and intermediaries would also reduce the appearance of agency or contractor favoritism. By objectively choosing the best technology commercialization mechanism and entity, a third-party intermediary shields the agency.

Strategic Challenge Number Eight: How Can We Develop a Tolerance for Failure and the Accommodation of Problems in This Significant Social Experiment Called Federal Technology Transfer?

Unfortunately, many policy makers and managers do not perceive federal technology transfer as a significant social experiment. Most will admit that we are not sure of the correct methods of accomplishing the government's goals and guidelines. This acknowledged uncertainty however, is not accompanied by a tolerance for the unsuccessful initiatives inherent in experimentation. Perhaps this intolerance is derived from a sense of urgency. Given the size of federal R&D expenditures, the derived technology is certainly a substantial, if relatively untapped, asset. Even though the management of technology from its generation to its utilization, is a difficult task, most U.S. institutions that create or apply technology do not acknowledge the extent of this challenge. As a result, innovative (and risky) practices are not being encouraged.

Compounding the challenge is another phenomenon which inhibits experimentation—the bureaucratic process of government and large private organizations. The dysfunctional aspects of bureaucracies are well understood by all. Federal employees know that rewards will not be forthcoming to those who cause their superiors to appear before Dingell's committee or who invoke investigations by the GAO. Novel approaches are especially risky to one's career when the effects of success are difficult to measure but the observation of rules violations is obvious.

If incentives are developed and awarded to those who do succeed in commercializing federal technology, we can assume that investigations will follow. Extraordinary rewards are not part of the typical bureaucratic process. Although a primary objective of the commercialization process is to create wealth, if any part of it is done within the public sector, jealousy is certain to instigate concern. Those who attempt novel methods, succeed and prosper are certain to attract attention, not all of which is likely to be positive.

Strategic Response Number Eight: Programs to Experiment with and Evaluate Different Technology Commercialization Processes

To increase the prospects for success in experiments designed to improve the technology commercialization process, there must be explicit programs to evaluate experimental elements. Such efforts should start with the lowest-risk experiments, perhaps those affecting the use of management technology which can improve the process. An example is the use of data bases which disclose industrial technology needs and interests and group-decision support systems and electronic meetings to initiate transfer processes.[18] Such innovations are less likely to be perceived as presenting potential conflicts of interest.

Additionally, incentives must be designed to reward these innovations just as royalties are shared with inventors whose patents generate revenues for their laboratories. Lastly, those members of Congress who have been foremost in the sponsorship of the legislation fostering technology transfer must now use their political influence to protect those innovators whose "smart" failures were well intended and without conflicts of interest.

Strategic Challenge Number Nine: How Do We Identify the Desired Socio-Economic Impacts of Technology Transfer?

A number of issues regarding government technology transfer can only be resolved if adequate models of economic and social impact can be developed to evaluate alternatives. Perhaps the most pressing issue to be illuminated by improvement in such models is the commercial effect of spending on defense-oriented research. Some critics, such as Laura Tyson, feel that the U.S. emphasis on defense research has handicapped our international competitiveness position because of limited opportunities for commercialization of resultant technology.

> Defense oriented research cannot replace civilian R&D. Past evidence suggests that America's reliance on defense as a technological engine can hurt rather than help competitive outcomes in commercial applications. In computerized machine tools, for example U.S. suppliers concentrated on unusual aerospace and defense applications, thereby losing a dominant share in commercial markets in such critical areas as robotics. Many observers now argue that civilian and defense needs are becoming even more divergent and that the current boom in defense research is diverting resources away from industrial pursuits.[19]

In addition to the issue of defense versus commercial research spending, other technology transfer issues require better measurement of social impacts. For example, investments in local facilities, job creation, domestic-supplier utilization, employee training, locations in "pockets of poverty," and community contributions, as well as negative impacts such as environmental infractions, might all be considered in a comprehensive model, with factor weights dependent upon the local needs. The potential impacts could be judged by scrutinizing commercialization plans from competing private bidders for the technology. However, expertise to judge the validity of such plans is likely, at least initially, to be lacking in public-sector decision makers. Also, if awards are made on the basis of the best promises, who would perform the monitoring in order to exercise any march-in rights which might be imposed to reclaim the technology if promises are not met? Perhaps a greater exchange of personnel between public and private sectors (a concept frequently espoused but inadequately implemented) would increase mutual understanding and improve the respective experience bases.

As an aside, the government bureaucracy is placing too much reliance on the usefulness of march-in rights in case of lack of performance of technology commercialization. First, the authors were able to find only a few instances in which the government exercised those rights. One explanation could be inadequate understanding and monitoring of the commercialization process by the responsible agency. Secondly, in those limited instances in which the rights were reclaimed, so much time had elapsed (usually several years) that, in any field of dynamic technology, the value of those rights was seriously diminished. Thus, if a private concern found that the government had proprietary technology which represented serious competition for its own, it might adopt a posture of licensing the rights, with minimal advance payments but ample promises of future royalties, with the intention of keeping public-sector technology from competitors. Such a strategy might not be viable if such action precluded future access by the firm to valuable government technology. However, the difficulty with this approach is that the government bureaucracy often has a limited memory and there is very little communication between agencies.

Strategic Response Number Nine: Developing the Model

In the short term, development of a comprehensive model of socio-economic impacts from public-sector technology commercialization can start with the formulation of a set of objectives or the criteria to be used to evaluate alternatives. Such a set of objectives should be uniformly adopted by all agencies. Perhaps, therefore, its formulation should be the responsibility of an interagency entity such as the National Technology Transfer Center, with assistance from such entities as the Office of Technology Assessment or the National Science Foundation. The resulting set of uniform criteria for all technology transfer RFPs or RFQs would greatly facilitate the evaluation of results and the eventual development of a complete model.

If federally-sponsored technology represents a sufficiently dramatic advancement over industry practice, awarding rights on an exclusive basis could significantly affect domestic as well as international competition. Realizing that such an industry restructuring could be caused by the manner in which awards of federal technology are made, the Bush administration was reluctant to establish a technology policy or an industrial policy that substantially increased federal involvement in critical industrial technologies. However, at the implementation level, decisions are made daily by federal employees and contractors to award federally-sponsored technology, usually with restricted exclusivities such as field-of-use, but occasionally without restrictions. There is more than an ideological issue at stake here. If the Clinton and future administrations desire to act as an economic King Solomon, models must be developed which give some insight into the socio-economic impacts of changes in the competitive structure of an industry. As difficult as it may seem to develop adequate models of this kind for established industries, the problems are compounded many times when a radical technological innovation is to serve as the cornerstone of an entire new industry.

Since most government bureaucrats have little experience judging the probable results of business ventures, new personnel skills must be acquired by the agencies or their contractor laboratories, most likely from the private sector. If the skills for such an endeavor do not exist in the executive branch, review boards of experienced private-sector managers can be organized to judge the probable impacts of commercialization plans from bidders for public technology. One thing is clear: if promises of commercialization results from successful bidders for public technology are to be monitored for performance, additional resources will be needed in terms of management systems and experienced personnel.

Strategic Challenge Number Ten: What Mechanisms Can Be Developed to Promote Industry/Government Interaction?

Every agency and government contractor has experimented with different technology-transfer mechanisms, such as exchange of research materials and samples, entrepreneurial spinoffs, collaborative R&D, licensing intellectual property, and technical assistance contracts and short-duration problem solving. Each technology source has its own unique array of preferred mechanisms. Because it is a more structured mechanism and a low consumer of resources, licensing intellectual property is the most common among many agencies. Launching spinoff ventures with departing employees is not encouraged by very many agencies or their contractors, because of the appearance of conflict of interest, the inordinate amount of resources required to assist a startup, and lack of knowledge and experience relative to incubation functions. (The ARCH organization which couples the University of Chicago with DOE's Argonne National Laboratory is a relatively unique experiment in the promising mechanism of entrepreneurship.) Legislation which establishes priorities for technology transfer to small and disadvantaged businesses, state and local governments, and universities and not-for-profit firms has been largely ignored by some agencies and contractors.

Because of the experience and preferences for certain mechanisms of most federal technology sources, industry would be well advised to review the records and historical transfer activities of potential government partners. Some, for example, have no compunctions in awarding exclusive rights to intellectual property, while others demonstrate great hesitancy. Because the expected royalties from licensing can pay for the next round of technology transfer activities and reward employee inventors, there are strong incentives in the system to encourage licensing. Because the returns are uncertain and the precedents nonexistent for equity sharing with technology recipients such as employee spinoffs, government bureaucrats and employees of large contractors are often loath to explore the possibility of this exciting arrangement.

Strategic Response Number Ten: Development of Novel Mechanisms

If the government truly wishes to facilitate the transfer and commercialization of its technology, it must find ways to encourage its agencies and contractors to adopt whatever mechanism is preferred by potential industrial recipients in each circumstance. The novelty required to invent new mechanisms may come from unique combinations of institutions. For example, experienced venture capitalists could not only bring funding for the maturation of early-stage technology but also apply their expertise to the development of new organizations. Just as industry is currently experimenting with new structures for strategic alliances, federal laboratories should consider themselves partners in the commercialization process as well as in the performance of mutual research efforts.

Technology pull is the current fad. Yet some studies suggest that technology push is most effective for radical innovations. Recognizing the current difficulty of consummating federal technology transfer to industry, combinations of push and pull mechanisms are frequently needed. Several good examples of successful combinations of mechanisms are provided in a study by the DOE Pacific Northwest Laboratory (PNL).[20] In those circumstances in which the government is unable to collaborate effectively with industry in defining R&D projects, technology push may be the only mechanism possible.

Strategic Challenge Number Eleven: How Can the Government Ensure Fairness of Opportunity for Access to Federal Technology?

One of the concerns of the federal government is the issue of ensuring fairness of opportunity to potential recipients of technology. The federal government has gradually moved away from the position that it should place all knowledge in the public domain through publication. Most agencies now recognize that more value is created if proprietary positions are established and protected and awarded on an exclusive basis. This exclusivity compounds concerns that some version of a "fairness" construct be applied in the transfer process.

Some attempts at dictating a fairness doctrine in technology transfer have been perpetrated by Congress, but sufficient ambiguity exists to ensure complexity in the implementation process. From the technology transfer acts of the early 1980s to the current ones, special considerations and priorities have been expressed for small and disadvantaged businesses, universities, not-for-profits and state and local governments. President Reagan further confused the intent of Congress by issuing an executive order stating that large businesses should have the same special considerations. If everyone has special consideration, obviously no one does. The interpretation and implementation of these departures from equal treatment as expressed by Congress have been slow by most agencies, and inconsistencies are prevalent, not only across agencies but also among contractors of the same agency.

The obvious dilemma in ensuring fairness is determining the most deserving party among those who request the technology. Less obvious is the issue regarding the appropriate ways to communicate the availability of the technology to those who might have an interest and a capability to commercialize it. The practices of government agencies and their R&D contractors with respect to communicating opportunities for technology commercialization vary from advertisements in the *Commerce Business Daily* to directed announcements to firms with whom the laboratory, agency or contractor has established relationships. Some agencies, such as the National Institutes of Health and the Environmental Protection Agency, have established electronic bulletin boards through which opportunities are announced. The practices of most agencies, however, leave some room for doubt that all potential recipients who might have legitimate roles to play in the commercialization of the technology have been made aware of the opportunity.

A complication to the fairness issue is the trend toward increased cooperation by sources of technology as well as potential recipients. Consortia of technology developers are evolving and maturing in their operations with considerable rapidity. As co-owners of intellectual property, co-inventors have occasionally found that disagreements and miscommunication (or lack of communication) can lead to serious consequences, such as different owners awarding exclusive rights for the same application to different recipients. In the federal government, it is not unusual to find several agencies responsible for the dissemination, including licensing, of a pool of technology. As more intermediaries (such as the National Technology Transfer Center and the Federal Laboratory Consortium) are given formal responsibilities from Congress, and other agents are selected by specific agencies, the potential for confusion is increased.

Strategic Response Number Eleven: Centralized Decision-making and Information on Awards of Rights

Ensuring fairness of opportunity is at least a two-step process: first, communicating the opportunity to acquire technology as broadly as possible, and then applying consistent selection criteria in an objective fashion. The outreach function for communicating opportunities can increase the probability of reaching all potentially interested parties by employing a variety of directed and broadcast methods. Using a variety of methods of communication for each opportunity can be prohibitively expensive. However, some methods such as electronic bulletin boards, can be highly efficient by promoting the use of the electronic board rather than each individual opportunity. These passive methods place the onus of communication upon the potential recipient, which some small business advocates believe to be unfair. Many small businesses do not have the capabilities for monitoring these passive channels.

One method of ensuring consistency in both the outreach and decision processes is to centralize the processes, at least at the level of individual agencies. NASA, for example, has multiple contractors operating regional technology transfer centers, any of which could be simultaneously working with different industrial clients to gain access to technology developed at one of NASA's field centers. Since final agreements are only consummated at a centralized patent counsel's office, the ultimate embarrassment will be avoided, but if the situation is confused with more intermediaries and poor coordination, much effort could be wasted by industry before the ultimate winner is selected.

Ensuring fairness in the selection process entails the development of consistent selection criteria and objective application to each proposal from a potential technology recipient. To accomplish this, it is necessary to develop models of socio-economic impact (see challenge number nine), coordinate a universally consistent process across agencies (see challenge number six), and evaluate and improve the models and the processes as more experience is acquired (see challenge number thirteen).

Strategic Challenge Number Twelve: How Can We Commercialize Public-Sector Technology Without Jeopardizing National Security?

Because the agencies and contractors of the Departments of Defense and Energy perform a tremendous amount of basic and applied research under the defense programs, the subject of so-called "dual-use" technologies has been debated with much-deserved fervor. Can we find commercial uses for and convert technology originally intended for military purposes? The early evidence suggests that we can if national security constraints are eased.

With the easing of international military tensions between the superpowers, the Department of Defense and the Department of Energy are reviewing their lists of critical technologies and improving the opportunities for U.S. businesses to compete by freeing previously restricted technology. In addition, these agencies have recently been invoking technology restrictions for national security purposes at the level of applications instead of on basic technologies. This means that for dual-use technologies, the military applications will be classified but applications with commercial potential will not. Concerns that lists of critical technologies have been too restrictive have also caused substantial reappraisal by the Coordinating Committee for Multilateral Export Controls (CoCom) and the U.S. Department of Commerce.[21] Given the relative greater importance of dual-use technology from the U.S. defense research base, as compared to our most serious international competitors, this peace dividend is likely to have substantial impact.

Strategic Response Number Twelve: Close Working Relationship between Government and Industry

Indeed, convincing arguments have been made that national economic security will be increased with the advent of closer cooperation between the federal government and the private sector.[22] Most experts would agree that national security, in a world with a declining threat from communism, should be redefined to broaden the security concept beyond military power. The importance of economic power to national security is becoming more widely recognized. In some areas of technology such as electronics and specialty materials advances in the civilian sector have outstripped the military. In these areas, technology transfer from the private sector to the government can be improved through the relationships and processes developed from the reverse flow.

The commercial application of dual-use technology is one very positive development through which the government is creating value in its technology (as well as in private-sector technology) with little additional monetary expenditures. The liberalization of export controls and national security classifications means that a peace dividend is truly accruing to U.S. technology users, especially those with economic impacts in foreign markets. In addition, new mechanisms must be found to raise the rate at which commercial applications are identified for technologies advanced within the defense research establishment.

Strategic Challenge Number Thirteen: Can We Develop Evaluation Systems and Learning that Will Facilitate the Evolution of an Effective System of Federal Technology Commercialization?

Improvement of a complex organizational process requires the development of learning through objective assessment of results and comparison to plans. Comparison of results to standards or baseline performance requires enough understanding of the process so that realistic expectations can be established. Because the possibilities for federal technology transfer processes are not well understood, few plans have been developed by the agencies. Congress dictated that those federal laboratories with sizeable research budgets should establish offices of research and technology applications (ORTAs), which in turn were required to develop five-year strategic plans for technology transfer. Although required to do so by law, very few laboratories undertook the planning effort, and even fewer developed a realistic plan.

One of the most difficult aspects of providing effective feedback on the effectiveness of a public/private partnership for technology transfer and commercialization is the need for access to the performance records of each partner. For example, the government by itself has neither the understanding nor the management capacity to monitor the commercialization efforts of its private-sector partners. The "enforcement syndrome" of many government/industry interactions make this measurement and communication process difficult. Managers on both sides must realize and act as though the greatest benefit from the feedback is learning, not the instigation of punitive actions to correct perceived or actual shortcomings from experimental activities.

Strategic Response Number Thirteen: Fund Objective Third-Party Evaluation System

Realistic plans by each laboratory for the commercialization of its technology must be developed with full recognition of the issues involved. Evaluation and learning depend on the baselines established by planning.

Congress should create and fund a program to facilitate the learning necessary for continuous improvement of the public/private technology commercialization process. The current authority and resources of the FLC or the NTTC could be expanded to perform this role. With additional resources, evaluation measures could be defined, expected outcomes planned in advance by each agency and contractor, information and data bases set up and used, and results disseminated for the benefit and learning of all parties concerned.

CONCLUSIONS

Increased cooperation between the federal government and industry to transfer and commercialize federally-sponsored technology must be a key aspect of any plan to increase U.S. international competitiveness. Policies and mechanisms for facilitating this cooperation are slowly evolving. Both sides must perceive these initial efforts as the first step in a great social experiment. Effective participation by both parties requires mutual understanding and a willingness to explore novel mechanisms and approaches.

Scientists and technology transfer personnel in the government's laboratories have been given new charters and incentives to cooperate with industry. Although the need and motivation to transfer federally-sponsored R&D have never been greater, the means for doing so are still uncertain.

Unfortunately, key policy makers and senior managers in critical institutions are often unaware of the complexity and scope of the public/private technology transfer and commercialization processes. A more comprehensive understanding of these issues is an imperative first step in the design of realistic partnerships. The incidence and effectiveness of these partnerships must be increased to affect our international competitiveness significantly. The Task Force on Management of Technology, National Research Council, recognized this shortcoming among many U.S. industrial and governmental leaders. In its 1987 report, the group listed as one of the eight most pressing technology-related issues for U.S. industry, "how best to accomplish technology transfer," including "assimilating externally developed technology and research results into the company's internal RDE&O activities."[23]

REFERENCES

1. National Science Foundation (1988). Real increase in 1988 research funds estimated at lowest rate in eleven years. *Science Resources Studies Highlights.* Washington: National Science Foundation.
2. Kline, S. J. & Kash, D. E. (1993). Technology policy: What should it do? In A. H. Teich (Ed.), *Technology and the future.* New York: St. Martin's Press.
3. Magnusson, P. (1991, April 1). Bush just might buy this plan—if no one calls it "Industrial Policy". *Business Week,* 27.
4. Hise, R. T. & McDaniel, S. W. (1988, Winter). American competitiveness and the CEO—Who's minding the shop? *Sloan Management Review* 49–55.
5. Hornig, D. F. (1984). The role of government in scientific innovation. In J. S. Coles (Ed.), *Technological innovation in the '80s.* Englewood Cliffs, NJ: Prentice-Hall, 55. It should be noted that Dr. Hornig recognizes one significant exception—the U.S. government role in the establishment of our world-class agro-industrial enterprise.
6. Council on Competitiveness (1991). *Gaining new ground: Technology priorities for America's future.* Washington: U.S. Government Printing Office.
7. Meyer, M. H. & Roberts, E. B. (1988, Summer). Focusing product technology for corporate growth. *Sloan Management Review,* 7–16.
8. Reich, R. B. (1990, January-February). Who is us?, *Harvard Business Review,* 53–64.
9. Kim, W. C. (1987, Spring). Competition and the management of host government intervention. *Sloan Management Review* 33–39.
10. Joiner, C. W. (1989, Summer). Harvesting American technology—Lessons from the Japanese garden. *Sloan Management Review* 61–68.
11. Flender, John O. & Morse, Richard S. (1977). The role of new technical enterprises in the U.S. economy. Committee Report of the Select Committee on Small Business, United States Senate *Small Business and the Quality of American Life* (Washington: 1987). Government Printing Office, 72–90.
12. Birch, D. L. (1987). *Job creation in America: How small companies put the most people to work.* New York: Free Press.
13. Radosevich, R. (1993). New business development from New Mexico's federal laboratories. In Kenneth Walters (Ed.), *Technology commercialization: Innovative alliances for economic development.* New York: Roman and Littlefield, forthcoming.
14. Botkin, J. W. & Matthews, J. B. (1992). *Winning combinations: The coming wave of entrepreneurial partnerships between large and small companies.* New York: John Wiley & Sons.
15. Staley, J. L. (1991, Fall). Global technology-asset management: A new survival skill for the 1990s. *Technology Transfer* 29–33.
16. Smilor, R. W., Gibson, D. V. & Avery, C. M. (1989, Spring). R&D consortia and technology transfer: Initial lessons from MCC. *Technology Transfer* 11–22.
17. Barber, J. (1990, June). DOE official's role in license examined. *McGraw-Hill's Tech Transfer Report* 4.
18. Kassicieh, S. K. & Radosevich, H. R. (1991, Summer). A model for technology transfer: Group-decision support systems and electronic meetings. *Technology Transfer* 43–49.
19. Tyson, L. D. (1988). Competitiveness: An analysis of the problem and a perspective on future policy. In Martin K. Starr (Ed.) *Global Competitiveness: Getting the U.S. back on track.* New York: W. W. Norton, 99.
20. Pacific Northwest Laboratory (1990). The technology transfer process: *Background for the U.S. national energy strategy.* (PNL-SA-17482).
21. Lewis, R. C. (1990). CoCom: An international attempt to control technology. In Robert W. Harrison (Ed.) *Technology transfer in a global economy.* Indianapolis: Technology Transfer Society, 67–73.
22. Kinoshita, H. (1989, August). Mutual security and dual-use technology. *Speaking of Japan* 20–24.
23. Task Force on Management of Technology, Commission on Engineering and Technical Systems, National Research Council (1987). *Management of technology: The hidden competitive advantage.* Washington: National Academy Press, Report Number CETS-CROSS-6, pp. 19–20.

TECHNOLOGY TRANSFER IN A TIME OF TRANSITION

Beverly J. Berger

Federal Laboratory Consortium for Technology Transfer
Washington, DC 20005

This is a time of extraordinary transitions for the nation's federal laboratories. The defense laboratories, of course, have been most profoundly affected by the end of the Cold War, which development alters their missions radically. But nondefense laboratories and the agencies that fund them also must adapt to a changing environment. Widespread concern about American competitiveness in a global economy places a premium on efficient utilization of federally funded technological resources. A large and growing federal deficit squeezes laboratory budgets and forces laboratories to look for customers and partners outside the federal government. An energetic new Administration seeks ways to "reinvent" government at all levels and is open to creative proposals to reduce costs and increase productivity, while also advancing American technology. Congress always plays an important role in reshaping federal programs and facilities, and many members of Congress have a strong interest in issues related to technology and industry; the largest Congressional caucus is the Caucus on Competitiveness.

Private corporations also are changing in ways that affect federal laboratories directly and indirectly. Spurred by recession and intense domestic and foreign competition, American companies have been adjusting and restructuring in many ways, including a shift toward the service sector that sometimes causes concern about the future of American industry. Within the growing service sector, information services have expanded dramatically. Increasing awareness of the importance of vast quantities of information to business and government, and especially of the central role of computers in collecting, exploiting, and transferring information, has led many to conclude that modern societies are moving from an "industrial age" into an "information age."

The information at issue, of course, is much more than data on accounts, markets, demographic changes, and the like, which often are immediately and directly useful to governments and the private sector. It also includes important aspects of technology, which now moves rapidly across borders, profoundly affecting the global economy and American competitiveness. This kind of information is valuable primarily to the extent that it translates into industrial capabilities and manufactured goods. During our transition into a "post-industrial" era in which information increasingly shapes the economy, it is useful to keep in mind that manufacturing is still where the rubber meets the road. Photons and electrons don't put potato chips on the platter or computer chips in the CPU. Somewhere along the line, someone—or some robot—still must mill parts or bend metal or inject plastic.

From Lab to Market: Commercialization of Public Sector Technology,
Edited by S. K. Kassicieh and H. R. Radosevich, Plenum Press, New York, 1994

The quality of the technologies behind processes and products is critical to success in the market place, but quality is relative and transitory, because technology is volatile. Rapid advances in science and technology combine with the increasing accessibility of information about those advances to accelerate the pace at which new technologies become part of everyday life, in the business world as well as in consumer markets. The reciprocal of rapidly advancing technology, unfortunately, is rapid obsolescence, as competitors quickly improve on technologies or create new ones.

The availability of technology in electronic form will increasingly drive the intensity of competition and the rate of obsolescence. For example, United States patents now are available on CD ROM. Quick and inexpensive availability of those patents will increase the number of firms that use or exploit them and will inspire further inventions and patents. Already, other companies, especially many foreign firms exploit inventions by patenting improvements on them, often before the inventor recognizes the necessity of improvements to achieve commercial application. Broader and faster access to patents can only accelerate this phenomenon of "picket" patents.

In the area of intense global technological competition, the interests of American businesses and federal laboratories converge in ways that are not yet fully appreciated in either the private or the public sector. The barriers to fuller utilization of federally funded technology are largely cultural. This paper addresses the nature of technology transfer, factors that inhibit it, ways to promote it, and means of measuring progress reliably without intruding on the technology transfer process to the point of interference. This discussion is by no means exhaustive. Rather, it is intended to promote creative thinking and energetic action by those interested in advancing technology transfer.

TECHNOLOGY AS A SOCIAL PHENOMENON

To say that technology has become volatile is not to say that it is a fragile, wasting asset. In the right environment, it is robust and dynamic, because it develops through an interactive process, like fermentation or fission. Technology begets technology through communication. Face-to-face and elbow-to-elbow communication is the most effective kind, as California's Silicon Valley, Massachusetts' Route 128, and other technologically oriented communities demonstrate.

Technology is a social phenomenon, an expression of culture. To thrive and to contribute its full value to society, it requires social interaction. It must inspire further development or competitive technology, and it must be brought to public attention and judgment via the market place. What it must not do (with the possible exception of certain weapon technologies) is to molder in the offices of an R&D facility. Whereas a few decades ago, a technology might be pulled off the shelf, updated, and incorporated in a successful product, today musty technology is usually fatal to a product. In personal computers and their peripherals, for example, a year or two can constitute a generation. Personal computers may be an extreme example of rapid obsolescence, but they make the point: "use it or lose it."

Like commercial technology, the technology available from federal laboratories is more than data in computers or drawings on a shelf. It is knowledge in the minds of creative and thoughtful persons. It, too is dynamic; if it remains within federal walls, much of its potential value to society is lost, and if it emerges years after its development, it may be irrelevant. The task at hand is to increase and improve the channels through which federal technology flows into the marketplace, and to devise incentives to accelerate the cultural process of technology transfer, the process that puts federal technology to work enhancing American quality of life and global competitiveness.

The communication link between the federal laboratory and the citizens in the marketplace is private industry, and the common language spoken over that link is Tech Transfer. Today, about as many people are fluent in Tech Transfer as in Esperanto. Our aim must be to make Tech Transfer a workable second language for everyone in the federal laboratories and the relevant federal bureaucracies—and in companies, universities, and not-for-profit institutions.

The success of technology transfer is especially important to the United States, as the federal budget crunch requires that we get more bang for the federal buck. Concern for our international competitiveness also requires that we benefit fully from our massive federal investment in technology. Moreover, the defense drawdown brings technology transfer front and center as a major contributor to defense conversion.

FEDERAL INVESTMENT IN TECHNOLOGY

Innovation has been one of America's great strengths, and American technology still leads the world in many important areas, even though a substantial portion of the technology that Americans have conceived and developed has remained within the confines of federal laboratories and the federal programs for which it was developed. A vast national asset has been commercially underutilized.

Hundreds of billions of dollars have been invested in the physical and human infrastructure of the federal R&D establishment, and about $25 billion more is invested annually. Many important technologies have spun off the programs funded by that investment. Whole industries owe their start to defense and aerospace programs. Development of computers was driven by the need to perform complex calculations within days rather than years. High-speed photography was driven by the need to study the performance of explosive materials. Metallurgy and machining were pushed to new frontiers by the demand for precision crafting of exotic materials; the science of nondestructive testing was developed and adopted by industry. The products and technologies that have percolated from the defense community into the commercial world over the last 50 years are innumerable and invaluable. From 1980–1990 alone the U.S. increased its investment in basic and applied research almost three-fold.[1]

One might argue that even more could have been achieved by direct investment in technologies for civilian purposes, but the fact is that a dangerous world demanded massive American investment in defense technology. The high motivation of the scientists and engineers in defense laboratories and industries, however, probably drove them to levels of creativity and productivity that they might not have reached in strictly commercial enterprises during peace time. In any case, Americans and people around the world have benefitted from many of the technologies developed by defense programs.

While the world is far from safe and tranquil, the pressures of World War II and the Cold War have eased, and the pressures of global economic competition have increased. Economic strength may well prove to be the most important aspect of national security. The Soviet experience certainly suggests that an unproductive economy undermines even the most determined and strenuous efforts to build military power. Massive diversions of human and material resources to military projects clearly hastened the collapse of the Soviet empire.

In the American economy, the costs of defense were defrayed substantially by the private sector role in defense R&D and production and by technological spinoffs from defense programs (the old form of domestic technology transfer before the term was invented). The challenge of defense conversion is to reduce the burden of defense even further without weakening our military strength dangerously. Spinoffs can no longer be eventual or incidental. Today, they are an important objective of defense programs.

During the Cold War decades, spinoffs from defense laboratories not only were incidental, but often were delayed by security concerns. Perhaps those concerns were exaggerated.

Perhaps, as Dr. Edward Teller argued years ago, greater openness of information would have accelerated technology development far more in the efficient American economy than in the cumbersome command economies of totalitarian societies. However, government caution in releasing information that might have endangered the lives of American troops and the nation itself was understandable. Today, the time for emphasizing classification concerns over technological dynamism seems to be behind us.

Except for a few, generally weapon-specific, technologies, most defense technologies can enter the commercial market immediately. We worry a bit less now about the availability of that technology, in part because we have learned how inefficiently totalitarian societies use their intellectual resources. Ironically, free capitalistic societies now are seen as a greater source of concern. While the Soviet Union was staggering toward collapse, our allies in Europe and Asia were challenging the once secure positions of many American industries as technological and manufacturing leaders across a wide range of fields. Although our position is not desperate, clearly we have slipped a bit, and there is general agreement that the United States should reassert itself in areas where traditionally it has been strong.

Many experts have cited management deficiencies as the principal cause of the relative decline of American industrial strength. In at least one area, technology transfer, it is clear that the managers of American companies do face major challenges. As the National Academy of Sciences noted, "Our conclusion, based on the available data and information, is that the U.S. performs relatively poorly in technology adoption, in contrast to our industrialized competitors."[2]

American industry, of course, is responsible for incorporating technologies into products, but the federal government can lend a hand by improving accessibility to the immense national resource of federal laboratories. From all quarters, we hear that federal laboratories are national treasures and that they should become more valuable to American companies. The report of the National Science Foundation's National Science Board Committee on Industrial Support for R&D recommended that the federal government "expand programs that directly support technology transfer activities in Federal laboratories."[3] A report of the private-sector Council on Competitiveness stated that "the Federal laboratories constitute a wealth of technical talent, and some represent world-class facilities that do not exist in industry...individual labs have great resources that industry should leverage."[4] One of the Council's nine recommendations is that "the Department of Defense should establish an outreach program to make the R&D and technical expertise in DOD more accessible to civilian industry."[4] Robert White, former Under Secretary of Commerce for Technology, has observed that "in our national laboratories we have the resources and outstanding talent, but with traditional cultures and missions that are not consistent with national industrial technology needs...we need to find ways to tap these resources more deeply, directly, and effectively."[5] Chairman George Brown of the House Science, Space, and Technology Committee, who is a strong proponent of technology transfer, has stated that, "U.S. science policy has traditionally focused on facilitating the performance of research, while neglecting...the ways that research results are integrated into society and translated into societal benefits."[6]

One of the best ways that the federal government can promote the societal benefits sought by Chairman Brown, along with many other members of Congress and Executive Branch officials, is to accelerate the process of technology transfer. Clearly, the time is right to focus on making technology transfer from federal laboratories work.

DEFENSE CONVERSION: TECHNOLOGY TRANSFER IN FATIGUES

Intense international competition, major changes in the strategic environment, pressures on the federal budget, and an American economy in the process of restructuring, combine to enlarge the role of technology transfer. While American companies are looking for ways to deliver more, faster, and cheaper, the government is looking for ways to stretch a shrinking defense budget and to bolster the economy. One step toward making ends meet is known as "defense conversion", which has been defined as "the process by which people, skills, technology equipment, and facilities in defense are shifted into alternative economic applications."[7] This process bears a striking resemblance to technology transfer. In fact, some of us in the Federal Laboratory Consortium (FLC) believe that the new language of defense conversion is really Tech Transfer spoken with a Pentagon accent.

I do not mean to suggest that technology transfer is the entire answer to defense conversion. Approaches to defense conversion come in many shapes and sizes, ranging from reorganizing the government to requiring increased corporate funding of laboratories to learn which have value to the commercial sector.[8] One organization suggested that a quasi-public corporation be established for a $5 billion investment in pre-commercial R&D.[9] Each of these approaches has merits. The Executive Branch and Congress will choose among these and other proposals to create a strategy for defense conversion. "It is time for the U.S. government to master a new and unfamiliar role in helping the private economy develop and diffuse technology *explicitly* for purposes of enhanced economic performance."[10] What technology transfer professionals bring to this project is proven tools and proposed tactics. Federal laboratories can provide technical assistance, cooperative research, patent licensing, facilities, and personnel exchange, the elements that make up successful federal laboratory technology transfer. As we see it, the primary laboratory role in defense conversion is to expand and expedite technology transfer.

REFINING THE PROCESS OF TECHNOLOGY TRANSFER

The purpose of this paper is two-fold: to encourage American companies to exploit an under-utilized American resource, the federal laboratories, and to discuss some of the ways in which federal employees, from government officials to scientists and engineers at laboratory benches, can improve the technology transfer process.

There are three basic parties to the process of communication: the government (especially research, development, demonstration, testing and evaluation [RDDT&E] facilities and their federal sponsors) private industry, and citizens in the marketplace. Government policies and practices have strong effects not only on federal facilities, but also on corporate decisions. Private citizens in the marketplace, however, are notoriously independent, and they will be the final judges of our success. Those citizens, acting through the mediation of private industry, will pass ultimate judgment on the success of our technology transfer efforts. Early returns in the form of highly successful products (a few of which are mentioned below) are encouraging. Guided by such experience, and with private industry as a partner, the federal government must continue to refine policies and practices to encourage an even greater extraction of federal technology and expertise.

A principal federal instrument for promoting technology transfer is the Federal Laboratory Consortium for Technology Transfer (FLC), a grass-roots organization formed in 1974 to promote the rapid movement of federal R&D into the mainstream of the American economy. Formally chartered by Congress in the Federal Technology Transfer Act of 1986, the FLC consists of over 600 member laboratories and research centers in 16 federal departments and agencies. A major purpose of the FLC is to help match these federal resources to the needs of private-sector as well as state and local government partners. To this end, the FLC develops and tests technology transfer methods, addresses barriers to the process, provides training, highlights grass-roots technology transfer efforts, and emphasizes national initiatives where technology transfer has a role.

By law, every laboratory with more than 200 scientists and engineers is required to have an Office of Research and Technology Application (ORTA). Most laboratories have an ORTA, though it may be called by another name. The ORTA (a term that is applied both to the office and to the technology transfer officer who heads that office) has the lead responsibility in the laboratory for technology transfer. In most cases, the laboratory's ORTA is also its representative to the FLC.

Supported by the FLC's nationwide person-to-person network, FLC representatives help initiate industry-laboratory contacts and make the necessary arrangements for technology transfer from federal sources to nonfederal users. Within the FLC, a legal network that consists of professionals from member R&D laboratories, departments and agencies, and industry works with technology transfer officers. The FLC legal network provides mechanisms for exchanging technology transfer ideas, sharing experiences, preparing summary issue papers, and supporting FLC training and education efforts.

The FLC, working with other technology transfer organizations, is a major force in accelerating the cultural changes necessary for effective domestic technology transfer, and benefits all who participate. The Administration gets input from the FLC into its policy making processes. Congress gets the information it seeks as it considers new legislation and evaluates implementation of existing legislation. Laboratories and their supporting departments and agencies, of course, benefit directly in innumerable ways. Most importantly, American companies benefit from the ability to match their needs to laboratory capabilities and from interaction with one another and with federal laboratories, and this benefit is tangibly manifested as competitive success in the international marketplace. Matching is done through the FLC Laboratory Locator Network which helps users identify and access federal R&D laboratories and centers by matching requests for technology, expertise, and facilities with the appropriate federal laboratory capabilities.

If wisdom may be defined as knowledge tempered by experience, then the "old hands" of the Federal Laboratory Consortium have a modest claim to wisdom in technology transfer, and they willingly share it with newcomers. Other practitioners seeking to improve their technology transfer skills, or just beginning a technology transfer program, can network within the Federal Laboratory Consortium, thereby drawing on the experience of others while making contributions of their own.

The FLC has learned that technology is transferred most effectively when potential users of federal technology can deal directly with the persons who have the needed expertise and capability. As discussed below, some important aspects of technology are not captured on paper or in computer files. For this reason, the FLC emphasizes networking both within the FLC and between the FLC and the private sector, to convey the rich information of human experience that must supplement recorded data in order to achieve genuine technology transfer. Almost twenty years of FLC networking has produced a wealth of insights into the dynamics of technology development and transfer throughout the United States.

For those new to defense conversion and technology transfer from federal laboratories, there is one very clear message:

> Try to avoid ... reinventing the wheel. Ensure that those engaged in defense conversion draw on the experience of professionals who have been working technology transfer for years. By tapping into our wealth of knowledge and expertise, they can accelerate the process and increase the efficiency of defense conversion. Building on FLC experience will be to the benefit of the Department of Defense laboratories, as well as of the laboratories of other departments and agencies and of the nation itself.[11]

Accepting the Challenge

Technology transfer professionals understand that there are cultural mismatches between industry and government, and that most companies have considered the benefits of working with federal laboratories to be outweighed by the burdens of dealing with the government. Times have changed, however, and old cost-benefit calculations may no longer apply. Company representatives will find it worth the effort to reach out to federal laboratories, where technology transfer professionals will help them bridge the cultural gap.

A Transition for Technology Transfer

The Executive Branch has struggled with technology transfer for years. The priorities of the Cold War, acute sensitivity to conflict of interest, and perhaps sheer weight of habit have combined to maintain a cultural style of holding technology close. This inertia seems likely to be overcome soon. The Administration, with the support of influential members of Congress, not only avows commitment to technology transfer but shows enthusiasm for fresh approaches. A technology report issued a month after President Clinton's inauguration stated that:

> This Administration will modify the ways federal agencies do business to encourage cooperative work with industry in areas of mutual interest...at the level of technology development, the fundamental mechanism for carrying out this new approach is the cost-shared R&D partnership between government and industry. All federal R&D agencies (including the nation's 726 federal laboratories) will be encouraged to act as partners with industry whenever possible. In this way, federal investments can be managed to benefit both government's needs and the needs of U.S. business.[12]

The report continued: "Agencies will make it a priority to remove obstacles to Cooperative R&D Agreements (CRADAs) and to facilitate industry-laboratory cooperation through other means."[12]

The Administration also made a commitment to "greatly increase nondefense Cooperative Research and Development Agreements (CRADAs) at the national labs."[13]

In March, 1993, the President announced the $472 million Technology Reinvestment Program, which is part of the President's Defense Reinvestment and Conversion Initiative. An October White House press release announcing matching TRP grants for 41 projects stated that:

> By encouraging the development of dual-use technologies in the post-Cold War economy, the TRP will help guarantee a strong national defense, support the growth of technologically sophisticated American firms, and create good, new jobs. The TRP will also provide technology assistance to small firms, including defense firms making the transition to commercial manufacturing.[14]

The largest of the TRP grants announced in October was for a project called "Technology Access for Product Innovation" (TAP-IN), which will help small businesses dependent on defense contracts make the transition to producing products for the private sector. TAP-IN was proposed by a consortium of the U.S. Chamber of Commerce's ChamberTech, NASA's six regional technology transfer centers, the Industrial Designers Society of America, the FLC, and three of the nation's largest accounting firms.

Secretary of Defense Les Aspin has strongly endorsed technology transfer in the context of defense conversion:

> Technology transfer is an important part of this Administration's plans to help revitalize the American economy. We all know that defense research and development has been immensely successful in providing the men and women of our Armed Forces with the highest quality weapons and equipment.

> Unfortunately, our nation has not been as adept as realizing the potential for private sector uses of the technologies developed in our military laboratories. . . Transferring technology can pay dividends to both defense and the commercial sector. Our military wins by continuing to have the benefit of defense laboratory research. Through the success of "dual-use" projects, our economy wins as well, providing private sector jobs and a higher standard of living for many more of our fellow citizens.[15]

Among practitioners of technology transfer, these statements have created a great sense of optimism and expectation. If we correctly gauge the intent and resolve of the Administration, the policies will be followed by demands for institutional changes to produce substantial results soon. Discussions with laboratory representatives of the FLC and with laboratory technical directors, departmental and agency technology transfer officials, and others lead to the conclusion that nothing less than broad cultural change is needed to achieve the full potential of technology transfer. The main purpose of this paper is to encourage discussion of ways to gauge and accelerate the cultural change that is under way.

THE ROLE OF LEGISLATION IN CULTURAL CHANGE

Legislation and executive orders already provide impetus and flexibility to technology transfer. Only minor legislative refinement is needed to smooth the processes. Passing legislation, however, can be the easy part. The hard part is creating within the government and the laboratories an environment in which the purposes of that legislation are achieved. As a recent report noted, "We are not fundamentally limited by legislation; we are limited by our will."[16]

Because of cultural impediments, full implementation of existing laws and executive orders challenges the Executive Branch. Technology transfer legislation encourages people to engage in practices that would have been improper or illegal not long ago. Government and laboratory employees are learning new behavior within these new laws. Caution is the conditioned, or acculturated, reflex of government officials and technologists asked to take actions to benefit industry. Industry, of course, has the mirror-image of that reflex. Both reflexes will change as the cultures change. We cannot expect individuals to risk getting too far ahead of their cultures. Fortunately, the cultural change that has been under way for a few years is accelerating.

MEASURING PROGRESS AND SUCCESS

Whenever a major national objective is established, the need arises for measuring progress toward that objective. Having established the objective of increased technology transfer, the government is interested in discriminating between those organizations that are effective in technology transfer and those that are not. The executive and legislative branches of the federal government must monitor performance for budgetary, political, and other reasons. In the case of technology transfer, private companies also will monitor performance, if only to determine whether they are missing an opportunity.

Measuring progress in technology transfer will require creativity. How do we know whether cultural change is really taking place, both in the government and in the private sector? What objectives will provide both standards and incentives for technology transfer? How do we measure the process of technology transfer without distorting or impeding it? This paper addresses three categories of measures. Measures of *output* of technology transfer and of quantifiable technology transfer *activities* are commonplace. *Intermediate indicators* of cultural change may also prove useful.

The Defense Conversion Commission has suggested measuring output in terms of "the number of jobs created, the number of new companies formed, and increases in productivity, revenues, and profits of private companies," in addition to the number of patents issued.[17] Although it is tempting to focus on such measures of technology transfer output as the value of federal laboratory technology and expertise to industry, these are the most difficult measurements to make. Value may be added at each of the many steps from the laboratory to marketing a commercial product, and it would be difficult to determine what portion of that value to assign to the underlying technology. Companies are not accustomed to making such measures, let alone publicizing them. More importantly, such measures are not available until a new technology is incorporated into a marketable product, which may be years hence. Similarly, revenues from patent licensing agreements (PLAs) provide good, quantitative lower-bound measures of value to companies, but do not provide timely information. Only with technical assistance programs can the value to the private sector ever be estimated quickly. Clearly, such historical insights are less valuable than near-real-time measures that can guide improvement of today's technology transfer processes.

The second approach is to measure, not output, but level of effort of technology transfer activities.[17] Such measures customarily include the total number of CRADAs[18,19] (about 100 through FY 1988 and almost 1500 to March 1993[20]), the number of consortia, and the total number of patent licensing agreements (PLAs). (It must be recognized that measures of CRADAs and PLAs, while useful, will always be lower bound estimates of technology transfer.)

Less common are normalized activity measures, such as the per capita measures at each laboratory of invention disclosures, patents applied for, patents issued, PLAs, and CRADAs. (A current benchmark is at least one CRADA for each 10 scientists and engineers.[21]) Another useful per capita measure in the future may be number of personnel exchanges, as these become more common. Another simple measure is the per cent of time each laboratory researcher spends on technology transfer.[16] These quantitative measures would provide good, current annual measures of activity, especially when normalized and computed per capita based on the numbers of scientists and engineers at each laboratory. They might prove even more revealing when normalized for funding. Such normalized activity or level-of-effort measures would be useful to the private sector, as well as to departments and agencies seeking to make appropriate comparisons among laboratories or across agencies.

The proposed new category of measure, intermediate indicators of cultural change, is less quantitative than output or level of effort but may prove more useful. We need intermediate indicators to gauge the rate of improvement in the technology transfer process and to provide guidance on how the process might be improved. (Some possible intermediate indicators are described below in association with personnel practices, licensing strategies, and foreign intellectual property protection.) Because we are seeking cultural change, we need real-time feedback on how the culture is adapting. Technology transfer performers also need to know how they are doing so that they can adjust their performance before being judged by the quantitative standards of their funding organizations and the Congress. This is particularly important because we have been asking those performers—the directors and staff of federal laboratories, their federal sponsors, and the private sector—to adopt a new way of doing business that conflicts with many aspects of the old way. They must be able to recognize when they are making progress and to convince others of that progress.

In the future, we will add to and refine these measures as contributions are made by practitioners—government, companies, universities, not-for-profit organizations. We will need new ways to measure and improved ways to motivate and speed cultural change across institutions.

LABORATORY-AGENCY ISSUES

The FLC and its network of technology transfer professionals is a shared national resource that is constantly enhanced by the experience of its member laboratories and their FLC representatives. Some laboratories have a strong history of successful technology transfer and have much to contribute to those just getting started. Although laboratories may be at different stages in their cultural transition, each can contribute. Useful measures can be derived from these experiences.

Each of the issue discussions that follow is drawn from experiences at FLC member laboratories and provides insight into the complex process of technology transfer and measures of success. Sharing what might be called "best practices" among member FLC laboratories substantially speeds domestic technology transfer.

Protecting Intellectual Property

> Intellectual property is a generic term which applies to any invention, discovery, technology, creation, development, or other form of expression of an idea that arises from research activities whether or not the subject matter is protectable under the patent trademark, or copyright laws.[22]

Appropriate protection of intellectual property is an integral part of any federal laboratory technology transfer program. It is a fact of business life that private firms develop or acquire new technology for the purpose of gaining or maintaining a competitive edge. They are unlikely to make the investments in product development, manufacture and marketing necessary to bring that technology to market if they can't protect their intellectual property. Federal involvement used to create a problem; the old attitude toward government-funded R&D, was that, since the public paid for it, it should remain in the public domain. However, when government-developed technology was equally available to all, it was used by none, or almost none. To paraphrase the familiar comment about responsibility, when everyone owns something, no one owns it.

Intellectual property is the currency of technology transfer. It also is a renewable national resource. Technology transfer both draws on and helps to replenish that resource. However, intellectual property is not inexhaustible. Like other renewable resources, it must be managed properly and not squandered. Unfortunately, all too often federally funded intellectual property is disseminated with little consideration of the consequences for the nation's competitive position. Although some laboratories capture almost all inventions, managers at many laboratories confide informally that 50–90% of inventions are not disclosed. (Preparation of a written disclosure of an invention is a necessary first step toward getting a patent.) Private-sector visitors to some laboratories confirm that they find much of interest that could have been patented (and sometimes still could be) but was not.

Scientists and engineers at all laboratories need to understand what constitutes an invention, how to make a proper disclosure, and the general process for obtaining a patent.[23] Scientists, whose career advancement traditionally has been linked to publication, especially need to be motivated to file patent applications on inventions that form the basis for their publications. Early publication in the scientific literature can destroy the ability to protect intellectual property, thereby reducing the likelihood of corporate investment in those inventions. Quick publication of results should be preceded by even quicker disclosure and filing for patents to avoid depriving American taxpayers of returns in the form of patent royalties or, more importantly, jobs.

Laboratory technology transfer professionals and their counterpart intellectual property attorneys, working as teams, can perform valuable service by enlisting laboratory scientists, engineers, and managers in treating intellectual property as the national resource it is.

Measuring Intellectual Property Protection

Intermediate indicators of increased protection of intellectual property may provide means of assessing the effectiveness of technology transfer efforts at federal laboratories. Useful per capita indicators (based on numbers of scientists and engineers) could be number of invention disclosures, patent applications, and patents granted. These also become indirect measures of the adequacy of the intellectual-property legal support provided by the laboratory and/or agency.

Unfortunately, most federal laboratories have not been very successful in getting inventors to make disclosures routinely. Some professionals engaged in the development end of R&D may be so intent on completing a project or meeting a deadline that they overlook opportunities to patent.[24] According to a GAO report, the provision for royalty sharing for federal laboratory scientists and engineers is inadequate motivation to increase interest in patenting.[25] This report recently was misinterpreted by a normally thoughtful and reliable columnist, Michael Schrage, as evidence of the failure of the federal laboratories "to spur commercial lab innovations," rather than as indicating that the existing monetary incentive alone is inadequate to create the necessary cultural change.[26] In fact, cash awards for making disclosures and filing patent applications at some laboratories are so modest that many employees find it not worth their time to fill out the forms necessary to obtain patents. In contrast, American companies often provide more generous rewards for making invention disclosures and getting patents. Companies can identify patents that contribute to corporate profitability, and inventors may be well rewarded by advancement.

In addition to financial incentives available from laboratories, the federal government now provides its inventors with a minimum of 15% royalty sharing. However, the odds that a single patent will earn an inventor significant royalties are slim. The chances of earning royalties increase dramatically when the inventor holds many patents.[27]

The laboratory intellectual property attorney and technology transfer professional, working as a team, are the natural purveyors of information about intellectual property, its importance in technology transfer, and writing disclosures and patent applications.

This is one of the areas where more laboratory Office of Research and Technology Applications (ORTA) staff and intellectual property professionals need to work together. As the chairman of the FLC Legal Networking Subcommittee has observed: "If the goals of the Technology Transfer Act are to be achieved, then lawyers and ORTAs must form a partnership based upon mutual respect. Each must recognize that the other is an equal member of the Laboratory Director's team."[28]

Achieving Foreign Patent Protection

Regarding the importance of world-wide patent protection, an October 1992 report by the United States Patent and Trademark Office (USPTO) observed that:

> Trends in patent activity also point to the increasingly international character of patent activity in the U.S. The fact that nearly half of all utility patents are now granted to foreign inventors is a testament to increasing levels of inventiveness abroad and to the importance of international trade, rather than a decline in U.S. inventiveness.[29]

Just as foreign inventors patent in the United States, it is important for American inventors to patent abroad. Technology that has patent protection only within the United States has much less commercial value. In a global market, foreign patents may be appropriate for any technology of significance. However, foreign patents may be precluded by premature public disclosure of inventions. Unfortunately, laboratory scientists and engineers are accustomed to gaining recognition through publication and rarely consider foreign patents. Consequently, much federal technology cannot be patented abroad.

To be a full player in technology transfer, agencies and laboratories need a strategy for foreign protection of intellectual property, including guidance on publication to prevent compromise of intellectual property rights and assistance in the complex and expensive process of applying for foreign patents. Such strategies also would provide guidance as to which countries should be considered for foreign patents of specific technologies. Existence of such strategies within an agency or laboratory is an indicator of sensitivity to the role of intellectual property protection in a global marketplace. The ability to implement a strategy is another indicator of adequacy of intellectual property legal support. The government has not yet developed a unified, comprehensive strategy on foreign patent protection.

Developing Good Licensing Practices

Related to the issues of both patents and copyrights is the matter of licensing practices. The government may grant exclusive patent licenses either separate from or in conjunction with CRADAs. To foster the widest possible use of technological advances throughout the economy, it may be wise to consider licensing a technology only for a specific field of use. Exceptions could be made for companies that manage and sub-license intellectual property very aggressively, thereby accomplishing the purpose of maximizing benefit to the society.

A good argument for this approach to licensing is made by the now classic example of Biobarrier™, a root barrier developed at Pacific Northwest Laboratory to protect low-level radioactive waste-disposal sites from root intrusion.[30] Biobarrier™ uses controlled-release chemicals to inhibit the roots of erosion-control plants from penetrating into such areas as uranium-mill tailing piles. A patent was filed in 1981 on the chemical technology, and dozens of firms were contacted. To date, three companies have been issued separate exclusive licenses for different fields of use: in sewer gaskets and irrigation lines, as caulk, and as a geo-textile placed under sidewalks and against structures. Nine other applications are under development and more may be found. Had the first interested company obtained an exclusive license for all uses, the other applications might have been missed.

Licensing arrangements, including the means by which exclusive licenses are granted, may be another possible intermediate indicator of the effectiveness of a laboratory's technology transfer program.

Accelerating Cultural Change Throughout the Federal System

Cultural change in the laboratories is critical to protecting intellectual property and advancing technology transfer. A recent NASA report describes the current situation. About a year ago, the NASA Administrator assigned a team drawn from NASA staff to "assess the NASA technology transfer processes and to make recommendations for improving those processes." In December 1992, that team submitted a thoughtful report that is well worth reading.[16] It noted, inter alia, that:

> Unfortunately, all too often, NASA employees, managers contractors, and grantees do not feel technology transfer is part of their job. Researchers are more inclined to view their job as technology development, and that the effort required to transfer technology is time and resources taken away from their 'real' job. NASA management has fostered this view, first by not actively encouraging and rewarding technology transfer activities, and second by discouraging some types of technology transfer activities, especially secondary targeted activities.

NASA should be commended for commissioning and releasing this candid report, which many other federal agencies and departments could take to heart. Fortunately, the means for accelerating the necessary cultural change are at hand.

The Federal Technology Transfer Act (FTTA) of 1986 requires that: "Each laboratory director shall ensure that efforts to transfer technology are considered positively in laboratory job descriptions, employee promotion policies, and in evaluation of the job performance of scientists and engineers in the laboratory."[31] The purposes of the law would be served much better if the requirement for evaluation against technology transfer standards reached up to the laboratory director and beyond. (At the September 20–23, 1992, meeting of the 46th National Conference on the Advancement of Research [NCAR-46] in Ruidoso, New Mexico, company representatives independently suggested using performance standards as a means of motivating change within government.)

Active participation of laboratory directors (that is, genuine leadership) is critical to effecting cultural change.

There are many ways for a laboratory director to exercise leadership beyond simply advocating technology transfer. One of the most effective ways to energize scientists and engineers and accelerate cultural change is to comply fully with the FTTA requirement to use technology transfer standards in evaluating and promoting laboratory staff. Highly motivated, mission-oriented government laboratory employees need reassurance that technology transfer will actually enhance their performance, not detract from it. Skepticism abounds regarding the effect of technology transfer activities on laboratory performance and careers.

Unfortunately, many laboratory directors have yet to implement this provision of the FTTA, even minimally. In some cases, this is because the director has not been unequivocally told by the funding agency that technology transfer is an integral and important part of the laboratory's mission. Without such guidance, laboratory directors sometimes conclude that technology transfer undermines the execution of the laboratory's central missions. For example, some have expressed concern that private companies will use technology transfer to skim the cream of laboratory talent. An active technology transfer program, however, can help to retain or recruit talented persons. Measures of other benefits of technology transfer to the laboratory are needed so that, when the day comes that a fine scientist or engineer is lost to the private sector, the laboratory director has convincing evidence of the many accrued benefits that outweigh the loss.

A laboratory director who believes that technology transfer is not in the best interest of the laboratory is unlikely to be persuaded by assertions that technology transfer serves wider interests, such as American competitiveness and prosperity. Exhortation rarely prevails over self-interest, and self-interest rarely extends much beyond an office or an institution. The FTTA points the way to a time-tested tool for shaping a federal employee's perception of self-interest: the performance evaluation. Like the scientists and engineers who report to them, the laboratory directors and their funding organizations can be evaluated against standards of performance that make technology transfer an integral part of each laboratory's mission.

It might be instructive to examine the personnel practices at laboratories with exceptionally good records in technology transfer to determine the extent to which they include technology transfer in performance evaluations for managers as well as scientists and engineers. Dr. Dwight Duston, a well-regarded federal R&D manager from the former Strategic Defense Initiative Office (SDIO), now Ballistic Missile Defense Organization (BMDO), requires that all his subordinate managers have technology transfer as part of their performance evaluation criteria and expects them to devote ten percent of their time to technology transfer. Duston's program has an excellent record in technology transfer. Out of a 125 million dollar R&D budget, he sets aside 2.5 million dollars for technology transfer.[32]

Application of technology transfer standards to the entire chain of command is important not just to promote technology transfer, but also to treat laboratory staff equitably. If any link in the chain of command is omitted from these requirements, a conflict is established between those below who are promoting technology transfer and those above who see it as a distracting side effort. Such possible conflicts can be resolved in advance by performance standards throughout the chain of command that establish the proper role of technology transfer. If everyone whose job is related to technology—whether at a laboratory, in a field office, or in a headquarters office of program management, procurement, or legal counsel—were rated in part according to success in promoting domestic technology transfer, the result would be a genuine team effort to achieve technology transfer. At NASA's Marshall Space Flight Center in Huntsville, Alabama, where technology transfer has been included in performance standards of laboratory scientists and engineers since 1990, the ORTA reports that the center's technology transfer efforts have been aided considerably by conforming with the law.[33] The center has responded with a high degree of customer satisfaction to an increasing number (more than 400 in fiscal year 1993) of requests from companies as a consequence of vigorous outreach which includes technology transfer agreements with the governors of seven states and initiatives with local Chambers of Commerce. The center also includes technology transfer in job descriptions and performance criteria of managers; new managers are required to complete technology transfer training before being placed permanently in management positions. The center's Senior Executive Service members soon will be treated similarly.

At the Hanscom Patent Prosecution Office, technology transfer is included in performance appraisal plans of those attorneys and other personnel whose duties include technology transfer.[34]

Consistent, pervasive use of motivational devices would provide another indirect but quantifiable intermediate indicator of the progress of laboratory technology transfer efforts.

Training to Accelerate Cultural Change

Training of personnel is a critical element in promoting technology transfer within an agency or federal laboratory. One of the recommendations that came out of NASA's study on technology transfer is that "Each center (laboratory) should provide technology transfer training for all employees."[16] Some federal agencies are putting training programs in place. For example, the Department of Energy, supported by the Oak Ridge Institute for Science and Education, is developing training programs for technical managers, technology transfer personnel and others involved in the transfer process.[35]

To support all federal agencies, the Office of Personnel Management (OPM) has made a module on technology transfer part of its two-week training program, "Science, Technology, and Public Policy" for government managers and executives. Recognizing the importance of technology transfer, OPM also has offered for several years a separate two-week course on managing federal technology transfer, "Technology Transfer and the Management of Federal R&D."[36] For both courses, OPM brings to its classes expert practitioners to share their experience.

The FLC is Congressionally chartered to provide training, and does so in conjunction with every one of its semiannual meetings. A full day is now devoted to "Fundamentals of Technology Transfer," which include such topics as ORTA responsibilities, outreach, and intellectual property law. Laboratory, agency, and company representatives attend these sessions in ever-increasing numbers. FLC regional meetings often have training sessions. For example, a recent meeting of the FLC Southeast Region included a half-day workshop for laboratory representatives, regional technology transfer staff, and personnel from state affiliates. The workshop featured a visiting entrepreneur with a product already on the market.

Attendees spent the morning brainstorming additional market applications for the underlying technology and ways that laboratories in the region could help with product development.[37]

The FLC Washington DC office also offers opportunities for two to four weeks of on-the-job training for laboratory technology transfer professionals interested in the Washington perspective on technology transfer. Longer professional development opportunities also are available through a competitive FLC Fellows program, which provides for special projects to be carried out over 6–12 months in the FLC Washington office.

Laboratory and agency managers committed to doing a first-rate job of technology transfer find ways to train their employees, often employing innovative approaches. Such training becomes another intermediate indicator of cultural change and enhanced technology transfer.

Making Awards for Excellence in Technology Transfer

The Federal Technology Transfer Act of 1986 (PL 99–502) directs the head of each federal agency making a significant R&D investment in government-operated laboratories to develop and implement a cash-awards program to reward scientific, engineering, and technical personnel for exemplary activities that promote the transfer of knowledge and technology. The United States Forest Service, which has a long tradition of working well with American industry, established the Chief's Award for Excellence in Technology Transfer in 1988; the first award was made the following year. This award may be made to as many as three Forest Service employees or groups of employees. Recipients are eligible for the unique Chief's Technology Transfer Plaque and a cash award of up to ten thousand dollars. This cash award is made when the winner has provided "exceptional value" with a "general benefit" applicable nationwide. The Chief's Award is just one of the motivational tools employed by the Forest Service to promote technology transfer.

The FLC has three national awards for service, contributions, and support of the FLC—the Harold Metcalf Award, the Representative of the Year Award, and the Outstanding Service Award. Each of the six FLC regions has four annual awards: Regional Laboratory Award Industry/Nongovernment/University Award, The Regional Coordinator's Excellence Award (to an FLC laboratory representative), and the Regional Appreciation Award. The winners of the latter two awards are nominated automatically for the corresponding national awards. The FLC also gives up to thirty awards a year to laboratory scientists and engineers who have demonstrated exceptional creativity and initiative in transferring technology with significant benefits to companies or state and local governments. The FLC awards usually are accompanied by letters of appreciation from department and agency heads, presented by their representatives at the FLC awards ceremonies. Some federal laboratories further honor FLC award winners with ceremonies, cash awards, and recognition in centrally located, permanent displays.[38]

Other professional societies also give awards for outstanding contributions to technology transfer, such as the Technology Transfer Society's Thomas Jefferson Award for individuals and the Justin Morrill Award to organizations.[39] Increasingly, recognition is being sought for those who make significant contributions to technology transfer. Recently, the Institute of Electrical and Electronics Engineers-United States Activities (IEEE-USA) approved the creation of the IEEE-USA Electrotechnology Transfer Award "to honor an individual(s) whose contribution(s) in a key government or civilian role led efforts to effectively transfer/apply federal or state sponsored developments in advanced electrical, electronic, and computer technologies to successful commercial sector opportunities."[40] This award was developed by the IEEE-USA's Defense R&D Policy Committee under the chairmanship of Dr. John M. Walker.

Positioning the Technology Transfer Officer

One way to speed cultural change is to position the technology transfer office for high, positive visibility, such as in the offices of a laboratory director or a Secretary or Assistant Secretary. Some laboratories have designated associate directors as technology transfer leaders. Examples are DOE's Ames Laboratory and FDA's National Center for Toxicological Research. At USDA's Southern Regional Research Center, the center director himself has taken the lead. Such prominent positioning of the technology transfer function is an indicator of the importance of technology transfer to the organization and of progress in achieving the cultural change necessary for successful technology transfer.

Making Efficient Use of Laboratory Outreach Personnel

Laboratory scientists and engineers tend to be introverts. Their nature is to focus on their work and interact chiefly with other scientists and engineers in furtherance of that work. More gregarious or extroverted persons gravitate to jobs associated with laboratory outreach, such as education, technology transfer, and Small Business Innovation Research (SBIR). Laboratories with vigorous outreach programs usually aggregate these programs to take advantage of opportunities for interaction and synergy. This can be beneficial, as one program's contact in the world outside the laboratory gates may lead to contacts for other programs. Cross-fertilization also can occur within the laboratory. For example, SBIR grant winners can be brought into the technology transfer program, to the benefit of the companies and the laboratory, as CRADA partners during SBIR Phase III.[41]

Conversely, SBIR program solicitations can be written to stimulate broader application and faster commercialization of a laboratory's already patented inventions. (A resourceful laboratory, aware of its own limitations, can also engage a university licensing office to sublicense laboratory inventions licensed to it.)[41] The Armstrong Laboratory (AL) has an SBIR project developing a technology transfer workstation (TTW) customized to meet the requirements of the AL ORTA. The TTW will provide a means of consolidating already developed tools and adding some new ones. Among the existing tools likely to be incorporated is the software known as the Invention Tracking System (ITS) developed under contract to the National Institutes of Health (NIH), which can be used to track inventions through the patenting and licensing processes. Upon Phase II award, a "tool kit" will be developed for use by programmers to customize the TTW software for use by ORTAs at other laboratories.[42]

At Los Alamos National Laboratory, the technology transfer program and the laboratory's economic development program are run out of the same office.[43] Agencies also can increase effectiveness by grouping related activities, as is done by the Defense Department's Advanced Research Projects Agency (ARPA), in combining technology transfer and SBIR.[44]

How outreach programs are organized can serve as another intermediate indicator of an organization's commitment to accelerating cultural change.

Making Laboratory Inventors Accessible

Support by the entire chain of command may enable the laboratories to meet another challenge: that of making the federal laboratory inventor available to a company that wants to license a federal laboratory patent. In many cases, regular assistance from the inventor is necessary for quick and efficient completion of the additional work necessary for commercialization. At the Lincoln Laboratory of the Massachusetts Institute of Technology (MIT), inventors are made available to licensees one day a week for a maximum period of 18 months. They are paid at their usual rate by MIT's Technology Licensing Office, which is reimbursed by the licensee.[45] Many federal laboratories have yet to find a comparable solution to this critical problem.

One way to make the inventor available is to develop a CRADA in conjunction with a patent licensing agreement so that the necessary bench-level collaboration may take place between the inventor and the licensee.[46] The private company may reap additional benefits because a CRADA may provide that any patentable invention growing out of the collaboration be assigned or exclusively licensed to the private-sector collaborator who licensed the original federal laboratory patent. A nonexclusive license for a government laboratory patent may suffice for a company that can work with the inventor under a CRADA as it develops a product and makes new, patentable inventions to which the company has exclusive rights.

Making Laboratory Expertise Accessible as Technical Assistance

As impressive as the technology, equipment, and facilities at federal laboratories are, by far the greatest resource is the experience and knowledge of the employees. As a laboratory professional said, "it is my belief that the greatest asset the (laboratory) has is the talent pool which is trained and experienced in forming multi-disciplinary teams to attack new problems."[47] Since federal technology travels on two legs, facilitating interaction between federal scientists and engineers and private-sector companies is critical. Technology transfer requires that laboratory staff meet and develop relationships with the private sector. Travel funds pay large dividends when laboratory scientists and engineers attend technical meetings, visit other laboratories, and generally expand their technical interactions beyond laboratory walls. Yet, travel budgets are among the first cut when funds are tight.

Enlightened laboratory management dedicated to technology transfer understands the importance of customer satisfaction. Private companies seeking federal technology always should be satisfied with the service they receive, even if the expertise they seek is not available in the first laboratory to which they turn. Good service will ensure repeat business, which is essential to successful technology transfer. To this end, the FLC provides the Laboratory Locator Network. Every federal laboratory is an entry point to that network, so every federal laboratory should be able to satisfy its customers either directly or by referring them to other laboratories with the needed expertise.

These continuing relationships involving technical assistance sometimes lead to more complex, cooperative relationships; they certainly provide federal laboratories with opportunities to benefit the American economy.

Working with Universities and Not-for-Profit Organizations

Although federal laboratories currently receive about two and half times as much federal funding as universities, universities conduct a disproportionately high amount of the nation's basic research, and American companies often turn to universities and other not-for-profit research institutions for technology. More often than not, the links among technologists at companies, universities, not-for-profit research institutions, and federal laboratories are weak. As more connections are made and collaborations become more frequent, American companies will be in a better position to use all of the federally funded technical resources to improve their products and processes and to contribute to the American economy.

The Pacific Northwest Laboratory (PNL), operated for the Department of Energy by Battelle Memorial Institute, is unique in its ability to work with companies, in part because the contract with DOE also allows PNL employees to do contract research for corporate sponsors in addition to their work for the Department of Energy and other federal agencies. Hence the expertise gained in doing work for the federal government can be transferred readily to the private sector.[48]

Assessing and Marketing Laboratory Capabilities

Before a laboratory can begin to market its technologies, it must perform thorough assessments to understand the laboratory's technical strengths and to identify which technologies are most likely to be commercially valuable. One highly marketable service a laboratory can offer is problem solving. Laboratories that provide technology assistance to companies in solving product and process malfunctions often find that it builds a confident client relationship that can lead to greater utilization of the laboratories' resources and to more effective technology transfer.[49]

Development of a marketing plan is the next critical step. Who is the market? How can that market be reached? These are questions that often are not asked properly or answered fully, even in the private sector; some businesses simply assume the answers, then fail when they prove wrong. Good answers to these questions are equally important to the federal organizations that deal with the private sector on technology transfer. The government is hardly full of experienced marketeers with refined intuition about the market; the closest most laboratories get to marketing is when requesting laboratory funds from headquarters.

How can federal laboratories learn about market demands? Answers are emerging from the experience of hundreds of laboratories and, we expect, will be shared broadly. Some laboratories have found that promotional packages and brochures that detail laboratory capabilities, facilities, and expertise are useful for informing the private sector. Specific technologies available for licensing can be advertised. "Industry days," participation in expositions and other occasions to showcase laboratory technology also help spread the word while also fostering marketing skills. One laboratory issued a video tape as part of a press release about a CRADA signing. The video featured embedded computing, an area where the laboratory is seeking more corporate and academic partners.[50] The opportunities these gatherings provide for gaining feedback from the private sector are valuable. Through advertising, meetings, and word of mouth, industry and academia can be invited to work cooperatively with the laboratory.

How well the laboratory carries out technical assessments and markets its technology provides another intermediate indicator of progress. One measure of marketing success is the number of incoming telephone calls from potential laboratory customers and the number of referrals from other laboratories and organizations (a measure of "connectivity" with the technology transfer network). Getting an estimate of the number of technical meetings attended and the number of corporate-laboratory visits, both at the laboratory and outside it, is simple. These measures can be normalized for funding or number of employees. Another way of increasing private-sector and community awareness of a laboratory is to donate excess equipment to local educational institutions, using the new legislative authority.[51]

The Role of Industrial Designers in Technology Transfer

Another approach to determining the market for a laboratory's technology is to engage industrial designers in evaluating applications and markets for that technology. In the laboratory science and engineering are brought to bear to accomplish a (usually) well defined task. Turning a raw, task-focused technology into a well-rounded commercial product that is marketable, user-friendly, and cost-effective to mass produce requires a thought process very different from the one that creates the technology in the first place. Industrial designers specialize in that thought process.

As nontechnical inventors, industrial designers are always looking for a better way to get something done. They value technological innovation and are quick to see commercial applications of technology. Also, because they often consult, they may be involved within a short time in dozens of product areas, some of them highly technologically sophisticated. Hence, they promote cross fertilization.

Industrial designers consider first and foremost how a user will interact with the product. That perspective and constant contact with the market produce an unusual combination of inventor and marketeer who can provide valuable service to a laboratory as a technology sleuth. As a popular business magazine noted, "Another stubborn problem that design can help solve is how to transfer new technology out of the labs and into the market...Because they can identify customer needs, designers can also do a better job than engineers of picking out technologies, hidden away in the R&D labs, that can solve real-world problems."[52] Moreover, trained to view problems from numerous perspectives, industrial designers can fill the key role of integrating R&D, engineering, manufacturing, marketing, and finance into a product development plan.

Industrial designers have been responsible for some remarkable commercial successes. An industrial designer in Long Beach California, utilized an electronic defense technology to create a popular commercial product, BeeperKid, that links parents and children to keep errant children from wandering beyond a certain radius. Production of the BeeperKid created 700 new jobs. A Boston industrial designer applied medical technology (air bladders) to athletic footwear to help Reebok create a spectacularly successful product.

Laboratories can locate industrial designers through the Industrial Designers Society of America.[53]

Recognizing That Technology Transfer Is a Two-Way Street

Technology transfer often is discussed as if it were a one-way process from federal laboratories to American companies. This is true only with isolated patent licensing agreements, which can occur at arms length. In fact, technology transfer is an interactive process. The mutual stimulation that occurs with the exchange of ideas, the synergisms that spring up—these benefits accrue to both parties. It is useful and necessary to quantify the benefits of technology transfer to federal laboratories as well as to private companies.

Los Alamos National Laboratory has found at its Superconductivity Technology Center that the requirement for industry partnership in all but the most basic R&D projects has been valuable in guiding R&D toward commercial uses.[54]

Few laboratories have worked on developing measures of the internal benefits of technology transfer, including leveraging federal resources. Benefits to a mission-oriented project could include, for example, avoided R&D costs.[55] Another benefit might be early achievement of objectives as a result of the contributions of private companies, both directly and through synergism.

A more direct measure of technology transfer level of effort or interaction is the number of days worked on site by persons not employed by the laboratory. This number could be obtained from laboratory visitor records. The ratio of collaborator days to employee days would reflect a laboratory's interaction with the private sector.

Closing the Cultural Gap: a Role of the FLC

Industry's most frequent complaint about federal laboratories is that they lack the sense of urgency that pervades the business world. The concept of return on investment is foreign to federal employees, and most federal employees have not needed to worry about their employers' going out of business. In the federal work place, there's always next year for the things that don't get done this year. At least, that's the way it used to be. Federal laboratories now are getting a taste of the pressures under which the private sector operates and are learning how to work on a private-sector time scale.

The government has two sources of instruction in this effort. One is the private sector, and the other is those federal laboratories that have learned how to work with the private sector. The FLC provides good access to experienced technology transfer professionals and their legal counterparts. The FLC Legal Networking Subcommittee attracts both attorneys and non-attorneys from the federal and private sectors.

Government attorneys must make a particularly difficult transition. Whereas industry was seen as essentially an adversary in contractual matters, now it is also a partner in facilitating cooperation in technology transfer. In FLC Legal Networking Subcommittee meetings one can see this transition process under way, as government attorneys seek new modes of interaction and work to meet corporate schedules. If corporate attorneys can match that transition, the technology transfer process should work smoothly.

Meeting the Management Challenge

Laboratory managers who decide to engage vigorously in technology transfer take on a challenge. They will need good technical assessments to determine what can be marketed, a strong public relations effort about laboratory capabilities to attract customers, and readiness to handle inquiries so that potential customers are not driven away by poor service. Also, invention disclosures and patent applications must be handled properly; if the administrative and legal staffs are not prepared for a surge, the educational program will be damaged by the negative experience of backlogs, and scientists and engineers will conclude that the laboratory isn't serious about identifying and handling intellectual property.

Another laboratory challenge is to find ways to give staff enough time and resources to make and sustain contacts. An industry survey found that person-to-person contact is important to successful interaction in technology transfer.[56]

This requires adequate travel budgets, not just for technology transfer professionals, but also for scientists and engineers. Adequate travel budgets, however, run counter to normal practice when budgets are tight. The usual practice is to maintain staff while reducing travel. This reduces *per capita* productivity by inhibiting the interactions that promote two-way technology transfer.

Innovative use of other resources may be called for. The Air Force's Phillips Laboratory contracts with the State of New Mexico, which contributes on an equal basis to support the Phillips Laboratory's technology transfer activities. The State subcontracts with the University of New Mexico for the latter's business, law, and engineering students to assess technologies and capabilities at the laboratory. University of New Mexico students also do preliminary work on Phillips Laboratory patent applications.[57] Phillips Laboratory also has a brochure that clearly explains licensing policy to potential licensees and laboratory employees.[58]

Another practical question at many laboratories is: when private companies have been persuaded to seek out a laboratory, how does that laboratory accommodate them? Whole areas of a laboratory may be designated for classified access, when only parts of them need to be. Could unclassified activities be grouped in easily accessible areas? If not, how should numerous uncleared visitors be handled to enable them to work productively at a laboratory?

Many of the challenges facing laboratory management are mirror images of challenges facing companies. 3M is well known for successful internal technology transfer, but most companies readily admit that they have a lot of room for improvement. Companies, too, can benefit from the approaches that federal laboratories develop to foster technology transfer. For these and other reasons, companies should take a strong and active interest in how the federal laboratories meet these challenges.

NATIONAL ISSUES

Although technology transfer legislation and executive orders are in place, there continues to be a need for both the Executive and Legislative Branches to exercise leadership to ensure achievement of the full potential of domestic technology transfer.

Congress has enacted legislation that provides an excellent framework. Very little additional legislation is necessary. Congress can assist the Executive Branch through oversight and by giving departments and agencies making excellent progress opportunities to testify in hearings and to serve as examples.

There is a need to develop consistent policies in implementing legislation, as American companies are hampered by disparities between laboratories and agencies. However, such policies must be developed very carefully to make sure that the flexibility built into technology transfer legislation is not lost. The Executive Branch can exercise leadership by using the interagency policymaking processes to tackle issues that span many agencies, a few of which are discussed below.

Conflict of Interest

Concerns about real or apparent conflict of interest may be the single greatest obstacle to technology transfer. Although the government is grappling with this issue, it is far from settled. As federal laboratory employees transfer technology to the private sector many issues arise, such as: may laboratory employees consult on commercial projects that they worked on as CRADAs? May laboratory employees hold equity positions in corporations whose technology they helped develop? Should laboratory employees disclose their financial holdings before working closely with companies? May laboratory employees be part of personnel exchanges with companies without being vulnerable to conflict of interest charges?

Congress anticipated these issues, and a House of Representatives committee report required departments and agencies to develop "employee standards of conduct for resolving conflicts of interest."[59] While some departments and agencies have made progress in this area, it remains a thorny issue. There is now an interagency effort under way to try to deal with conflict of interest and its appearance.

Copyrighting Software Created by Federal Employees

Although software is an area in which American companies excel, we have not taken all the steps necessary to protect that important intellectual property. While government employees of most of our trading partners can copyright software, our federal employees cannot.

In 1992, Congress considered legislation to allow limited copyrighting under H. R. 191 and S. 1581.[60] (H. R. 523, a bill identical to the original H. R. 191, was introduced by Congresswoman Constance Morella on 21 January 1993; no companion bill has been introduced in the Senate as of September 1993.) It was not enacted during the last session of Congress because of concerns expressed by the information industry that the flow of information from the federal government to the public might be obstructed. If that concern is valid, the bill can be amended to accommodate it. However, it may turn out to be unfounded, as was the concern expressed by American companies when legislation was proposed to mandate royalty sharing with inventors employed by the Federal Government. Some private firms worried that their inventors would demand similar sharing. Six years after the legislation was passed, that seems not to have been the case.

CRADAs and Fairness of Opportunity

A highly visible issue related to CRADAs is the question of fairness of opportunity to companies to enter into CRADAS. Federal procurements are controlled by the Federal Acquisition Regulation (FAR). CRADAs however, do not fall under the FAR. The CRADA instrument is intended to break new ground, to enable corporate and government laboratories to work together in new and creative ways. (The Navy has a brief document on FAR creep that explains what a CRADA is and is not.)[61]

In a recent court case on a challenge to an EPA CRADA by Chem Service Inc., the court dismissed the case due to Chem Service's lack of standing [62] because one of the purposes of the enabling legislation is to "improve the economic...well-being of the U.S.," rather than to benefit individual companies.

The latitude that technology transfer law intentionally provides for creativity causes uncertainty and confusion. Fairness of opportunity is likely to be an area where the Administration takes a significant step forward.

Multiple CRADAs

Laboratories may have multiple CRADAs in place in the same technical area and even involving the same laboratory employees. This is possible if 1) the companies are focused on commercializing different products, and 2) laboratory employees are careful to avoid any "cross fertilization" between CRADAs. Agencies and laboratories can achieve these conditions if they are very careful in setting up the CRADAs.[63]

Recently, the Naval Air Warfare Center, Weapons Division (NAWCWPS) at China Lake, California, signed a CRADA for a joint project at NAWC's newly established Embedded Computing Institute (ECI). The ECI is the focal point for cooperative ventures into embedded computing, the integral use of computers in larger systems. NAWC has been grappling with the problems of embedding computing for almost a decade and has sought help in the private sector. The goal of the ECI CRADA is to develop stand-alone work stations that can look nonintrusively into embedded computer systems. Additional companies and universities are expected to join in working with the ECI. The NAWC expects the payoff from the CRADA to be "a more efficient military research-and-development system, a strengthened commercial sector better able to compete in world markets, and academic institutions training scientists and engineers in the cutting-edge technologies that will shape tomorrow's world."[64]

Giving FFRDCs CRADA Authority

Rapid progress also is likely with respect to technology transfer from Federally Funded Research & Development Centers (FFRDCs) and other federally funded institutions not considered traditional R&D laboratories. (FFRDCs are R&D performing organizations exclusively or substantially financed by the Federal Government to meet a particular R&D objective or, in some instances, to provide major facilities at universities for research and associated training purposes. Each center is administered either by an industrial firm, a university, or another nonprofit institution.) Two examples of southwestern FFRDCs that have the potential for entering into CRADAs are the Inhalation Toxicology Research Institute, administered by Lovelace Biomedical and Environmental Research Institute, and the Institute for Advanced Technology, administered by the University of Texas at Austin.

The Defense authorization for FY 1993[65] gives authority to the Secretary of Defense through the Advanced Research Projects Agency (ARPA) and the Service Secretaries to permit directors of FFRDCs to enter into CRADAs. This extends CRADA authority beyond that provided in the original 1986 legislation, to FFRDCs that are not government owned. The contracting officer for each FFRDC contract now determines, case by case, the technology transfer issues related to each FFRDC such as:

- What costs associated with technology transfer are allowable under the FFRDC's contract?
- Are there any restrictions on what an FFRDC's CRADA partner may do with either the technology or the CRADA partner's name for the duration of the CRADA?
- Can an FFRDC that has been invited by the Chairman of the FLC be prevented from becoming a member of the FLC, even though benefits accrue from association with other technology transfer professionals and attorneys?

New policy guidance on these and other issues may be necessary to enable FFRDCs to be more effective agents of technology transfer.

Expanding the Federal Technology Transfer Resource Base

Other federal institutions that are not considered traditional R&D laboratories also may have significant technology. Consider, for example, the Central Intelligence Agency's (CIA) developing capability to teach foreign languages using interactive multimedia technologies. In July 1991, the Director of Central Intelligence approved the creation of the Federal Language Training Laboratory (FLTL), pursuant to the FTTA. The FLTL signed a CRADA in February 1993 to incorporate industry expertise and promote broader use of the technology in the private sector.[66] The CIA may define other facilities as laboratories to create opportunities for more CRADAs.

There are many other federally funded entities with resources suitable for technology transfer. An important category is test and evaluation facilities, which are likely to have more near-term value to companies than R&D facilities have. As the Council on Competitiveness stated "many of the Federal labs have excellent testing facilities and can play a major role in validating technology prior to the introduction of commercial products."[67] Test and evaluation facilities have particular value to small companies that do not have the resources to build comparable test equipment. The Air Force Materiel Command has recognized the potential technology transfer value of such facilities and has delegated CRADA authority to its laboratories and large field manufacturing complexes (Air Logistics Centers) following Command certification. A case in point is the Advanced Manufacturing Center at McClellan AFB, California, which is supported with research development, testing, and engineering (RDT&E) funds. This facility was delegated CRADA approval authority by its headquarters at Wright Patterson AFB, Ohio.[68] Other facilities that are funded with other than R&D funds also have technology that is valuable to the private sector.

Specifically addressing this issue, a report prepared prior to the House of Representatives vote on legislation establishing CRADAs states:

> The term "government-operated laboratory" in this Act is intended to be generic and to apply to all organizations primarily sponsored or owned by the Federal Government and engaged in research development, engineering, test and evaluation, or otherwise developing or maintaining technology. This includes, but is not limited to federal laboratories, national laboratories, centers, directorates institutes, federally-funded R&D centers and depots.[69]

Informal communications with originators of the FTTA indicate that they intended that CRADAs be used whenever federally funded technology, equipment, facilities, etc. are of potential use to the private sector. The inclusion of "depots" in the committee report language was meant to indicate that the type of federal institution is immaterial; CRADAs should be available to companies as a mechanism for tapping into federal technology resources wherever those resources may be useful. Executive Branch formalization of this principle for government-wide application would create many more opportunities to make federal-technology-related investments available to American companies.

PRIVATE-SECTOR ISSUES

Congress has enacted legislation, the Administration vigorously supports technology transfer, and agencies and laboratories are geared up and ready to go. What remains is for more companies to use the resources of federal laboratories. One cannot over-emphasize the point that Administration policies can be successfully implemented only if American industry seizes the opportunities presented.

Making Contact with a Federal Laboratory

What is the best way for a company to benefit from more than 700 Federal laboratories funded at about $25 billion per year, with over 100,000 scientists and engineers who have an estimated two million years of technical experience, and with thousands of government licenses available?

The technology search should begin with a look inside the corporation to determine what technological advances could make the greatest difference in the bottom line. Companies should think big, because the resource they are tapping is big. In the best of all possible worlds what technologies would be useful? What technological gap could be filled? In particular, what technology could most improve the bottom line?

A large American corporation was highly successful with this approach even without the benefit of CRADAs. A person with a high-level champion was charged with extracting from each of the corporation's operating divisions one-page descriptions of technological advances that would make the largest impact on the bottom line. Armed with thirty or so of these statements, this corporate prospector mined two large national laboratories for useful technologies. He estimates that, over two to three years, his company got the benefit of $10 million of R&D and gained a three-year lead on competitors for one major product. This success was achieved even though the laboratories were far from an obvious match for the corporation. One of the major products of the corporation is diapers. For decades, the principal research of these laboratories—Sandia National Laboratories (Albuquerque) and Los Alamos National Laboratory—has been dedicated to nuclear weapons.

Clearly one cannot predict where valuable technology will be found within the vast federal laboratory system. No laboratory should be ruled out. However, the approach described above, successful as it was, should not be taken as a model in all respects. The corporation might well have gained even greater value if it had approached not two but hundreds of federal laboratories.

When a company can identify a specific technological need, a gap to be filled, or a technical challenge outside the expertise of the company, we encourage it to contact the FLC. The FLC representative will use the FLC Laboratory Locator Network to match requests to the capabilities of one or more federal laboratories and will provide the company a name and telephone number for a point of contact at each of those laboratories. Each point of contact will be a technology transfer professional who will have checked with the appropriate technical specialist to verify a technical match and the availability of the specialist. The likelihood of a productive match using the FLC Laboratory Locator Network increases with the detail of the information provided on the technology needed. Among 100,000 federal scientists and engineers, there almost certainly is someone who can be valuable to a company.

The technology transfer professional also will "facilitate the collaborative relationship," as we say in the trade. That means guiding both the private-sector partner and the laboratory technical specialist in identifying and using the appropriate technology transfer mechanisms.

Thinking small also can be productive, and not only for small businesses. Companies of all sizes should tap laboratory expertise to solve technical challenges outside their expertise. Federal scientists and engineers have surmounted more difficult technical problems than most companies ever encounter. Two examples of companies that took advantage of laboratory trouble-shooting capability are:

- A company technologist had difficulty annealing a particular alloy. He discovered that Sandia had relevant expertise, so he spent an afternoon there at a laboratory bench, working with a Sandia employee. He left with the knowledge he needed in his head and the skills in his hands. The cost of solving his company's problem was one pleasant afternoon at Sandia.[70]
- A small company in Washington state ran into technical difficulties in trying to fill a $1 million order. The design of the product was sound, but the product coming off the line was defective because of corrosion, and no one at the company could discover why. Two members of a laboratory's staff visited the company, found the cause of the malfunction, and tracked its source to a stage of production. Within a day or two, the problem was solved, and the company was able to fill its order. The cost was only a modest charge for laboratory tests.[71]

Not only small companies can benefit from consulting federal laboratories. General Motors received help from a federal laboratory to find and fix an intermittent power problem in a new model vehicle that already was in production.[72]

Many small companies can access federal laboratory expertise through state and local organizations that customarily provide technical assistance, such as Small Business Development Centers (SBDCs) and Economic Development Centers (EDCs). In nineteen states, the FLC has carried out two-day workshops to enhance the technical assistance capability of the existing infrastructure supporting small businesses. Participants learn how to turn to the FLC if their other resources don't provide the answer. (On average, 90% of inquiries can be satisfied using published materials and databases without engaging the FLC.) After engaging the person-to-person laboratory network, the FLC provides the local organization a suitable contact at a federal laboratory well suited to providing the needed assistance. This FLC Laboratory Locator Network is invisible to the small businessperson who sees only that help is forthcoming from the local organization. This use of the capabilities and networks of existing organizations to reach smaller businesses is representative of the FLC's approach in general, which is to get good leverage on FLC resources and to provide maximum benefit to small companies.

The FLC works with professional societies, trade associations, and the Chamber of Commerce to get the word out to companies about the resources available at federal laboratories. The FLC recently began a joint program, ChamberTech, with the U.S. Chamber of Commerce to reach out through local chambers to thousands of businesses of all sizes and in all sections of the economy.[73] The program is designed to help businesses use the expertise of federal laboratories to develop and commercialize new technologies, processes, and products.

Companies needing further inspiration to draw on the federal technology bank might consult *Mining the nation's brain trust*[74] and choose one of the strategies described there. They also should become familiar with the full spectrum of mechanisms available for working with federal laboratories. The FLC document, *Mechanisms for accessing federal resources*, describes the use of equipment and facilities together with such technology transfer mechanisms as CRADAs licensing, and Space Act Agreements.[75] This FLC document, cleared by the relevant member departments and agencies, including DOD and NASA, is also useful to companies with IR&D funds, as it states clearly that IR&D funds may be used as a company's

CRADA contribution, provided that the costs would have been allowable as IR&D, absent a CRADA.

The FLC Locator Network is not the only way to find a federal laboratory resource. An FLC publication, *Tapping federal technology* [76] is designed to help companies find their way to federal laboratory facilities, technology, and expertise. It identifies and describes over 150 federally funded resources, including clearinghouses, information centers, catalogs, bulletin boards, and databases. Among these are NASA's Regional Technology Transfer Centers (RTTCs) and National Technology Transfer Center (NTTC), and the Department of Commerce's National Technology Information Service (NTIS), which publishes a valuable directory.[77] All of these have specialized capabilities to help technology transfer. The FLC has arrangements with each of these organizations so that, when a laboratory resource is needed, the FLC network is available to find an appropriate match. Similarly, when a requestor can appropriately use the resources of another organization, the FLC makes the referral. The FLC is developing a protocol to make sure that these "handoffs" always are made successfully. As our information infrastructure expands into an "information superhighway" it should be increasingly easy to tap into federal laboratory resources to advance American competitiveness and economic strength.[12]

Cultural Adaptation in the Private Sector

The biggest contribution to accelerating cultural change in the private sector will come from the companies themselves. Companies that have worked with federal laboratories can, by publicizing the benefits they have received, encourage others to follow their lead.

The FLC provides a forum where companies can share experiences regarding the process of incorporating new technologies in their products and processes. Richard Marczewski of General Motors, chairman of the FLC Large Business Committee, has committed to facilitating such exchanges, with the idea that, through sharing, cultural changes in American industry will be accelerated. The activities of the FLC Large Business Committee can also prove valuable to the federal laboratory and agency professionals as they become increasingly familiar with how companies approach technology transfer opportunities. Such knowledge will be particularly useful to agencies and laboratories as they refine their efforts to reach out to companies.

The Council on Competitiveness has noted that "all too often, industry has moved too slowly in harnessing the labs' potential for developing new technology with commercial applications and has failed to take full advantage of new legislation."[4] One reason for this is the NIH (Not Invented Here) syndrome, which is as common in the private sector as in the government. Companies cannot afford to restrict themselves to technologies developed in-house. To prosper in the international marketplace, they must take full advantage of American technological resources, including those in the federal laboratories. Corporate management must find ways to rehabilitate (through appropriate incentives) corporate researchers crippled by the NIH syndrome so that they begin to tap into federal laboratories. As a DuPont employee has observed: "Our job is to shorten the time to market using the best technology available, not the best technology DuPont can create."[78] There are encouraging signs of progress in this area. According to a recent report, "there has never been livelier private sector interest in accessing the abilities and resources of the (federal) laboratories."[79] This bodes well for the United States, as "the countries that best integrate the generation of new knowledge with the use of that knowledge will be positioned to be the leaders of the twenty-first century."[80]

Practices of Large Companies

How do big companies usually access federal laboratories? Some have individuals dedicated to scouting laboratories. For example, Jim Nottke of DuPont tells how his company has assigned an employee full time to work at a federal laboratory on a government project. This person is becoming DuPont's expert on tapping into federal laboratories. Nottke also says that DuPont recently has hired six employees from national laboratories, and nine former DuPont employees now work for national laboratories. Such exchanges of personnel make the capabilities of federal laboratories more accessible to DuPont.

Other large companies invest resources to involve their suppliers or vendors in cooperative work with federal laboratories to obtain the latest technology. The large company may not want intellectual property rights for itself, because the technology is outside its domain and licensing is not part of the company's business, but it benefits from helping vendors and suppliers obtain those rights through a CRADA.

A recent sample of high-level managers of large companies investing heavily in R&D provided some useful generalizations about interactions with federal laboratories.[81] Person-to-person contact was regarded as crucial to successful interactions. That technology transfer requires personal interaction should come as no surprise. Our most deliberate technology transfer process, graduate education, finds students traveling across the world to learn at institutions and with professors of their choice. If technology transfer really took place just by reading technology papers, students would not need to travel. Similarly, throughout the Cold War, Russian engineers repeatedly learned how much essential technology was omitted from American plans obtained through espionage. Those same Russians achieved some of the most successful technology transfer in history by exploiting captured German scientists. As these and many other cases suggest, critical elements of technology do not translate well from the human mind to paper; they must be passed along person-to-person.

The amount of professional interaction between laboratory and corporate personnel has increased significantly over the past four years, with the most informal methods being the most frequent. Companies with at least moderate experience with federal laboratories have increased their interaction. As David Roessner of the Georgia Institute of Technology observed, "familiarity breeds collegiality." By far the dominant incentive given by industry respondents for interacting with a federal laboratory was "access to unique technical resources."

In the end, it comes down to human beings communicating with each other. Just as companies depend on human creativity and judgment, their ability to mine the federal laboratories will depend on the human interaction between a firm and the federal laboratories. The firm must send the right person in search of technology, and the laboratories must help find the right specialist or team.

Measuring Value to Companies

How do companies measure value received from interaction with federal laboratories? There doesn't seem to be a systematic approach or effort, although some federal laboratories now ask CRADA partners to provide dollar estimates of savings or earnings that result from CRADAs. DuPont's Nottke says that DuPont doesn't add up the resources.[78] For university-funded work, adding up the contract dollars is easy, but for DuPont, there are no quantitative measures for federal laboratories. Roessner refers to the high value provided by some of the less tangible payoffs from interaction with federal laboratories and suggests that industry "work to develop evidence of these kinds of payoffs that will have as much credibility as the more tangible forms, such as expected profits and business opportunities." [81] One articulate employee of an energy company says that licensing government patents is not critical and unfortunately seems to be greatly over-emphasized by some technology transfer professionals and their societies. He says that, for now, we should accept benefits from interaction on faith, assuming that adequate measures will not be developed for some years. In the meantime, industry has access to existing federally supported facilities. He suggests that, in the short term, we measure benefits as avoided cost of capital. A slightly different approach is "whether the work could have been done better or faster through other mechanisms such as in-house R&D or contract with another technology provider."[82]

American companies may wish to consider developing for internal use sets of measures for accessing federal technology, expertise, facilities and equipment. Some may indeed be end point measures, such as reduction in R&D cost, time to market, market share, and increased profit. Others may be activities, such as CRADAs, PLAs, and numbers of contracts with federally funded technology people. Companies may also want to consider intermediate measures similar to those suggested for federal laboratories to promote internal cultural change. In fact, both companies and federal laboratories would probably benefit from collaborating on developing intermediate measures of cultural change. Although the two sets of measures might have only a few identical elements, some collaboration might make sure that their individual efforts were fully complementary and mutually supportive.

CONCLUSION

With respect to federal laboratories, defense conversion *is* technology transfer. Many persons previously uninvolved in technology transfer take it very seriously as defense conversion. A remark made to the FLC Vice Chairman that "You've been doing defense conversion for a long time" illustrates the new attitude.[83] Now that general awareness of the value of technology transfer has been achieved, the challenge is to find ways to increase its rate exponentially so that it reaches its full potential in contributing not just to defense conversion, but to American international competitiveness.

As discussed above, choosing sound goals and appropriate measures of progress toward them is critical. Creative and disciplined effort is required to develop new and innovative measures for evaluating technology transfer progress and, above all, ways to foster the necessary cultural changes. It bears restating, however, that, while broad objectives can be mandated by government, the most effective goals and measures will be those established through the cooperation of the persons and institutions most directly involved.

To borrow an image from another cooperative venture, the mountain climber can only use pitons within reach; a piton set on the summit of the Eiger by someone who ascended the easy back slope is of no use to the climber struggling up the north face. Technology transfer goals set by a remote authority that are irrelevant or beyond reach will have little constructive effect on laboratory scientists and engineers. Technology transfer goals, like pitons, must lead inevitably upward. Also like pitons, they should be set by those who use them so that they will follow the best track from any given location to the peak. The most useful and effective goals and measures will be those set by each organization, focused on its needs, and embraced and used by its management and staff.

For these reasons, this paper describes an illustrative range of measures and goals, but by no means pretends to present the full spectrum. Output measures of productivity, intermediate indicators of cultural change, and normalized activity measures can help in assessing progress and allowing comparisons within and between organizations, provided that these measures and indicators are applied with full understanding of the complexity of the social process of technology transfer and of the limits of the measuring instruments. Poorly chosen measures can be destructive.

In particular, we must choose our measures carefully to avoid creating useless artifacts of the measuring process. We must avoid the well-known phenomenon of fostering more of the measured quality or quantity at the expense of genuine productivity. It happened once in a distant land that, when management of a certain remote facility became aware that green trees were especially favored by a visiting VIP, certain steps were taken to curry favor. Trees were planted along the barren approach to the facility, then painted green after they withered.[84] Thus was long-term cost incurred for short-term praise.

If what we are seeking is more communication, greater involvement, openness to fresh approaches, and fewer barriers to technical collaboration, we need to be sure that we don't unintentionally foster easily measured but secondary phenomena at the expense of genuinely productive activities. It is all too easy to encourage hasty transplants and spray painting with inappropriate measures.

Ultimate success in technology transfer will be recognizable by disappearance of the matter from topical discussions. When we no longer question the value of technology transfer, when it has become such an integral aspect of work at federal laboratories and U.S. companies that it blends into the background, technology transfer will have arrived.

Acknowledgement

In the three and a half years I have served as the Washington, D. C. Representative of the Federal Laboratory Consortium, I have had the privilege of associating with many fine people engaged in domestic technology transfer. The observations made here are drawn from these associations. Only a few of the contributors to this paper are cited directly, but all deserve credit for their help and for their larger work with the FLC in promoting technology transfer.

REFERENCES

1. U.S. Congress, Office of Technology Assessment (1991 May). *Federally funded research: Decisions for a decade* OTA-SET-490. Washington: U.S. Government Printing Office.
2. U.S. Cong. Senate Committee on Government Affairs (1992 July 21). *The government role in civilian technology.* Statement of John Sullivan Wilson. 102d Cong., 2d sess.
3. National Science Board (U.S.) Committee on Industrial Support for R&D (1992). *The competitive strength of U.S. industrial science and technology: Strategic issues.* Report of the National Science Board, Committee on Industrial Support for R&D (NSB-92-138). Washington: National Science Foundation, p. v.
4. Council on Competitiveness (1992, September). *Industry as a customer of the federal laboratories.* Washington: Council on Competitiveness, 1,4.
5. White, R. M. (1992, Winter). In search of a technology strategy. *The Bridge 22* (4), 10.
6. U.S. Congress House of Representatives Committee on Science, Space, and Technology (1992, July). *Report of the Task Force on the Health of Research.* (Chairman George E. Brown, Jr.), 102d Cong., 2d sess., Report 102.966 Washington: GPO.
7. Defense Conversion Commission (U.S.) (1992, December 31). *Adjusting to the drawdown.* Report of the Defense Conversion Commission, Washington, 1.
8. Committee on Science, Engineering, and Public Policy (U.S.), Panel on the Government Role in Civilian Technology (1992). *The government role in civilian technology: Building a new alliance.* National Academy of Sciences, National Academy of Engineering Institute of Medicine. Washington: National Academy Press, 3.
9. Council on Competitiveness (1991). *Gaining new ground: Technology priorities for America's future.* Washington: Council on Competitiveness.
10. Committee on Technology Policy Options in a Global Economy, National Academy of Engineering (1993). *Mastering a new role: Shaping technology policy for economic performance.* Washington: National Academy Press, v.
11. Schmid, Loren C., Chairman, Federal Laboratory Consortium (1993, April 21). Public testimony before the Committee on Armed Services, Subcommittee on Research and Technology, U.S. House of Representatives, 12.
12. Clinton, W. J. & Gore, A. Jr. (1993, February 22). *Technology for America's economic growth: A new direction to build economic strength.* The White House: Office of the Press Secretary, 8.

13. Clinton, W. J. (1993, February 17). *A vision of change for America.* Report accompanying address to the Joint Session of Congress. Washington: U.S. Executive Office of the President, 54.
14. White House (1993, October 22). Office of the Press Secretary.
15. Aspin, L. (1993, April 21). Letters of congratulation to defense laboratory scientists, and engineers who won FLC awards for excellence in technology transfer.
16. Creedon, J., Abbott, K., Ault, L. et al. (1992, December 21). *NASA technology transfer.* Washington: NASA Institutional Team on Technology Transfer, 12.
17. Defense Conversion Commission (U.S.) (1992, December 31). *Adjusting to the drawdown* Report of the Defense Conversion Commission, Washington, 28.
18. United States Department of Commerce (1989, July). *Federal Technology Transfer Act of 1986: The first 2 years.* Report to the President and the Congress from the Secretary of Commerce. Washington: U.S. Department of Commerce, 7.
19. United States Department of Commerce (1993, January). *Technology transfer under the Stevenson-Wydler Technology Innovation Act: The second biennial report.* Report to the President and the Congress from the Secretary of Commerce. Washington: U.S. Department of Commerce, 1 Appendix B.
20. Mullins, R., Acting Deputy Director, Office of Technology Commercialization, U.S. Department of Commerce (1993, March). Personal communication.
21. Stern, R., Technology Transfer/Small Business Manager Electronics and Power Sources Directorate, Army Research Laboratory Ft. Monmouth NJ (1993, March 17). Personal communication.
22. Federal Laboratory Consortium for Technology Transfer (1992, November). *ORTA Handbook, No. 2 of Federal Laboratory Consortium Handbook Series* Sequim WA: FLC Publications, 3–1.
23. Schmid, L. C. (1992, March 12). *Federal laboratory and industry relationships.* Statement to U.S. Senate, Committee on Governmental Affairs, Subcommittee on Government Information and Regulation, 102d Cong., 2d sess. Washington: GPO, 50–58.
24. Linsteadt, G., FLC Administrator (1993, March 15). Personal communication.
25. U.S. General Accounting Office (1992, December). *Technology transfer: Barriers limit royalty sharing effectiveness.* Report to Congressional Committees (GAO/RCED-93–6). Washington: n. p. 3.
26. Schrage, M. (1993, February 19). Slashing national labs' budgets might really get them thinking. *The Washington Post* (Final ed.), Sec. Financial, 3.
27. Erlich, J. H., Chief, Hanscom Patent Prosecution Office Air Force Legal Service Agency, Waltham MA (1993, March 5). Personal communication.
28. Elbaum, S., Chief, Intellectual Property Law Division Army Research Laboratory and Chairman, FLC Legal Networking Subcommittee (1992, November 3). *Enhancing our intellectual property resources: The roles of ORTAs lawyers, and laboratory directors.* Paper presented at the FLC Fall 1992 Meeting, Scottsdale AZ, 2.
29. U.S. Department of Commerce, Patent and Trademark Office (1992, October). *Highlights in patent activity.* n. p., 18.
30. Pacific Northwest Laboratory (1988, October). *Putting science and technology to work: A casebook of transferred technologies.* Prepared for the U.S. Department of Energy under Contract DE-AC06–76RLO 1830, PNL-SA-16279. Richland WA: Pacific Northwest Laboratory.
31. *Technology Innovation: Chapter 63, United States Code Annotated Title 15, Commerce and Trade: Sections 3701 to 3715 (As Amended Through 1990 Public Law and with Annotations)* (1991). Prepared for The Federal Laboratory Consortium for Technology Transfer. St. Paul MN: West Publishing Company, 24.
32. Duston, D., Director, Innovative Science and Technology Office, SDIO, The Pentagon (1993, June 30). Personal communication.
33. Akbay, I., Director, Technology Utilization Office George C. Marshall Space Flight Center, AL (1993, September 30). Personal communication.
34. Erlich, J. N., Chief, Hanscom Patent Prosecution Office Air Force Legal Service Agency, Waltham MA (1993, September 27). Personal communication.
35. McKinley, T., Associate Division Director, Oak Ridge Institute for Science and Education (1993, October 5). Personal communication.
36. Houston, S. R., Associate Director/Program Manager Western Management Development Center, Office of Personnel Management Denver (1993, April 20). Personal communication.

37. Wright, H. B., Manager, Technology Transfer & Commercialization, Tennessee Valley Authority, Knoxville, and FLC Southeast Regional Coordinator (1993, October 12). Personal communication.
38. Dudick, C., Manager, Technology Commercialization Pacific Northwest Laboratory, Richland WA (1992, May). Personal communication.
39. Hayes, K., Coordinator, Technology Transfer Information Center, National Agricultural Library USDA, Beltsville MD (1993 October 4). Personal communication.
40. Brantley, C. J., Manager, Government Activities, IEEE-USA (1993, October 5). Personal communication.
41. Stern, R., Technology Transfer/Small Business Manager Army Research Laboratory, Electronic & Power Sources Directorate, Ft. Monmouth NJ (1992, July). Personal communication.
42. Blair, D., Chief, Office of Research & Technology Applications, Armstrong Laboratory, Brooks AFB TX and FLC Mid Continent Regional Coordinator (1993, October 5). Personal communication.
43. Stark, E., Program Manager for Economic Development and Technical Assistance, Los Alamos National Laboratory, Los Alamos NM (1993, April 20). Personal communication.
44. Jacobs, M. C., Small Business Innovation Research Program/Industrial Liaison, ARPA (1993, April 6). Personal communication.
45. MacLellan, D. C., Assistant Director, MIT Lincoln Laboratory, Lexington MA (1993 February 17). Personal communication.
46. Erlich, J. N. (1993, June). *Mechanisms for transferring federally owned technology.* Presented at the Licensing Executives Society, Washington, 2.
47. Richards, G., Assistant Laboratory Counsel, Lawrence Livermore National Laboratory, Chairman of the FLC Legal Networking Working Group on Liability and Indemnification (1993, March 15). Personal communication.
48. Bagley, J. F., Vice President, External Affairs, Battelle Memorial Institute, Washington (1993, March 5). Personal communication.
49. Wright, H. B., Manager, Technology Transfer and Commercialization, Tennessee Valley Authority, Knoxville TN, and FLC Southeast Regional Coordinator (1993, October 12). Personal communication.
50. Wunderlich, E., Head, Commercial Applications & Transfer Office, Naval Air Warfare Center, China Lake CA (1993, March 18). Personal communication.
51. *American Technology Preeminence Act, PL 102–245, Title III, Sec. 303. amending 15 U.S.C. 3710.*
52. Hot products: How good design pays off, (1993, June 7) *Business Week* 54.
53. Berube, K., Program Manager, Industrial Designers Society of America, Great Falls VA (1993, October 5). Personal communication.
54. Stark, E., Program Manager for Economic Development and Technical Assistance, Los Alamos National Laboratory (1993, October 5). Personal communication.
55. Lanham, C. E., Manager, Army Domestic Technology Transfer Program, Army Research Laboratory, Adelphi MD (1993, March 9). Personal communication.
56. Roessner, D. J. (1993, March). *Patterns of industry interaction with federal laboratories.* Report of the School of Public Policy, Georgia Institute of Technology, Atlanta, 7.
57. Rodriquez, P., Air Force Contract Manager, Phillips Laboratory/State of New Mexico (1993, March 30). Personal communication.
58. USAF Phillips Laboratory (1993). *Licensing of Phillips Laboratory patented technologies.* (Prepared by the Patent Prosecution Office, Air Force Legal Services Agency, Office of the Judge Advocate General). University of New Mexico.
59. H. R. 3773, 99th Congress, 2nd Session, 4 (1986). Committee report accompanying P.L. 99–502.
60. Federal Laboratory Consortium for Technology Transfer (1992). *Issues related to copyrighting federal software.* Prepared for the Committee on Science, Space, and Technology, United States House of Representatives. Washington: The Consortium.
61. Erickson, R., Staff Patent Attorney, Office of Naval Research, Arlington VA (1993, January 25). Personal communication.
62. *Chem Service, Inc., v. Environmental Monitoring Systems Lab-Cincinnati of the U.S. EPA et al.* , Civ. Action No. 92-0989 (E. D. Pa 11 January 1993).

63. Lanham, C. E., Manager, Army Domestic Technology Transfer Program, Army Research Laboratory, Adelphi MD (1993, March 17). Personal communication.

64. Naval Air Warfare Center, Weapons Division (1993 February). *ECI advances technology sharing.* Press release, China Lake CA.

65. U.S. Congress House of Representatives (1992, October 1). *National Defense Authorization Act for Fiscal Year 1993.* Conference report to accompany H.R. 5006. 102d Cong., 2d sess., Report 102–966. Washington: GPO, 40–41.

66. U.S. Central Intelligence Agency, Office of Public and Agency Information (1993, February 23). *Cooperative research and development agreement between Analysas Corporation and CIA.* Press release.

67. Council on Competitiveness (1992, September 24). *Industry as a customer of the federal laboratories.* Washington: Council on Competitiveness.

68. Phillips, G., FLC Far West Regional Coordinator, on behalf of Brigadier General Paul, Deputy Chief of Staff for Science and Technology, Wright Patterson AFB (1993, October 5). Personal communication.

69. H. R. 3773, 99th Congress, 2nd Session, 4 (1986). Committee report accompanying P.L. 99–502.

70. Thompson, O., Program Manager, Technology Transfer Sandia National Laboratories (1991, March). Personal communication.

71. Schmid, L. C., Manager, Office of Technology Deployment Pacific Northwest Laboratory, Richland WA (1990, March). Personal communication.

72. Simon, J., Manager, Technology Leveraging, General Motors (1993 January). Personal communication.

73. U.S. Chamber of Commerce, Office of Media Relations Department (1993, February 8). *U.S. Chamber announces Chamber tech initiative.* Washington: Press release.

74. Grissom, F. E. & Chapman, R. L. (1992). *Mining the nation's brain trust: How to put federally-funded research to work for you.* Reading, MA: Addison-Wesley.

75. Federal Laboratory Consortium for Technology Transfer (1992, June). *Mechanisms accessing federal resources.* Sequim, WA: Federal Laboratory Consortium Publications. This document was approved by the relevant federal departments and agencies, including DOD and NASA.

76. Federal Laboratory Consortium for Technology Transfer (1993). *Tapping federal technology: Inventions, expertise, and facilities.* No. 3 of *Federal Laboratory Consortium Handbook Series.* Sequim, WA: Federal Laboratory Consortium Publications.

77. Center for the Utilization of Federal Technology National Technical Information Service, U.S. Department of Commerce (1990). *Federal laboratory & technology resources: A guide to services, facilities, and expertise.* Reading, MA: Addison-Wesley.

78. Nottke, J. E., Director Technology Acquisition, DuPont Central Research & Development (1993, March 16). Personal communication.

79. U.S. Congress, Office of Technology Assessment (1993 May). *Defense conversion: Redirecting R&D* OTA-ITE-552. Washington: U.S. Government Printing Office, 103.

80. Committee on Science, Engineering, and Public Policy (1993). *Science, Technology and the Federal Government: National Goals for a New Era* (National Academy of Sciences, National Academy of Engineering, Institute of Medicine.) Washington: National Academy Press, 48–49.

81. Roessner, D. J. (1993). *Patterns of industry interaction with federal laboratories.* Report of the School of Public Policy Georgia Institute of Technology, Atlanta, GA, 7.

82. Center for Strategic and International Studies (1993). *National Benefits From National Labs.* Washington: Center for Strategic and International Studies, 48.

83. McNamara, M., Technology Transfer Manager, Naval Undersea Warfare Center Division, Newport CT (1993, April 20). Personal communication.

84. Belenko, V. (1980). *MiG pilot: The final escape of Lt. Belenko.* New York: McGraw-Hill, 105.

DOING TECHNOLOGY TRANSFER IN FEDERAL LABORATORIES

Robert K. Carr

Technology Consultant
Arlington, VA 22201

PART 1

Why should a federal laboratory worry about how to improve technology transfer? Primarily because there is a widespread perception in industry and government that the U.S. is not realizing an adequate return from its substantial investment in the federal laboratory system. Furthermore, in the wake of a serious deterioration of the U.S. trade balance, the federal government has been looking for new ways to increase U.S. competitiveness. New technology is widely considered a critical element in improving productivity, and such improvements are, in the long run, the only way to improve the nation's competitiveness and standard of living.

The annual federal research and development (R&D) budget is $71 billion, of which the federal laboratories spend over $25 billion. Annual industrial R&D expenditures are $72 billion. A shift of some of the value created by federal-laboratory R&D to industry could add significantly to the latter sector. Finally, federal laboratories have certain facilities, capabilities, and technologies that are not available in the private sector. The only way some of these technologies (e.g. supercomputing) can be brought quickly to bear on the competitiveness problem is to transfer them expeditiously from the federal laboratories to industry.

The federal laboratories' potential to contribute to the nation's technology development becomes more evident when one considers the magnitude of human talent in these institutions. The Federal Laboratory Consortium (FLC) estimates that there are 100,000 scientists and engineers working in federal laboratories, with an average of 20 years' experience in their fields.

There is a continuing interest in Congress in increasing the commercial spinoff from the federal laboratories, and after a decade of technology transfer legislation, there is a growing demand for measurable results. Indeed, there are indications that much more could be done by the federal laboratories to increase rates of technology transfer.

Although university technology transfer programs operate in a somewhat different environment, they nonetheless provide an interesting comparison to the federal laboratories. For example, MIT, one of the more active technology transfer universities, granted nearly the same number of licenses in 1990 as the entire Department of Energy (DOE) laboratory system.

From Lab to Market: Commercialization of Public Sector Technology,
Edited by S. K. Kassicieh and H. R. Radosevich, Plenum Press, New York, 1994

MIT's royalty income from licensing activities in the same year was over twice that of the entire DOE laboratory system. (Some other major research universities such as Stanford and the University of California have licensing, royalty, and R&D figures in the same range as MIT's.) For comparison, the MIT research budget (including the Air Force's Lincoln Labs, which MIT operates) for that year was $700-million (approximately 80% U.S. government sponsored) while the DOE laboratories spent over ten times that amount [1,2]. The head of the office of technology licensing at MIT estimates that an order of magnitude increase in technology transfer from the federal laboratory system is probably feasible at current levels of R&D. Unless one assumes that federal technologies are significantly less relevant to industry than those from universities, the explanation of the gap in the rates of technology transfer must lie elsewhere. The gap is probably caused by differences in the technology transfer process. That is the focus of this paper.

While much is known about how to do technology transfer, not all of the best techniques are diffused throughout the federal system, and there are new areas of technology transfer (e.g., cooperative research and development agreements [CRADAs] and cooperation with industry associations) that have yet to be fully exploited, particularly by the DOE laboratories. This paper, based on a report for senior management at the Los Alamos National Laboratory, is based on a review of available technology transfer literature and interviews with technology transfer professionals in federal laboratories, primarily DOE laboratories, as well as universities. It examines the technology transfer process and catalogs some "best practices" (see Part 2) in technology transfer (as defined by the institutions that use them) noting that universities generally have much higher technology transfer rates than federal laboratories, and suggesting some reasons for these higher rates.

THE TECHNOLOGY TRANSFER PHENOMENON

What is technology transfer? The FLC defines it as "the process by which existing knowledge, facilities or capabilities developed under federal R&D are utilized to fulfill public or private domestic needs."[3] There are many other forms of technology transfer, including across national boundaries. However, this paper focuses on technology transfer from federal laboratories and, to a lesser extent, from universities to the private sector for the purpose of increasing national competitiveness.

It is a tired cliché that technology transfer is a "person to person" activity, or a"body contact sport." Tired or not, the cliché is accurate. Another author has characterized the technology transfer process as being less like a relay race where a baton is passed from runner to runner on the way to the finish line and more like a basketball game where the ball is passed back and forth between team members, leading up to a score.[4]

Inventions spring from and reside in the minds of men and women. Written descriptions, samples, or even working prototypes rarely convey all there is to be known about a new phenomenon, substance product, or process. Further development, which is almost always essential for commercialization, must be based on the entire body of knowledge about an invention. The inventor's knowledge, as well as intuition about further potential, must be transferred via personal contact between individuals on both sides, making it a critical part of technology transfer.

Much of the study of technology transfer focuses on the creation capture, transfer, development, and commercialization of intellectual properties. This paper is no exception. But while the transfer of an intellectual property is often thought of as the essence, the sine qua non, of technology transfer, such a view is myopic. Signing R&D agreements and transferring intellectual property are among the few elements of technology transfer that lend themselves to measurement—thus they tend to be measured. Though they may be seen as the engine of technology transfer, personal contacts are the lubricant that allows the engine to run. Furthermore, nonpatented know-how, ideas, and suggestions often constitute information of considerable value, however difficult it is to measure and evaluate this sort of technology transfer.

FEDERAL LABORATORIES AS TECHNOLOGY TRANSFER AGENTS

Do all federal laboratories have the same capabilities for technology transfer? Probably not. Some are handicapped by the nature of their principal mission and by their lack of experience in the technology transfer process. Some laboratories have missions to develop technologies for government consumption, technologies that are not generally useful in the private sector. Transfer from these laboratories tends to be horizontal (developed as an offshoot of the principal mission) rather than vertical (developed as direct result of the principal mission). Most of the technologies commercialized from the weapons laboratories, for example, are transferred horizontally. On the other hand, other federal laboratories have missions to develop technologies specifically for transfer to the private sector (although the government may also be customer). Laboratories such as the National Renewable Energy lab (formerly SERI) can and do transfer technologies developed in pursuit of their principal mission, as well as those which are offshoots of that mission.

Weapons laboratories, in particular, are handicapped by a historical modus operandi that might be characterized as 1) accomplish the mission, 2) figure the cost, 3) bill the government. This modus operandi was appropriate to the era in which it was developed. It is increasingly difficult to apply in a era of decreasing budgets, and it is completely out of step with the mentality of the commercial sector. It tends to make collaboration between weapons laboratories and commercial firms more difficult than that between industry and civilian laboratories.

Some federal laboratories have been in the technology transfer business for a long time and have a particularly successful record. NASA's 1958 enabling legislation required that the agency engage in technology transfer. Demonstrably the most successful technology transfer organizations in the federal government are the Agricultural Research and Extension Services. The Department of Agriculture has been in the technology transfer business for most of this century. It spends nearly half of its R&D budget on dissemination and transfer of agricultural technologies. The technology transfer budgets of federal latecomers to the process are tiny fractions of this percentage.

TYPES OF TECHNOLOGY TRANSFER

Most technology transfer programs tend to focus on technology licensing and cooperative R&D. This is understandable, since these two activities generally involve the transfer of intellectual property and hold the promise of royalty income for the originating institution. They are also the most measurable forms of technology transfer. However, there are a number of other forms of technology transfer which can bring significant benefit to industry.

According to a recent survey, most responding companies thought that promising payoffs from interaction with federal laboratories would not come from licensing, but from visits to laboratories, information dissemination by laboratories, technical consultation, workshops seminars, and cooperative research, in that order. Licensing contract research, and employee exchange were ranked as least likely to have future payoffs. This study, completed in 1988, surveyed R&D executives in large companies. It was to be repeated in 1992.[5]

These attitudes, which are somewhat surprising and warrant further investigation, nonetheless suggest that an optimized technology transfer programs should promote activities that are highly ranked by industry as well as by the federal laboratories. Thus, a laboratory may want to become more proactive in its technical-information dissemination efforts as well as in its programs for industrial consulting, technical seminars, and visits by industrial personnel.

Consulting, in particular, is an excellent technique for bringing the expertise of laboratories to the attention of industry, and can introduce industry to the possibility of additional interaction with laboratories, such as collaborative R&D or licensing opportunities. Increased consulting by federal-laboratory scientists and engineers also has great potential for increasing technology transfer. If all federal scientists were permitted to consult one day per week, as many university scientists are, this practice would make over five million consultation-days per year available to industry.

TECHNOLOGY TRANSFER TO SMALL AND MEDIUM-SIZED FIRMS

The United States is a nation of small businesses. Small and medium-sized enterprises (SMEs) are critical to the U.S. economy, and the many new SMEs are the result of America's unique entrepreneurial spirit. They are responsible for most of the new jobs created in the United States over the past decade. They are also responsible for much of the high-technology innovation that is critical to productivity growth. Since they are often supplier firms, providing services and intermediate goods to larger companies producing a final product, their efficiency is critical to the competitiveness of the entire U.S. economy.

Except for high-technology firms, SMEs usually do not have their own research facilities and often have difficulty solving technical problems that limit increases in their productivity. Thus, these firms are potentially important clients for technology transfer and other assistance. However, reaching SMEs is a fundamental problem for federal laboratories. SMEs barely know that the laboratories exist, are unclear about what they have to offer, and usually have no idea how to gain access to them. Contacting and transferring technology to SMEs can be a labor-intensive activity for a laboratory. Nonetheless, Congressional intent is very supportive of increased interaction between laboratories and SMEs, and the 1986 Federal Technology Transfer Act (FTTA) requires that laboratory directors give preference to small business in choosing CRADA partners and in licensing patents. Therefore, laboratories need to find new, efficient ways to deal with SMEs. One promising way is the use of intermediaries such as industry associations, or state and regional economic-development or technology-assistance networks.

The FLC and the new National Technology Transfer Center also have a number of programs to facilitate contact between federal laboratories and SMEs. The FLC has cooperative relationships with a number of state business-technology-assistance networks, the National Institute of Standards and Technology (NIST) manufacturing centers, the NIST industrial extension services, and a number of small business associations. The focus of the FLC efforts is to help SMEs take advantage of the federal technologies available to them by increasing their knowledge of the federal laboratory system and how to reach it. Since SMEs often do not have the resources to enable their employees to travel widely, the FLC efforts are concentrated on local and regional efforts to link SMEs with nearby laboratories. Following training efforts carried out in cooperation with state and local organizations, the FLC has found that the number of requests by SMEs for federal technologies tends to rise dramatically.

NTTC has a variety of programs to facilitate government-industry transfer. It opened an 800 telephone service, enabling SMEs to reach appropriate federal researchers rapidly with problems and needs. NTTC also has several other new projects for bringing federal, state industry and university groups together in innovation teams for identifying issues and dealing with them, and for training federal and industrial innovation managers.

Because of the difficulty of dealing with SMEs, technology transfer from laboratories tends to focus on larger firms, which have the wherewithal to seek out and take advantage of the federal laboratories. While there is nothing inherently unproductive about such relationships, laboratories should consider whether large firms are always the best vehicle for commercialization of a given technology. Large firms are often less dynamic and innovative than SMEs, and laboratories need to be sure they are not dealing with technological dinosaurs.

INFORMATION-DISSEMINATION AND TECHNOLOGY-SEARCH PROGRAMS

Information dissemination was one of the earliest forms of federal technology transfer. It was used in a major way by NASA, which was required by its enabling legislation to transfer useful technologies (called spinoffs) to industry. These programs are an excellent way of publicizing technical information that does not constitute intellectual property. Furthermore, they are a good introduction to the capabilities of the federal laboratories and are a way of sparking the interest of firms in further interaction with the laboratories. Information-dissemination programs are not a substitute for other types of technology transfer, but they are a useful supplement.

There are also technology search programs, which enable industry, state and local governments, and the general public, to locate information, expertise, and technology in the federal laboratory system. The laboratories are usually passive participants in these programs, although they are the ultimate source of the information that resides in a number of databases and is searched by technology search organizations. Laboratories could increase their accessibility by establishing proactive relationships with technology search organizations and providing up-to-date information about their expertise and R&D programs.

MARKET PULL VS. TECHNOLOGY PUSH

Market Pull. The current focus in federal technology transfer is on industry-led or market-pull transfer. It is appealing to business, since that sector has a significant role in defining the R&D underlying any resulting transfer. Since the business sector is in the lead, it is almost a foregone conclusion that the resulting technologies will find commercial markets. There are often fewer problems with conflict of interest and fairness of opportunity in market-pull programs, since they are usually initiated by the private sector, not by a laboratory or its employees. The new emphasis of market-pull technology transfer is beginning to change the historic technology-push bias of the federal technology transfer systems.

Technology Push. In spite of the current enthusiasm for market-pull technology transfer, one should not lose sight of the fact that most new technologies, particularly the break-throughs, have emerged through technology push—for example, the use of nuclear energy.

Most major scientific discoveries are unexpected, serendipitous events, not the result of direct research. As scientific knowledge accumulates, and as the tools available to scientists become more powerful, the future may see major discoveries emerge from R&D efforts focused on particular problems. Historically, technology push has been the major path for new technologies, dramatic and mundane, to enter the market, and is likely to remain so for the indefinite future. Thus, laboratory technology transfer programs must remain structured to handle technology-push as well as market-pull transfers. Successful handling of transfers based on technology push requires a significant market effort.

Modified Technology Push. Modified technology push occurs when a research institution promotes technologies for which a commercial market is known to exist. Environmental clean-up technologies are an excellent example of this type of transfer. The demand for these technologies is well known. Thus, laboratories may develop technologies to deal with environmental problems in the absence of direct interest from a commercial firm, but with the knowledge that there is almost certainly a customer out there somewhere.

COOPERATIVE RESEARCH AND DEVELOPMENT

Cooperative research and development is in vogue as the newest paradigm for laboratory/industry technology transfer. Many laboratories are actively pursuing collaborative arrangements under the CRADA authority contained in the 1986 and 1989 technology transfer acts. While some agencies have been executing CRADAs (other agencies prefer the acronym CRDA) for years, the experience is relatively new to DOE. The department and its laboratories are still developing new administrative techniques, such as model and umbrella CRADAs, which are designed to speed processing and approval.

Within the DOE system, there is considerable diversity in processing CRADA agreements among the laboratories and the area offices. Some laboratories have had joint work statements and CRADAs processed by their area offices in days. Others experience significant delays. CRADAs are now receiving high-level attention at DOE. Thus, the current wide variation in how area offices treat CRADAs will probably diminish as standard procedures and provisions come into general use.

However, a number of industrial partners of DOE laboratories complain that current delays in approving CRADAs and, particularly, funds for laboratory R&D efforts, are too long for the CRADA process to be of value. They contrast DOE laboratories' delays with other agencies where approvals are decentralized and agreements can be put into place quickly. They insist that it is essential to seize temporal advantage with new technologies. This requires rapid approval of agreements funding, and commencement of work.

The CRADA provisions of the 1986 and 1989 technology transfer acts are legal vehicles for cooperative R&D between laboratories and industry. CRADAs do not automatically create market-pull technology transfer since laboratories can develop proposals on the basis of existing technologies in order to push them to market. However, in most cases, the opportunity exists to bring industrial partners into the process of defining continuing R&D and therefore making the resulting work responsive to market forces. Whether a CRADA is market-pull or technology-push transfer, it is nonetheless an opportunity for both sides to leverage their investment in the R&D program, and the industrial investment validates the market's interest in the technology.

Codevelopment of technologies is another form of cooperative R&D, as exemplified by the superconductivity pilot centers at Los Alamos, Argonne, and Oak Ridge. In these centers, the laboratories and cooperating companies are developing a technology that did not exist on either side before the cooperative work began. Each participant brought unique capabilities and strengths to the project and the technologies were developed through joint efforts.

While cooperative R&D offers considerable promise for increased cooperation with industry, it should not eclipse other valuable forms of technology transfer. Nor should it be thought of as a fiscal salvation for the laboratories in an era of declining federal budgets. Big numbers are being reported and projected for cooperative R&D programs. However, industry contributions to joint R&D projects are unlikely to replace reduced federal funding. Furthermore, government funding for CRADA activities will have to grow substantially before they can substitute for other programs in decline.

Nevertheless, CRADAs and other forms of joint R&D do have the potential to jump-start meaningful cooperation between industry and the laboratories. CRADAs, in particular, break down barriers between two important sectors of the national R&D community and may be the vehicle by which industry fully realizes the value of the federal laboratory system. Successful experiences with CRADAs may also generate new industry interest in other forms of technology transfer.

TECHNOLOGY LICENSING

Licensing is the classic method of technology transfer from laboratories and universities. It is a way of conveying access to intellectual property, usually that which arises from technology push, although cooperative R&D programs also generate licenses. Much is known about licensing, and the associated techniques are well diffused. There are some variants of licensing policy among the institutions surveyed, but they are less important than the differences in the way technologies that might be licensed are captured and then marketed.

Since the passage of the technology transfer legislation of the past decade, federal technologies have been more frequently licensed on an exclusive basis. Such exclusive licensing is usually a requirement by which the licensee must invest in considerable additional R&D to bring the technology to market. (These developments costs have frequently been estimated at as much as ten times the cost of the research that created the technology.) Exclusive licensing is problematic in that it increases fairness-of-opportunity and conflict-of-interest possibilities which must be managed. Indeed, exclusive licensing is not always the best way to ensure the commercialization of federal technologies. For a specific field of use or geographic area, it is one way to grant exclusivity while maximizing access. Software and some other technologies are frequently diffused more widely if licensed on a nonexclusive basis—e.g., two major university technologies, the Cohen-Boyer gene-splicing technique and the X-Windows software system.

START-UP OR SPINOFF COMPANIES

A small number of important new technologies (or groups of related technologies) are most quickly and effectively commercialized by new firms, generally called start-up or spinoff companies. Such technologies are usually identified by their revolutionary rather than evolutionary nature. A second method of identifying start-up technologies is by their fit into the ongoing activities of existing companies. If the technology does not fit into any established firm's activities, it is unlikely (performance requirements notwithstanding) to receive the same high-priority treatment it would from a start-up.

Start-up companies come into the world in a number of ways. They are often founded by the inventor(s), who may be using their life savings and working out of a garage, or who may be well organized and well capitalized. In the latter cases, the inventors constitute the new firm's ownership and management as well as its technical inspiration. Some institutions, particularly universities, give start-ups by staff members a wide range of support, from locating venture capital to providing business assistance and space in local business incubators. On the other hand, some federal laboratories take a hands-off approach to in-house start-ups to avoid fairness-of-opportunity problems.

In other cases, the inventors' institution is the agent of establishment. In these cases, the inventor may own equity or have a salaried advisory relationship, but usually has no management function or control over the new company. Institutions which take charge of start-ups themselves are responsible for locating venture capital and a management team, and may themselves take equity in the new venture in lieu of royalty payments. The institution-founder approach reduces fairness-of-access and conflict-of-interest problems, since the institution, not its employees, is the agent of establishment. Institutions engaging in starting new businesses sometimes use venture-capital firms or other technology transfer intermediaries to assemble the start-up firm.

A number of new companies (estimated at more than fifty) have spun off from the federal-laboratory system, particularly the DOE laboratories in the past ten years. For the most part, these companies have been started by former employees commercializing ideas or technologies developed in their research in the laboratory. Most of the cases involved no formal transfer of intellectual property—employee-entrepreneurs simply took ideas or know-how with them when they left the laboratory. These start-ups were usually of the shoestring variety. Many failed, but a few succeeded and flourished. Among the better known are Amtech (Los Alamos) and EG&G-ORTEC (Oak Ridge). In a very small number of cases, federal laboratory start-ups have generated controversy, usually concerning fairness of opportunity. Although wrongdoing has not been established in any of these cases, the controversy has diminished the enthusiasm in federal laboratories for start-up ventures.

A potential growth area exists for laboratory technology transfer programs to encourage and/or manage start-up firms. Where they are technologically and commercially appropriate, spinoff companies have several advantages for laboratories. They generally locate in the area around the laboratory and contribute to local development more directly than do distant firms. Start-ups have the potential for considerable growth. The new companies and the jobs they create generate local economic development which, in turn, provides support for the laboratory's technology transfer program. Future controversy over fairness and conflict of interest may be mitigated if start-ups are initiated by the laboratory or an intermediary acting on its behalf. Nonetheless, uncertainty surrounding the fairness and conflict-of-interest issues constitute a significant barrier to a potentially fruitful path of federal technology transfer.

SPECIAL CASES: BIOMEDICAL TECHNOLOGIES AND SOFTWARE

Technology transfer in the area of biomedical technologies is qualitatively different from that in the physical sciences. Discoveries made in basic biomedical research can often be commercialized without additional development costs. (However, these technologies often face regulatory hurdles that make commercialization difficult.) Transfer from medical schools and biomedical R&D faculties constitutes an important part of the technology transfer portfolio of the parent institutions.

Software is perhaps the most easily transferable technology. It is normally very close to being a commercial product at the time it is transferred, and constitutes a significant percentage of technologies transferred from laboratories and universities. Software is sometimes not licensed exclusively because a) expensive further development is not usually required, and b) the software may be diffused more widely if it is made available to all.

TECHNOLOGY TRANSFER FOR LOCAL DEVELOPMENT

Many federal laboratories and the locales surrounding them believe that technology transfer can be a significant stimulus to local economic growth. Examples of such growth are found in many high-technology centers in the United States, foremost among them Silicon Valley in California and the Route 128 corridor around Boston. Spinoff companies from Stanford and MIT are now among the high-technology giants of the U.S. economy.

Several studies of Silicon Valley phenomena, as well as of the effects of federal laboratories, universities, and technology transfer on local economic growth, show that technology transfer from a major R&D facility is not the only element required to stimulate technology-based economic growth. Other key factors are the presence of support services, a favorable business climate, a technologically savvy work force, availability of venture capital, and innovative policies on the part of the local economic-development authorities and the federal laboratory. Furthermore, other studies indicate that established agglomerations such as Silicon Valley and Route 128 exert very powerful attractive forces on new high-technology firms. These areas provide all the critical support services needed by the firms, as well as opportunities to network with other companies working in similar area. Laboratories close to such high-technology agglomerations are well situated to take advantage of them. Distant laboratories, such as Los Alamos, may find that firms that have spun off locally eventually move to high-technology areas where related firms are already established.

A review of two areas similar in many ways to northern New Mexico (Oak Ridge and Huntsville) indicated that most of their high-technology firms were there primarily because of the procurement practices of the laboratories, not because of technology transfer. One off-the-cuff estimate given during this review indicated that firms established pursuant to transfer of technology constituted no more than ten percent of all the local high-technology companies. Instead, it appears that the laboratories' practice of subcontracting a significant part of their R&D provided the major incentive for high-technology firms to locate in the area. However, the high-technology contract firms may have created an environment which made the surrounding area an attractive location for subsequent firms based on transferred technologies.

TECHNOLOGY TRANSFER MODELS

The nature of technology transfer programs in federal laboratories and universities can be described in terms of an evolving model developed by Jon Sandelin of Stanford University's technology licensing office. Traditionally, technology transfer has been organized according to the legal or administrative model. The model adopted by an institution was usually determined by where the technology transfer program was located organizationally. Those programs located in the legal divisions usually followed the legal model, and those in research or support division followed the administrative model. No existing technology transfer programs began in the marketing model. All of the latter have been created through conscious transformation of legal or administrative programs.

Legal-model technology transfer programs (often called patenting programs) were usually run by the organization's legal staff and focused exclusively on patenting inventions. Before the passage of technology transfer legislation, minimal emphasis was placed on commercialization by government patenting offices; most technologies went on the shelf and stayed there. Licensing public discoveries was required to be nonexclusive. Thus, federal technologies were of little interest to firms, since the significant investments required to bring them to market were an imprudent undertaking without the protection of an exclusive license. Universities with legal-model technology transfer officers also have low rates of technology transfer.

Administrative-model technology transfer programs in universities usually were created as part of an administrative or support organization or were attached to a school or department with an active R&D program. After the technology transfer legislation of the 1980s federal laboratories began to move toward the administrative model. Technology transfer offices were enlarged and were usually staffed by administrators (although lawyers still played a prominent role). More attention was paid to the commercial possibilities of disclosed technologies, and an effort was made to evaluate technologies for commercial potential before patenting. There was a new willingness to grant protection in the form of exclusive licenses for inventions requiring significant additional development before commercialization. Marketing efforts used by administrative-model offices tend to be limited to advertising in publications like the *Commerce Business Daily* and trade journals.

The marketing model is generally thought to have its origin in the Office of Technology Licensing founded by Neils Reimers at Stanford in 1969. Under the marketing model, the technology transfer office must accumulate and have on hand a large inventory of technologies to market to industry. Therefore, technologies are frequently patented without a potential licensee in sight, and scientists are motivated to increase disclosures of their inventions through a simplification of the patent process and by rewards in the form of royalty income and other incentives.

Marketing-model technology transfer offices actively market technologies available for licensing, with the objective of finding an appropriate licensee and concluding a license agreement. Technology transfer offices using the marketing model (others include the National Institutes of Health, the Argonne Chicago Development Corporation, the Massachusetts Institute of Technology and the University of California) have entrepreneurial staffs with experience in marketing as well as in specific technology areas. If potential licensees have not been identified by the inventor, staff members use their knowledge of specific industry segments to locate candidate firms, and can usually, in short order, identify a number of potential licensees. Marketing-model offices do not usually employ formal advertising of available technologies, in order to limit curious inquiries and conserve staff time for the most promising prospects.

Marketing-model technology transfer offices often fund part of their activity by using part of the institution's royalty income to cover operating costs. (Under the legal and administrative models technology transfer offices are usually funded exclusively from the institution's budget.) It is interesting to note that most of the leading technology transfer universities (measured by licensing and royalty volume) use some variant of the marketing model.

An opportunity for two types of marketing exists for federal laboratories. The first is marketing available technologies, carried on in much the same way as in marketing-model laboratories and universities. The second is increased marketing of the federal laboratories themselves to the large segment of U.S. industry that is unaware of opportunities or is reluctant to engage the federal laboratories in cooperative technology ventures.

TECHNOLOGY TRANSFER—RISKY BUSINESS

Technology transfer is more like a business than a normal government activity. Successful technology transfer involves risk-taking at many levels. At the business level, technology transfer may fail. Licensing or cooperative R&D agreements may fail to transfer anything and may not produce marketable technologies. Start-up companies can (and often do) go down the drain, taking invested funds and institutional resources with them.

At the bureaucratic level, active technology transfer programs often run the risk of criticism or censure from their agencies, the General Accounting Office, or Congress. At the legal level, the risk of suit is present over such issues as fairness and U.S. preference. In fact such gray areas in policy are often delineated and a body of case law builds up. However, the National Institute of Standards and Technology (NIST) has a 40-year history of collaborative R&D, usually negotiated on an exclusive basis, without a single legal challenge. This suggests that the risks of dealing with private firms on an exclusive basis may be overestimated.

Whatever the level, risk is an inevitable part of successful technology transfer programs. Virtually all knowledgeable observers point out that the only way to eliminate risk is to eliminate technology transfer, and, if the level of transfer is increased, the level of risk must increase as well. These observers also note that federal technology transfer programs are, of necessity, more concerned with limiting risk than are those in the university sector, and are therefore more bureaucratic and cumbersome. The challenge for laboratories is to focus on managing the business, legal and bureaucratic risks of transfer in an informed manner, consciously deciding where to position themselves on the transfer/risk continuum. Government policy makers can also assist technology transfer efforts by reducing legal and regulatory obstacles to prudent risk-taking and by clarifying the legal and regulatory boundaries to acceptable conduct.

LIMITS TO TECHNOLOGY TRANSFER IN THE FEDERAL LABORATORIES

Technology transfer invades the scientific culture of most federal laboratories. It intrudes on the time and interest of laboratory scientists and tests the legal framework in which the laboratories operate. Several factors limit technology transfer. Institutions wishing to carry on active programs need to understand and mitigate these factors, which fall into two groups, cultural and structural.

Cultural limitations

Cultural limitations include the barriers caused by cultural differences between the groups on each side of the technology transfer process (i.e., the laboratories/universities on one side, and the business sector on the other). These cultural gaps cause misunderstanding of the needs and motives of technology transfer partners and sometimes result in aborted efforts. For example, businessmen may not be able to tolerate the bureaucratic procedures that federal laboratories recognize as mandatory to the process. (Laboratories may also be critical of such procedures.) Government staff members may not fully appreciate the business side's need to minimize procedures (and therefore costs) and to move quickly before market opportunities disappear.

Conflicting priorities between a laboratory's primary mission and technology transfer can limit the laboratory's technology transfer effort. In spite of legislative and policy changes that have recently elevated the status of technology transfer in the mission of the federal laboratories, the required attitudinal changes at the program-manager and bench levels will come more slowly.

Information gaps between the laboratories and industry are another limitation. Most industries have only the vaguest idea of the range of technologies and capabilities available in the federal laboratories, and the laboratories generally do not have a clear idea of what industry wants and needs.

Since technology transfer is a relatively new phenomenon for many laboratories, adaptive change is required on all sides. Institutional inertia and normal resistance to change make this difficult. The government side may realize the need to move nimbly and to take sensible risks. Industry must free itself of the "not-invented-here" syndrome and end its suspicious view of the federal government.

There may be a significant role for intermediaries in bridging the government/industry culture gap. The technology transfer programs of a number of industrial nations with significant R&D efforts often include intermediary organizations which in some way bridge the gap between government laboratories, universities, and industry. Some such institutions are the subject of experimental programs here in the U.S. (see the best practices menu in Part 2).

Structural Limits

There are four structural limits to technology transfer.

National Security. Particularly at laboratories doing defense work, the classification of some technologies limits technology transfer. The procedures now in place, while lengthy, are reasonably well understood. However, given the new international situation, there will be pressure to ease the national security criteria for evaluating new technologies in order to commercialize some previously classified technologies to improve national competitiveness.

U.S. Preference. The 1986 FTTA specifies that preferences must be given to business units located in the U.S. and, in the case of foreign-owned companies, their governments must permit American firms to participate in cooperative R&D programs in their countries. The 1989 National Competitiveness Technology Transfer Act (NCTTA) strengthens the 1986 provisions and requires preference for any firm which will undertake further design, development, and manufacturing in the U.S. If a firm is foreign-owned, additional reciprocity requirements try to limit the movement of U.S. technologies to foreign countries. However, in some cases, there have been no U.S. -based firms which were competent partners for the transfer of a given technology, and thus transfer to a foreign-owned company was the only way to bring the technology to market.

Fairness of Opportunity. Congress also requires fairness of access to federal technology transfer programs. At present procedures for correct implementation of the fairness-of-opportunity or -access requirements have not been spelled out, and uncertainty over what is likely to pass muster in an inspector-general or congressional investigation only strengthens the normal caution exhibited by all bureaucratic organizations. Thus, technology transfer offices in federal laboratories are often understandably reluctant to adopt marketing strategies that speed technologies to the private sector and maximize efficiency of operation. They prefer extensive publicity and advertising, a time- and resource-intensive method.

While fairness is a normal feature of most government programs, it is not typical of the business world. Because successful technology transfer programs must operate more like businesses than government programs, technology transfer agents in the public sector must be creative in applying the fairness provisions. Many universities interpret fairness of opportunity as mandating equal access to the institution's transfer program rather than to each individual technology. These technology transfer offices publicize their programs to industry and maintain a roster of interested firms which they contact first when they are marketing inventions. Many transfer offices (particularly in universities) view widespread advertising of available technologies as not only nonproductive, but counterproductive, since it generates additional staff work and rarely, if ever, leads to a license.

Conflict of Interest. The potential for conflict of interest exists almost everywhere, and technology transfer is no exception. In the federal government, recent laws and regulations (in the procurement and ethics areas) have developed limits on employee activities; these seemingly try to render conflict of interest impossible. If these kinds of limits are applied to federal-laboratory technology transfer programs they will, among other things, severely reduce the involvement of scientists in the transfer of their inventions.

Such involvement is not only inevitable, but also highly desirable. Contacts between inventors and private-sector colleagues often lead to recognition of the commercial potential of a new technology. These contacts often provide valuable insights on the directions of further R&D which will increase the commercial potential of a discovery. After the formal transfer of an invention, the inventor's knowledge and intuition usually must be transferred to the receiving firm through consulting or some other involvement.

Many universities begin with the assumption that virtually all transfers involve the potential for conflict, so they take control measures. The university community has found that the disclosure of financial interests of all participants is a key tool in dealing with conflict problems. This information is used by department heads, technology transfer managers, or others to monitor potential conflict situations and to take steps to limit conflict if it arises.

Cases where fairness-of-opportunity and conflict-of-interest issues come together, as they do when inventors have preexisting relationships with potential licensees, require more careful handling. Often such firms have unique experience and may be the only logical licensees. Licensing to such firms should be treated with additional care, including an examination of the inventor's financial interests. This is a case where the risk can be managed by looking for specific conflict areas and making sure that people with conflict problems are kept at arm's length from negotiations, ultimately allowing the laboratory to license to the firm.

Working close to the four structural limits adds additional elements of risk to technology transfer. Institutions need to recognize and evaluate the degree of risk, and balance this with the level of technology transfer desired. Even high risks can be managed, but if an institution opts for little or no risk, it is de facto opting for little or no technology transfer.

A WORD ABOUT MEASURING TECHNOLOGY TRANSFER

As interest in technology transfer has increased, so has interest in measuring success (or, presumably, lack thereof). However, evaluating technology transfer presents a number of problems. Historically there has not been a serious constituency supporting technology transfer measurement and evaluation, nor has there been much research into the subject. As a result, it is a primitive science. Technology transfer effectiveness is difficult to conceptualize, and so is its measurement. Moreover, commercial success may depend on a combination of factors, of which the transfer of a federal technology is only one (the others being technologies generated internally or obtained from other sources).

Finally, evaluations are often carried out to justify existing programs. Such efforts tend to start with the desired outcome and work backward, using whatever supportive data can be found. In order to maintain credibility, technology transfer evaluators should avoid inflated claims, particularly in a case, as noted above, where a specific transfer of federal technology may have been only one of many factors involved in an industrial success.

Four models of technology transfer measurement have been postulated: the out-the-door model, the market-impact model, the political model and the opportunity-cost model.[6]

The out-the-door model measures transfer of some countable thing usually intellectual property or information assets (brochures, etc.). This model has the advantage of being in common use. Almost every technology transfer institution counts its licenses and CRADAs. What such numbers mask is the quality and magnitude of the counted event. A single license may transfer revolutionary new technologies or just a modest process improvement. A CRADA agreement may provide for ground-breaking R&D in revolutionary areas or a modest exchange of results in parallel R&D projects.

A variant of the out-the-door model, the in-the-door model, measures disclosures and patents. If these measures are low, resulting technology transfer is likely to be low as well. However, high numbers of disclosures may indicate simplified and user-friendly disclosure and patenting procedures, or the presence of effective incentives. High numbers of patents applied for and issued may be more closely related to the size of the institution's patenting budget than anything else. To be most meaningful, these measures must be linked to subsequent licenses and eventual royalties.

Measuring market impact is an attractive approach to evaluation, since an impact on the market is the ultimate goal of technology transfer. One common measurement, royalty income, reflects market success and also has the advantage of being in common use. However, it is not a very useful tool for technology transfer managers in the short term, since revenues usually lag behind licensing by many years and vary dramatically from license to license. Royalties tend to measure a) success five to ten years ago, and/or b) a small number of "home runs."

Half of Stanford University's royalty stream comes from just one (nonexclusive) license—the Cohen-Boyer gene-splicing technique.

Since most public-sector technology transfer programs have political origins, many evaluations cater to the needs of political sponsors. Reports to Congress tend to stress bureaucratic success and achievement in spending authorized funds. These are not necessarily trivial accomplishments, and may be precursors of actual transfers and market impact. At the minimum, they satisfy the need for a short-term evaluation of an inherently long-term process.

Opportunity-cost models examine technology transfer program expenditures and ask what else could have been done with the same funds. Evaluations based on this model might question whether a program is more valuable than other activities that could have been undertaken. The opportunity-cost evaluation may be most useful in comparing competing technology transfer programs within an institution rather than as a measure of absolute success.

The easily countable elements of technology transfer should not be viewed as all there is, as some significant transfers may not be adequately captured by conventional metrics. Less quantitative methodologies must be used to survey technology transfer successes adequately, particularly from the market-impact perspective. Some possible approaches are included in the following examples:

- Surveys to query technology recipients about applications of transferred knowledge may be the best way to identify market successes. Federal laboratory technology recipients may be able to provide information on such things as cost savings, value added, and jobs created. Surveys could elicit how else the problem might have been solved, or where else the technology might have come from, and why the firm chose to proceed with a federal technology as opposed to finding some other solution. A summation of many such company surveys could provide a fairly reliable picture of the effectiveness of a laboratory's or an agency's technology transfer program.
- Collections of testimonials and positive anecdotal information provide examples of successes that are particularly useful for justifying specific programs and/or activities in given geographical locations.
- Case studies of spectacular successes or failures in technology transfer offer insight into how future efforts can imitate past successes and avoid past failures. Since the case-study methodology is well established, this approach should be easy to implement.
- Studies of further diffusion of a transferred technology can provide an indicator of a truly revolutionary development representing a major contribution to the economy. The Cohen-Boyer gene-splicing technique is perhaps the best example of a widely diffused revolutionary technology developed from public-funded research.

It is unlikely that we will uncover any magic-bullet solutions to the technology transfer evaluation problem. Additions to the theory and practice of technology transfer evaluation and measurement will require more study and agreement among experts as to the meaning of these imperfect indicators.

CONCLUSIONS

Why is the rate of technology transfer (albeit judged by imperfect measures) so much greater from a few marketing-model research universities than from the federal laboratory system? Available data indicate that the differences are one or two orders of magnitude, a very significant gap.

Is university research inherently more appropriate to the needs of the commercial sector? Probably not. While an increasing part of university R&D funding comes from industry, investigator-initiated basic research is still the norm. Even if this type of research were significantly more transferable to industry (questionable), the difference would not seem to explain an order-of-magnitude lead over the federal laboratories.

Are university scientists inherently more inventive than their colleagues in the federal-laboratory system? Although the quality of scientists and inventors may vary from institution to institution, there is no evidence that the overall quality of inventors is higher in universities than in federal laboratories. However, universities have had incentives for technology transfer in place longer than have the federal laboratories, so university inventors may be more aware of and inclined to engage in technology transfer.

Are commercial firms inherently more comfortable working with universities than with the federal laboratories? To a certain extent this may be true. Almost every industry executive has attended a university, while very few have ever visited a federal laboratory. Nevertheless, some cultural gaps also exist between the commercial sector and the universities. But it seems unlikely that the different widths of cultural gaps can explain all of the vast technology transfer chasm.

Does the co-location of some successful technology transfer universities and high-technology industrial agglomerations, such as Silicon Valley and Route 128, increase the success of the universities' technology transfer activities? It almost certainly does, providing an industrial "vacuum cleaner" for technology push from the universities. However, no such effect occurs at the federal laboratories that also are in proximity to these high-technology industrial areas.

What, then, is the explanation for the dramatic difference in transfer rates? Several factors are suggested by interviews with technology transfer professionals. The different attitudes of industry toward universities probably explain a small part of the gap. The relatively late start of many federal laboratories in the technology transfer business may provide additional explanation, particularly for the much lower royalty income of the federal laboratories.

However, the explanation for the major part of the gap almost certainly lies in the technology transfer process, particularly in the use of the marketing model. Several factors indicate that process is the most important determinant of technology transfer rates. First, significant differences in rates of technology transfer exist between universities using the marketing model and those using more traditional technology transfer methods. Some well-known research universities with legal-model technology transfer offices have very low rates of transfer. The quality of researchers, the nature of the technologies, and the width of cultural gaps probably do not vary widely between these two types of universities, but the process does. Second, technology transfer officials in marketing-model universities note that they now have much higher rates of transfer than was the case before adopting the marketing model.

Should federal laboratories adopt a more aggressive marketing strategy to interest industry in their technological capabilities? Or should they continue the more traditional approach of waiting for serendipity to make the connections that lead to transfers? The marketing approach is strongly suggested. After all, industry expects to be marketed to. Almost everything that industry acquires, from earth-moving equipment to toilet paper, is the result of a marketing effort. In addition, internal company decisions are usually the result of internal marketing efforts by competing groups.

Thus, if the federal laboratories want to compete in the commercial market place of ideas, they must practice more aggressive marketing. In addition, they will need to target their marketing efforts carefully. For example, corporate R&D managers are usually not the most productive targets, since they often see the federal laboratories as potential competitors. Marketing efforts aimed as sales or product development managers may be more successful.

Can federal laboratories catch up with the MITs and Stanfords of the world by adopting the marketing model? Probably, but there are limitations. Federal laboratories do not operate in the same policy environment as universities. The former are constrained by the structural limits discussed above. In spite of this, the laboratories could still adopt many technology transfer processes used by the marketing-model universities. Shifting to more business-oriented marketing methods would probably make significant increases possible, as would adoption of selected technology transfer techniques that have been successful in other institutions.

Laboratories, their agencies, and the Congress continually need to reexamine the structural limits on federal transfer. The end of the Cold War may permit some easing in national security restrictions on technology transfer, but few observers believe that a flood of transfers are being held back by current security requirements. Neither do many technology transfer professionals want a green light for unlimited transfer to foreign firms. Indeed, most believe that they should transfer to U.S. firms whenever possible. However, the continuing confusion over precisely what constitutes a U.S. firm (a complex problem beyond the scope of this paper and not likely to improve) introduces restraints.

Areas where there is room for policy clarification and change include conflict of interest and fairness of opportunity. A clearly articulated policy and a more broad-minded interpretation of these two limits would permit federal laboratories to act more like businesses. Congress and the Administration have spoken with several voices on these issues, leaving laboratory technology transfer practitioners to feel their way in the dark. Even the most capable bureaucrats lean to cautious and defensive behavior. Uncertainty about the boundary between acceptable and unacceptable conduct can only induce more caution and hesitancy. If federal-laboratory technology transfer efforts are to move faster, government policymakers must reexamine clarify, and preferably liberalize the conflict-of-interest and fairness strictures surrounding these activities.

One solution to these problems has been a relationship between some laboratories and an intermediary organization that takes title to and/or markets intellectual properties on behalf of the laboratory. Such organizations, which are usually not-for-profit, are generally located in the private sector. They therefore have the ability to operate more like businesses and can be useful in bridging the gap between corporate and federal-laboratory cultures. These organizations may be catalysts or creators of start-up firms based on laboratory technologies before the latter are marketed to industry. There are several variants of this arrangement being tried at present that merit continued monitoring and evaluation for possible use by additional laboratories. Several such arrangements are discussed in the best practices menu with follows.

Acknowledgements

The author is grateful to Dr. Ted Maher of the USDA Extension Service and James Syckoff of US2LL Technology Exchange for many of the concepts described in this article.

REFERENCES

1. Association of University Technology Managers (1991 April). The 1990 survey of university patenting and licensing activities. In GAO report *Technology Transfer, Federal Agencies' Patent Licensing Activities.*
2. Preston, J. T. (1991 June). Success factors in technology transfer. In *Preparing the way: Technology transfer in the 21st century* proceedings of the Technology Transfer Society's 16th annual meeting, Denver.
3. The Federal Laboratory Consortium for Technology Transfer (1990 November). *Representative's Manual.*
4. Dorf, R. C. (1988). *Models for technology transfer from universities and research laboratories.* Technology Management Publication TML.
5. Roessner, D. J. & Bean, A. S. (1990 Fall). Industry interactions with federal laboratories. *Journal of Technology Transfer.*
6. Bozeman, B. (1991 June). Evaluating technology transfer success: A national survey of government laboratories. In *Preparing the way: Technology transfer in the 21st century,* proceedings of the Technology Transfer Society's 16th annual meeting, Denver.

Part 2: MENU OF BEST PRACTICES IN TECHNOLOGY TRANSFER

What follows is a menu of technology transfer best practices as defined by the institutions practicing them. It was developed through a series of interviews with technology transfer professionals at selected federal laboratories and research universities. Of course not all of the techniques below are suited to every institution transferring technology. In particular, universities are usually subject to fewer federal legal and regulatory restrictions on their technology transfer operations than are the federal laboratories. Even within the federal system, agency regulations and practices and specific laboratory cultures influence individual laboratory technology transfer practices. That said, the methods and techniques described in this menu are thought by those who use them to increase technology transfer execution and to be worthy of consideration by any laboratory wanting to adopt a full-court press in technology commercialization.

ORGANIZING THE TECHNOLOGY TRANSFER FUNCTION

A number of organizational models for technology transfer are used at the federal laboratories and research universities surveyed. The most active technology transfer laboratories tend to have the function organized at least at the division level (Argonne National Laboratory [ANL], Sandia National Laboratory [SNL], and Lawrence Livermore National laboratory [LLNL]) or at the associate-director or vice-presidential level (Oak Ridge National Laboratory [ORNL] and Pacific Northwest Laboratory [PNL] with policy and operations combined within the same organization. Usually these organizations include licensing and cooperative research and development (R&D) activities. Some technology transfer organizations also handle work for others and less frequently, the patenting process, which in most cases still resides in the office of the organization's general counsel. However, a few recent or planned reorganizations shift the patent activity into the technology transfer office (Brookhaven National Laboratory [BNL] and LLNL). Most observers feel that the patenting process should be organized under the senior executive responsible for the rest of the technology transfer process.

It is the author's observation that an increasing number of the surveyed institutions are moving in one way or another to bring together the formerly diverse functions that make up part of the institution's technology transfer effort. Judging from current trends, more and more institutions will consolidate technology transfer policymaking authority with operational elements (patenting licensing, cooperative R&D, work for others/contract research, promotion of start-up firms, institutional or individual consulting, and others) to achieve better coordination.

INVOLVING THE SCIENCE AND TECHNOLOGY STAFF

Since scientists and engineers are the source of technology that is transferred to industry, successful technology transfer organizations make sure these specialists are knowledgeable and enthusiastic players in a laboratory's transfer effort. They are kept informed of current policies and regulations, and they are encouraged to participate in whatever way they can. The following attitudes and techniques are frequently used to support this effort:

Management Encouragement and Support

One cliché has successful technology transfer efforts requiring support from the highest levels of laboratory management. This is certainly true, but several professionals have pointed out that strong support must also be manifested by middle- and lower-management personnel who make day-to-day decision about funding, program direction, and staff activities.

Formal Recognition of Technology Transfer Activities

Technology transfer is a relatively new activity at most federal laboratories, and is slow to be absorbed into their cultures. A few formal actions can accelerate this process. Modification of job descriptions and personnel evaluation systems to recognize technology transfer sends a clear message that management is serious about its support for the activity. These changes in job descriptions and evaluation systems not only raise awareness and provide recognition but are also legal and regulatory requirements.

Technology Transfer Networks

Several laboratories have created formal or informal technology transfer networks (ANL, BNL, PNL). In general, these networks link the central technology transfer office and the group level, and are made up of individuals who are interested and knowledgeable in the field. They may be volunteers in an informal network, appointees in a formal system, or jointly funded experts who serve both their group and the technology transfer office. In addition to their normal scientific work, they may be charged with a number of transfer functions. These include keeping abreast of national, agency, and laboratory technology transfer laws, regulations, and policies and briefing their colleagues on changes, keeping abreast of technological programs and developments within their group, identifying technologies with commercial potential, and assisting inventors in filing invention disclosures and presenting inventions to the laboratory's patent or technology transfer office. The individuals making up the network may receive some formal training in technology transfer theory, policy and practice (ANL).

CAPTURING INTELLECTUAL PROPERTY

A major part of technology transfer is the passing of intellectual property from a laboratory to a private-sector firm. Therefore, active institutions try to make sure they capture all the intellectual properties they develop by maximizing invention disclosures (and thus their inventory of licensable technologies) and by patenting as many promising disclosures as budgets permit.

Changing the Culture

The cultures of many federal laboratories do not encourage the identification and commercialization of inventions. In fact, many scientists and other employees may not recognize potential intellectual property. Training technology transfer officials, members of technology transfer networks, and scientists themselves to recognize their intellectual property will contribute to changing that part of the laboratory culture that works against technology transfer.

An Inventor-Friendly Disclosure and Patenting Process

The disclosure and patenting process is complex and time consuming. If too much of the burden is levied on the inventor, it will become a disincentive to disclose. Therefore, many laboratories strive to make the process as simple and effortless as possible for the inventor. Disclosure forms are minimal. If a network exists, the local technology transfer representative is often trained to help his inventor colleagues perform preliminary evaluations of inventions for patentability, complete disclosure forms, and present them to the technology transfer or patent office. Patent personnel at many institutions conduct interviews and visits to elicit information and data from inventors for use in preparing the patent application. (Note: implementing inventor-friendly patenting processes usually increases the cost of the institution's patent program. Most of the institutions surveyed employ outside patent attorneys to supplement or replace internal patent counsel.)

Incentives for Inventors

Even the most inventor-friendly disclosure and patenting program will cost inventors time away from their ongoing scientific work and/or their personal lives. Many institutions have found it worthwhile to provide incentives to inventors for participating in the patent process. Federal legislation and university practice grant significant percentages of any royalties to inventors. However, these are distant and often minimal rewards. Thus, many institutions also offer incentives to inventors for patents applied for and/or received. Significant monetary awards (either directly to the inventor, or to his group or research project or both) are thought to be the most direct and effective. Some institutions also employ honorary awards and publicity to reward patent inventors (National Aeronautics and Space Administration [NASA], ORNL, ANL, PNL).

Responsiveness to Inventors

Responsiveness by technology transfer officials to inventors is another way of motivating scientists to disclose. If the technology transfer office (or patent office) responds quickly to inventors' inquiries and encourages disclosures, and if disclosures are quickly evaluated and worthy disclosures (liberally defined) are patented, a message will be sent to scientific staff members that the office is there to serve them. Favorable publicity will be spread by word of mouth, the best form of advertising.

Technology Ferrets

The use of technology ferrets is a technique pioneered at government laboratories in the United Kingdom. Ferrets are technology experts who represent (usually) a consortium of firms and who have free access to government research facilities. Individuals serving as ferrets are familiar with the technology needs of the industry they represent and can identify laboratory technologies of possible interest to industry. When such technologies are identified, laboratories can take steps to make sure they are patented and may decide to devote maturation resources to move them in a commercial direction. Since only a small number of individuals are required to represent one or more industries, the problem of granting appropriate clearances is reduced.

EVALUATING AND PATENTING INTELLECTUAL PROPERTY

Not every invention, no matter how technologically stunning, has commercial value. The patent and technology-marketing process is resource-intensive and costly. Therefore, inventions are always evaluated and winnowed before a decision is made to patent and ultimately market them. There are many approaches to the patent-decision process spread along a continuum. At one end is the liberal patenting style of the Massachusetts Institute of Technology (MIT) (about 50% of all disclosures); at the other end is a process that postpones the patenting decision until an invention is thoroughly evaluated, the market researched, and a licensee located.

Most institutions have made a conscious choice about their position on the patenting-decision continuum, based on their technology transfer philosophy and the budget available for patenting and transfer. Institutions doing significant market research and even locating licensees before patenting usually do so because their patenting and technology marketing budget is limited. Institutions that make patenting decisions early and liberally usually argue that to be effective they need a large inventory of technology to market. Many of them can demonstrate high percentages of patents licensed and relatively high rates of royalty income.

Evaluation by Inventors

Nearly every organization involved in technology transfer relies to some extent on the inventor's knowledge of the technology area for a preliminary assessment of marketability. Often the inventor, through contacts developed with private-sector colleagues at scientific meetings, knows one or more firms working in related areas.

Evaluation by Technology Transfer Office Personnel

At some institutions, technology transfer personnel with experience in marketing and relevant technology areas make patenting decisions based on their knowledge of the field. At institutions where inventors have been well briefed about technology transfer and patenting (i.e. through a technology transfer network), the percentage of patentable disclosures submitted to the technology transfer office is usually higher than average (ANL, MIT, Stanford University).

Consultation with Cooperating Firms

A number of technology transfer offices consult with commercial firms in related technology areas when evaluating inventions. Often these firms have licensed the institution's intellectual properties in the past. Some institutions give cooperating firms the first opportunity to license the technology (MIT, Stanford, PNL).

Market Research

Most federal laboratories do not have the internal capacity to do market research and must employ outside organizations. One interesting approach is the synergistic relationship now developing between some federal laboratories and schools of business at nearby universities. In most of these relationships, the laboratory uses the services of the business school and its graduate students to perform market surveys on its inventions at low cost, and the business school gets real-life material to use in the education of its graduate students. In some cases, the students go on to work in the laboratory's technology transfer effort (ORNL, LLNL).

Patenting Committees

Many institutions use a patenting or intellectual-property committee for patenting decisions. Often, one or more of the above techniques provide input to the committee's decision-making process.

MARKETING LABORATORY TECHNIQUES

Marketing should be a key part of technology transfer from federal laboratories and universities alike. Before the 1980s, many institutions thought they only needed to put their technologies on the shelf and make known their willingness to transfer. This approach is now generally acknowledged to be ineffective. Institutions which are successful in technology transfer are generally characterized by a heavy emphasis on their marketing activities. Marketing strategies vary widely. The choice and implementation of a strategy is bounded, among other things, by the institution's interpretation of its obligations under fairness-of-opportunity strictures and by the human and financial resources available to the technology transfer office. The more effective strategies include:

First-Contact Marketing Strategies

Institutions using this strategy (many universities) try to locate a licensee from information provided by the inventor or from the technology transfer office's own knowledge of firms. Institutions employing this strategy often conclude a license, even an exclusive license, with the first firm demonstrating sufficient interest and capability. Proponents of this approach argue that it conserves scarce technology transfer resources. A simplistic explanation of their argument is that contacting and negotiating with only one company requires as little as "one unit" of technology transfer resources. Unless multiple licensees are intentionally sought (i.e. a nonexclusive license), contacting many firms uses many resource units to conclude the same license, transfer the same technology, and eventually produce the same royalty income. MIT, which uses the first-contact marketing strategy, produces a licensing volume equal to the entire Department of Energy (DOE) laboratory system with only seven technology transfer professionals and an equal number of support personnel.

Focused-Publicity Marketing Strategies

Some institutions (PNL, ORNL) use a focused-publicity marketing strategy, which usually consists of a press release announcing the availability of the technology, distributed to trade publications. In addition, some institutions use carefully targeted mailings. Laboratories and universities taking this approach believe that firms that are seriously interested in acquiring external technologies monitor the trade press, and feel this level of publicity is adequate to meet fairness-of-opportunity requirements. Institutions using focused publicity usually take simultaneous proactive measures to locate licensees.

Bulletin boards are a subset of the focused-publicity strategy. There are several computer bulletin boards that list new and available technologies, usually by type. Advertising technologies on one or more bulletin boards is a relatively simple matter, requiring minimal resources on the part of technology transfer offices. Institutions can advertise more widely (in the *Commerce Business Daily*, for example) that all their technologies are to be found on a particular bulletin board and assume that interested firms will monitor the boards. The National Institutes of Health (NIH) technology transfer program uses a bulletin board as its principal method of advertising technology transfer policies and methodologies, as well as licensing and cooperative-research-and-development-agreement (CRADA) opportunities (University of California and NIH).

PREPARING TECHNOLOGIES FOR COMMERCIALIZATION

Most federal-laboratory inventions are far from a commercial stage even though they may appear to have great potential. Much additional development is usually required before they become part of a commercial product or process. Furthermore, it is difficult to interest firms in immature technologies unless a proof of concept or prototype exists. Several techniques are used to make inventions more attractive to firms.

Internal Maturation Funds

Several federal laboratories have established maturation funds (ORNL SNL, NASA Marshall), which are used to continue the development of a technology in a commercial direction when the agency's funding has ended or when development in a commercial direction cannot be funded by the agency's program. Industrial partnering for the maturation program is usually sought if the project goes into multiple years.

External Maturation Funds

Some institutions (Georgia Institute of Technology) have associations with external technology transfer organizations that provide grants of maturation funds similar to those of internal programs. The external organizations usually have contracts to acquire and market technologies they support.

Maturation by External R&D Organizations

At least one university (Stanford) uses the services of external commercial laboratories or engineering firms to do maturation work on technologies, usually to build prototypes. In Stanford's view, much of the R&D required to make a technology commercializable is not appropriate for university research by Ph.D candidates, and thus is better performed by nonacademic organizations.

Cooperative R&D

Many inventions form a base on which a cooperative-R&D program with industry can be built, using the CRADA authority in federal laboratories. Since the CRADA process is viewed by many (particularly within the DOE system) as difficult and cumbersome, it is too soon to tell whether cooperative R&D based on CRADAs will be an efficient technology-maturation tool for any but very large efforts. However, experience is accumulating rapidly, and the coming years will further define the usefulness of the process for maturing technologies.

Transferring Technology Locally

Laboratories and state and local economic-development authorities in many areas have an interest in using federal laboratories to spur local technological and economic development. Several technology transfer techniques appear to have encouraged such investment:

Location Incentives in Licenses. Some federal laboratories (ORNL and Idaho National Engineering Laboratory [INEL]) offer favorable license terms for firms that agree to locate in the state or locality around the laboratory. In addition to the investment that could occur as part of the license-commercialization process, licensee firms may also move some of their scientists to locations near laboratories in order to be close to continuing developments.

Technical Assistance to Local Industry. An effective method for transferring technology to the local economy is through participation in local or regional technology-assistance or -extension networks. Often federal laboratories (ANL BNL, ORNL, and others) participate, along with local universities, as the back end of networks developed and managed by the state government or a regional organization. These networks connect local firms having technology-based problems with the technologies and know-how of the back-end institutions. One laboratory (ORNL) provides up to two no-cost staff-days of expertise to local firms through the state network.

One NASA technology utilization office (TUO) is said to have responded to a local technology-assistance network by collecting technology problems of local firms, printing them on cards entitles "Can you solve this problem?" and putting them on each table in the facility's cafeteria. The responses were said to have inundated the TUO staff. Scientists and engineers naturally respond to intellectual challenges a trait that can be positively exploited.

USING TECHNOLOGY TRANSFER INTERMEDIARIES

Several interesting models involving intermediaries have been developed to facilitate technology transfer from federal laboratories and universities. These models employ intermediary organizations, private-sector entities that are operationally independent of the laboratory or university and thus free of some of the restraints imposed on technology transfer in public institutions.

Argonne Chicago Development Corporation (ARCH)

ARCH is a not-for-profit [501(c)(3)] corporation affiliated with the University of Chicago, the DOE contractor for ANL. ARCH has a contractual agreement with the university under which it can take title to intellectual properties developed by the university and ANL. It focuses its efforts on starting up companies, but it also engages in joint ventures with private industry, technology licensing, and combinations of all three.

ARCH is an experiment designed to bridge the gap between the different cultures of the university/laboratory and the business sector. Its organizational home is in the Chicago School of Business, where its CEO is an associate dean. Graduate students work on ARCH commercialization projects, and, in fact, the corporation has hired a number of business school graduates. Its employees are compensated by salary and incentives for performance. In addition, ARCH is expected to cover its expenses from revenues within a few years. This arrangement is designed to make the organization think more like a business. Its board of directors includes both academic and business executives.

ARCH's approach to start-up companies is unique among the federal laboratories in that it believes that start-up ventures are the best way to commercialize revolutionary discoveries. In other laboratories, start-ups have been problematic, high-risk proposals often involving the extensive involvement of scientific staff, to the detriment of the institution's primary mission and raising difficult fairness and conflict-of-interest problems. ARCH, on the other hand takes the lead in the formation of start-up ventures. It decides whether a start-up is justified by the nature of the technology, selects a management team (almost always made up of business professionals, not the inventing scientists), and locates seed and venture capital. The continuing involvement of inventors with the new company is critical to success, and ARCH usually fosters a relationship between inventors and firms, sometimes giving equity positions to inventors as compensation for their services to the company.

Industry Associations

The U.S. Department of Agriculture's Agricultural Extension Service and the Technology Transfer Information Center at the National Agricultural Library are undertaking a project in which they are working with an industry association to foster technology transfer based on genuine market pull. The project began with identification of the association, the Hardwood Research Council, which represents a large number of relatively small firms. The HRC served as a forum in which to identify technology-based problems faced by the industry as a whole. Once the problems were identified, the services of a technology agent were employed to produce reports containing 1) a thorough technical discussion of the problem and the deficiencies of the available technologies, 2) the industry parameters within which the technologies must work, 3) an analysis and evaluation of the potential technologies compared to the industry-established parameters, 4) a recommendation on which technologies should be investigated further for possible commercialization, and 5) a business opportunity statement that discusses the size of the market for the commercialized technology, replacement cycles, cost constraints, and other pertinent factors. The third step of the process was to search for and match technologies in federal or university laboratories with the expressed needs of the industry. Interestingly, the hardwood-industry project turned up a promising cutting technology at the Department of Energy's INEL.

The industry-association market-pull process may or may not require new R&D, depending on the state of the potential technological solutions found in the laboratories. Once a promising technology is located (potential technological solutions might require further development through a CRADA, maturation funding, or other arrangement), the laboratory can license it, not to the individual firms in the industry, but to the industry's equipment and service firms which can quickly disseminate the new technologies to the end users.

The use of industry associations to identify generic technologies useful to an entire industry is an effective way for the federal-laboratory system to provide technology assistance to small and medium-sized firms without the need to strike a large number of small deals with individual firms.

A similar effort with industry associations has been undertaken by the new National Technology Transfer Center (NTTC). NTTC has created an 800 telephone system for linking companies needing technological assistance with appropriate federal laboratories free of charge. The center also is a point of contact between the federal laboratories and industry associations (trade and research) as a means of solving generic industry problems.

Venture Capital Firms

Some institutions use venture-capital (VC) firms as intermediaries to develop start-up firms. MIT is perhaps the best example of this approach. In the last five years, MIT has spawned forty start-up companies, in spite of a policy limiting the institute's direct role in the formation of start-ups. Its approach is to locate a VC firm interested in an MIT technology, and then to pass further start-up responsibilities to the VC firm (i.e., doing a market analysis of the technology, writing a business plan, locating management, and organizing the company). The invention is subsequently licensed to the start-up created by the VC firm. Involving a VC firm at the earliest stages of a start-up venture limits the involvement of the MIT technology licensing office and increases the influence of the market on the process of forming the new venture.

MIT does not have an in-house venture-capital fund and does not invest in its own start-ups, preferring to let the market make all the decisions about the commercial viability of such endeavors. However, it does take equities, often in lieu of up-front fees and initial royalties. The equities are managed by the MIT treasurer's office, which handles them as it would any part of the institute's portfolio, making further decisions (e.g., to sell) just as they would for any equity.

Battelle Memorial Institute

The Battelle Memorial Institute (BMI) operated PNL and has a unique relationship with the laboratory. Battelle is a not-for-profit corporation that has significant R&D operations beyond those undertaken by PNL at its Richland, Washington, site. In fact, PNL is a floating subset of the BMI Richland operation, which can grow or shrink almost on a weekly basis as BMI resources are shifted in and out of DOE contract work. The institute often uses its own funds for both technology transfer and scale-up activities based on government-funded technologies.

BMI employs an innovative concept in its technology transfer operation, using portfolio officers, who, unlike their counterparts in other technology transfer offices, function as idea managers. They take responsibility for ideas that come to their attention as invention disclosures or through their technology transfer network. They take a new idea all the way through to commercialization. The idea managers also are responsible for evaluating the commercialization potential of the technology, recommending additional R&D funding from the institute (approved by a panel), and carrying the idea through to licensing or, in a few cases, a start-up company.

Technology Brokers

There are commercial firms in the United States and elsewhere (Research Corporation Technologies, British Technology Group) that market technologies developed in laboratories and universities. It is interesting that very few of the surveyed institutions employ such brokers. Most technology transfer professionals feel that using these firms can be advantageous to smaller, less well-known R&D institutions, particularly those that do not have the internal resources for patenting, marketing, and licensing inventions. Generally, such firms relieve the institution of the costs of these functions in exchange for a percentage of the eventual royalty stream. They generally skim the cream of institutions' invention disclosures, effectively leaving the rest unpatented and unmarketed.

The California Technology Venture Corporation Proposal

The University of California's office of technology transfer is preparing a proposal to the university regents to establish a technology commercialization venture to be called the California Technology Ventures Corporation (CTVC). CTVC is visualized as a profit-making firm created to bridge the gap between the university and the private sector. The current proposal envisions the University of California with majority ownership of CTVC, and private high-technology firms (a number of such firms have already expressed an interest) with a minority interest. The university would not invest its own funds in the corporation, the private minority owners providing all the liquid investment.

CTVC's purpose would be to fund further development of UC inventions (i.e., provide maturation funds) in order to produce a prototype or otherwise enhance the commercial potential of the technology. The commercialization development funded by CTVC would take place in the university laboratory that made the invention or would be done by a commercial engineering or technology firm, as appropriate.

The corporation would not have any rights to university inventions nor would it ever be the licensee of UC technologies. All UC technologies with development funded by CTVC would be advertised to potential licensees in the same way as other UC technologies. Obviously, the private-sector firms that are CTVC investors would, by virtue of their proximity, have better insight into the potential value of CTVC-funded technologies. When CTVC-funded technologies are licensed, the firms would take a portion of the royalty streams as payment for their investments in commercialization development. Since CTVC development funding would increase the potential value of new technologies, licenses for these more advanced technologies should be worth more.

The corporation would also organize start-up companies to commercialize UC technologies. In these cases, CTVC would provide the initial seed money and would then co-venture the start-up with commercial venture-capital firms (the involvement of commercial firms is seen as essential to verify the market's interest in the technology). It also would locate the management and prepare a business plan for the new firm. The technology would be licensed directly to the new firm, in which CTVC would generally own equity as a way of recouping a return on its investment.

The CTVC proposal will be initially submitted to the University's Technology Transfer Council. Ultimate approval by the UC regents is thought to be likely, although the process may take the better part of a year. The UC office of technology transfer envisions that the UC-operated national laboratories could also participate in CTVC activities if they desire.

USING INFORMATION PROGRAMS AND TECHNOLOGY SEARCH ORGANIZATIONS

NASA publicizes its technologies through the magazine *Tech Briefs*. Originally an internal NASA activity, *Tech Briefs* has been privatized and is now published by a commercial firm, supported by paid advertising in the magazine. NASA provides the technical information to the publisher, who prints the information in the magazine and distributes it free of charge. Firms interested in a particular tech brief contact the appropriate NASA laboratory directly for additional information. *Tech Briefs* has a circulation of 200,000 copies per month, and NASA gets between 120,000 and 150,000 inquires annually about technologies described in the magazine. The commercial success of *Tech Briefs* suggests that this approach might be an appropriate way for other federal laboratories or groups of laboratories to publicize available technologies.

NASA is also responsible for a technology-search network consisting of NTCC and six regional technology transfer centers (RTTCs). The latter were recently established by NASA, replacing an older network of IACs (industrial application centers). NTTC's mission is to create a user-friendly hub permitting industry, government organizations, and the general public to locate information on research, expertise, and technology in the federal laboratory system. (Although NASA is the establishing agency, NTTC's focus is government-wide.) NTTC is the focus of an information system on federal R&D consisting of itself, the Federal Laboratory Consortium, the RTTCs, and state, local, and regional economic-development offices.

NTTC's first order of business was to establish a computer-based communications and information system (CIS). The CIS provides NTTC's technical and information agents with simultaneous access to numerous (its own and other commercial and federal sources) databases containing information on federally funded research and development. Users will gain access to the NTTC through its widely advertised 900 number.

An effort similar to that of NTTC has been under way for some time operated by the former NASA New England Industrial Application Center (NERAC). NERAC is now an independent, not-for-profit organization supported by fees and subscriptions for its services, paid by firms interested in gaining access to the federal laboratory system. Its information specialists assist firms having technology needs to find a federal laboratory or university that is working in the same area and that may offer a solution to the problem of the inquiring firm.

Finally, the Federal Laboratory Consortium (FLC) operates a locator service designed to link potential users in industry, government, or academia with a federal laboratory employee having expertise in a particular area of interest to the inquirer. The clearinghouse also locates existing technologies in federal laboratories that may solve an inquirer's technology problem. FLC, as mentioned above, collaborates with NTTC.

CONVENTIONAL WISDOM ABOUT TECHNOLOGY TRANSFER

While there are few definitive studies of technology transfer, much conventional wisdom has developed. What follows is a collection of conventional wisdom (CW) that is shared by many technology transfer observers:

- Barriers to technology transfer are mostly man-made (organizational resistance, laws and regulations, etc.). It requires persistent, vigorous activity by men and women to overcome them.
- Every technology to be transferred requires a champion on both sides of the process. The champion is required to slay dragons and sweep away obstacles to the transfer that are constantly erected by the system. The champion can be the inventor or a formal technology transfer agent. A champion is probably more important on the receiving side (i.e., the firm) than on the sending side.
- People are the key to success in technology transfer. (This CW is obviously applicable to many areas of endeavor.) One well-known technology transfer official said that if he were starting a new transfer operation, he would rather have a free hand in hiring his staff than in reorganizing the operation. Successful marketing-model universities emphasize the importance of hiring staff with backgrounds in selected technical areas, as well as in business and other entrepreneurial activities. These kinds of employees are more likely to have a business rather than a bureaucratic approach to technology transfer, and are essential to a successful marketing-model technology transfer office.
- Limit the role of lawyers (appropriate apologies). Nearly all technology transfer officials emphasize that transfer fundamentally is not a legal process, but rather a business activity. Therefore, it is important that technology transfer not be considered a legal function run by lawyers. As one technology transfer official put it, lawyers in technology transfer are like the brakes on a car. You wouldn't want a car without them, but neither would you want the brakes to control the car's movement. Many professionals suggest that lawyers should be an integral part of the technology transfer organization and report to the senior technology transfer official.

REFERENCES

1. *Agricultural Libraries Information Notes* (1991) *17* (8).
2. Maher, T., & Hayes, K. (1991 Summer). Technology transfer to the hardwood industry: Testing new strategies. *Journal of Technology Transfer.*

ACCELERATING TECHNOLOGY DEVELOPMENT FOR ECONOMIC COMPETITIVENESS

Kay Adams

Industrial Partnership Center
Los Alamos National Laboratory
Los Alamos, NM 87545

INTRODUCTION

The United States is emerging from the Cold War era into an exciting, but challenging future. Improving the economic competitiveness of our nation is essential both for improving the quality of life in the United States and for maintaining a strong national security. The research and technical skills used to maintain a leading edge in defense and energy now should be used to help meet the challenge of maintaining, regaining and establishing U.S. leadership in industrial technologies. Companies recognize that success in the world marketplace depends on products that are at the leading edge of technology, with competitive cost, quality and performance.

Los Alamos National Laboratory (LANL) with its Industrial Partnership Center (IPC), has the strategic goal of making a strong contribution to the nation's economic competitiveness by leveraging the government's investment at the laboratory, personnel, infrastructure and technological expertise. As part of the national economic investment plan, the laboratory's technical effort should be engaged in industry-LANL partnerships. More specifically, LANL proposes a three-pronged partnership approach which will systematically address the challenges facing this nation:

1. The first component is "National Challenges," requiring the development of new technologies and/or the massive integration of existing capabilities. Meeting such challenges should advance the national well-being. National challenges require broad and long-term (four to ten years) collaboration between industry and government to accomplish high-risk, high pay-off ventures. The national laboratories can play a seminal role in assuring technical success in such ventures.

2. The second component focuses on well-defined, near- to mid-term (one to five years) technology programs. These activities involve cost-shared partnerships with industry to co-develop technologies that are market-driven and may also directly benefit the U.S. national security.

3. The last component addresses the grass roots economic development of America through intensive collaboration between the national laboratories and small- and

From Lab to Market: Commercialization of Public Sector Technology,
Edited by S. K. Kassicieh and H. R. Radosevich, Plenum Press, New York, 1994

medium-sized businesses. This activity concentrates on the transfer of innovative technologies from the laboratories to emerging U.S. enterprises.

TRANSFORMING THE FUTURE: NATIONAL CHALLENGES

National Challenges are complex problems with both high-risk and high-payoff potential, the solution of which would advance the national well-being. To meet a National Challenge will require the development of multifaceted, revolutionary technologies (e.g., the Manhattan Project) and/or the extension of existing capabilities with massive engineering and systems integration (e.g., the Apollo program). Such programs are too large and too long-term to be addressed by any single company or industry sector. During pursuit of National Challenges, we expect many technology developments to have near-term industrial applications and advance the technology base of industry.

Meeting National Challenges in collaboration with industry would improve the nation's competitiveness in international markets and would serve the best interests of both the public and private sectors. Private investment can support development of near-term products for market consumption while public funds support research on innovative ideas. However, there is likely to be no unique funding source to support high-risk/high-payoff large-scale programs addressing National Challenges. Accomplishing these National Challenges would require the formation of broad long-term (4 to 10 years) collaborations between industry and government.

LANL's past defense-related missions have created a diverse, multidisciplinary science and engineering culture. This total technology competence has developed large, complex weapons and pursued scientific programs including large scale computation and simulation, rapid prototype-demonstrations, process scale-up, vertical integration with manufacturing facilities (DOE production facilities or private industries), large scale testing, diagnostics, servicing, and retirement of components or systems. These strong capabilities can now be exercised in cooperation with industry.

The Laboratory's National Challenges

A number of National Challenges have been jointly defined by LANL, government and industry. Identified below are several National Challenges that well match the lab's evolving core competencies. These challenges will significantly improve the nation's technical capability as applied to public needs and to US industrial competitiveness in the world marketplace.

- *National Information Highway.* Using the extensive LANL capabilities in networking and parallel computing, the lab will continue to improve the "National Information Highway" which can provide industry with affordable, real-time access to technological and market information. Information contained in this information highway would be as diverse as financial data, data banks for the Human Genome, medical data, public broadcasting to enrich leisure time, etc.
- *Transportation and Energy in the 21st Century.* LANL can collaborate with automobile companies to build a 21st century automobile that is light-weight, ultra-high-mileage productive, safe, recyclable and environmentally benign. Industry, government and the laboratory realize that future public transport will emphasize mass transit, and that future individual transport may evolve from the internal combustion engine to some version of a hydrogen-based or all-electric vehicle. To support this future vision, the laboratory will emphasize the development of advanced materials and manufacturing processes needed by the transportation industry, and will engage with the energy industry in the

development of new technologies for energy production, power transmission and energy storage (e.g., for fuel cells).

- *Manufacturing in the 21st Century.* Manufacturing techniques in the 21st century must use modern simulation techniques as an important tool in conceptualizing new materials, new designs, and entire new manufacturing processes. Bringing such agile and adaptive capabilities to the tooling, fabrication, and manufacturing industries would revolutionize the nation's industrial ability. The most immediate benefit to manufacturing would be the reduced cost and time needed in retooling for production of new products. Combining computer design capabilities with advanced materials and processes could support U.S. preeminence in many high technology manufacturing sectors and would make possible the efficient and rapid adaptation of environmentally sensitive manufacturing processes.
- *Biomedical Technology and Health Care*
  - Using technologies currently being codeveloped with industry, the Laboratory, with cooperating private companies, can create an integrated genetic-analysis system that will use advances arising from the Human Genome program. The result will be the capability to diagnose and prescribe individualized health care based on one's genetic predisposition and acquired environmental mutations.
  - Through collaboration with major pharmaceutical firms and health care providers, significant advances in structural biology and computational chemistry can be made so that truly "rational drug design" is a real possibility before the year 2010.
  - Technologies to revolutionize diagnostic medicine and to handle the requisite large data bases and information flow efficiently will be required to enable future health care/maintenance to be both affordable and universally available.
- *Restoring and Monitoring "Spaceship Earth"*
  - LANL can develop and improve technologies that enable complete *in situ* restoration of land and water sites contaminated by hazardous and mixed waste (liquids including chemicals and petroleum, and solids including radioactive debris). The laboratory can make prototypes and deploy clean-up technologies in actual "test-bed" sites; transfer of technologies to industry will be accomplished by integrating laboratory and industry expertise at these field tests and demonstration sites. Phased implementation will progressively treat more and more types of contaminants.
  - LANL can help create an entire industry sector devoted to developing, deploying, and maintaining a network of satellite and ground-based remote sensing platforms which assess the environmental integrity of land, water and atmosphere. Additional industry sectors would collect, assess, interpret and prioritize the assembled data. Such technology and capability would provide global, real-time data to evaluate pollution, quantify global climate change, and enable long range weather forecasting. The sensor suites on the remote platforms would also support defense and non-proliferation inspection/validation.
- *Total Life Cycle Accountability of Hazardous Materials*
  - LANL can create and lead an industry/national laboratory consortium to develop integrated technologies for total life cycle accountability and disposal of hazardous and energetic materials and their waste. These activities would be tightly integrated with the over-all design and operation of the envisioned "Weapons Complex of the Future." Petroleum, gas, mining, chemical and microelectronic industries would be direct beneficiaries of the technologies developed in this program. These developed technologies and industrial infrastructure would apply to all processes involving hazardous and energetic materials and their by-products/waste (raw materials, pro-

cessed materials, fabrication, handling, storage, deactivation, transmutation, reclamation, and remediation).

Recommendations for Implementation

1. The Administration and DOE should assign responsibilities for the National Challenges to officers at the senior management level. Since most National Challenges will cross-cut within and across several government agencies, a senior level office is required to ensure commitment in setting, prioritizing and maintaining the strategic direction for these efforts. Directions should emphasize industrial needs and benefit the "public good."
2. Each Laboratory should be given the incentive to use its unique capabilities and resources to solve the technical problems in these National Challenges. Industrial partners who will develop concepts and ultimately commercialize the innovations must be involved in these developmental activities. These efforts will provide the "seed corn" for later industrial partnerships and technology transfer to industry.
3. Industry should participate in setting national priorities for further development and commercialization of the technologies resulting from accomplishments of the National Challenges.

FOCUSING ON THE PRESENT: INDUSTRIAL PARTNERSHIPS FOCUSING ON NEAR TERM PROBLEMS AND OPPORTUNITIES

Cost-shared partnerships with industry will address the near-term (one to three years) or mid-term (two to five years) collaboration described in this section. In some cases this collaboration will be an important step in addressing National Challenges. The background for the collaboration and its tie to DOE National Laboratories is important to note.

Recognition that commerce must have the ability to harvest the fifty years of technological excellence residing in the national laboratories resulted in the National Competitiveness Technology Transfer Act (NCTTA) of 1989. For the first time, the national laboratories were provided with the ability to enter into cost-shared, market-driven collaboration with industry with the specific goal of commercializing government-developed technologies. The payback for industry on the high cost of investing in new technologies often lies in the competitive advantage of exploiting a technology through the rights provided by patents and copyrights. Bedrock to the NCTTA is the ability for companies to protect information and data generated through the collaboration and to gain full and exclusive rights to the resulting patents, software and other intellectual property.

Examples of Ongoing Industrial Partnerships:

- Deposit of diamond film on cutting tools and wear parts for the automotive industry that will increase the life of the part. The result is greater yield, less down time, and higher productivity.
- Development of new sensors for automobile engine systems that will provide improved performance. The result is lower exhaust emissions, greater fuel efficiency, longer engine life, and a cleaner environment at a lower cost.
- Use of neural networks for in-line process control in the manufacture of surgical instruments. The result is higher precision, increased yield, reduced need for human intervention of the production floor.

- Use of high performance computers to model reactors used in refining low grade crude oil. The result is improved yield of high-value fractions of refined oil, increased use of low-grade, less expensive oils found in the Western hemisphere and reduced dependence on Mid-Eastern oil.

A Prescription for Success in Technology Transfer Programs

Companies know that market opportunities cannot wait. While Congress and DOE have set the stage, exploiting current opportunities for industry collaboration continues to require bold steps and a sharp focus. The following recommendations are steps essential to building partnerships for shorter term success.

1. Streamline DOE's industry partnerships program review and approval process. Current DOE programs funding collaboration with industry are dramatically oversubscribed. Of the companies committed to at least a 50% cost share on a collaboration, only 5 to 10% can now be supported. Many excellent projects are rejected simply for lack of funds. For those collaborations that do receive funding, the process from the point of initiation of the proposal preparation to commencement of work often exceeds one year. Streamlining the proposal preparation, review, recommendation, authorization, and project implementation cycle will greatly reduce the growing perception that the process is the only product of working with DOE.

2. DOE should maintain oversight responsibilities and delegate accountability for strategic direction definition and implementation to the DOE national laboratories. DOE should continue its program commitment to technologies identified and defined by industry. The definition of strategic direction and the authority to implement procedures aligned with industry should be delegated to the laboratories, with full accountability for successful execution of resulting programs. Each DOE laboratory has the responsibility of defining for industry the laboratory's core competencies and how these capabilities align with industrial future technological needs. DOE should maintain oversight responsibilities and ensure representation/coordination of the Laboratory's activities with Congress and other government agencies.

LANL strongly supports the planned expansion of the industrial partnership programs to affect significantly the economic competitiveness of U.S. industry. It is anticipated that these programs will emphasize stronger links to LANL's core competencies, greater input on program directions, and closer tie-ins with major DOE programs. In addition, it is recommended that DOE seek ways to streamline the process of executing and implementing industry-laboratory partnerships, including the delegation of authority (even directing discretionary authority) to the national laboratories.

BUILDING STRONG TECHNOLOGY TRANSFER MECHANISMS THROUGH SMALL BUSINESSES: ECONOMIC DEVELOPMENT AND LICENSING

Small business drives the nation's employment and much of its commercial innovation. However, most small and minority-owned businesses lack the resources to seek and develop technological opportunities with federal laboratories. Building on past successes, LANL proposes initiatives to strengthen significantly its ties to small and medium-sized business.

A Track Record of Success

Since the 1970s, LANL has worked actively in local economic development and in encouraging small business cooperation and spin-offs from the laboratory. These efforts have facilitated:

- Los Alamos Economic Development Corporation (LAEDC), established in 1983. LAECD is now widely recognized both for local success and for its state and regional leadership in SBIR and business assistance promotion.
- Los Alamos Small Business Center, established in 1985 as one of the early "business incubator" facilities in the U.S. and the first associated with a federal laboratory. This center now operates three facilities.
- Over forty business spin-offs currently employing over 600 people. These spin-offs are in such disparate technology areas as computer security, controls software, machine parts inspection, shipping container identification, nuclear safeguard, instrumentation, and environmental clean-up diagnostics.
- Aggressive licensing activity. Twenty-one of twenty-two licenses (patents and software) granted by the laboratory as authorized agent of the University of California were to small businesses. The number of licenses is growing rapidly.
- Several successful technical assistance programs, often in collaboration with Sandia National Laboratory, including New Mexico's **STARS** (State Technical Assistance Resource System) program and LANL's Industrial Manufacturing Internship Program.

A Need for Directed Effort

Because of the stretched resources of every small business, new opportunities assistance must be available with convenience and efficiency. Similarly, the sheer number of small businesses which might need assistance or profit from new laboratory technologies can be overwhelming. These considerations indicate the benefit of using of existing intermediary organizations—small business development centers, various state, local and regional groups, and trade groups—to facilitate communication and identify specific needs and opportunities. LANL's natural constituency is in the Southwest, particularly in the laboratory's home region and state.

Success requires a broad set of cooperating organizations with multi-agency state and federal cooperation. Several existing organizations at the state, regional and national levels are poised to cooperate effectively in these initiatives. These include most state science/technology/economic development organizations, NASA's Regional Technology Transfer Centers, the Federal Laboratory Consortium and the U.S. Chamber of Commerce.

Initiatives

- Develop regional small business technology assistance programs through partnerships with local and regional economic agencies, state and local governments, educational institutions, and federal laboratories.
- Develop substantial, continuing dialogue with venture capital groups for rapid exploitation of major commercialization opportunities suitable for start-up, small, minority, and woman-owned businesses.
- Cooperate in the development of technology theme initiatives (e.g., in environment, biotechnology, computation) for synergism in start-ups and expansion.

- Create a process to identify small business technology needs, particularly those common to many small businesses (such as environmental compliance), and match them with the laboratory's capabilities and technologies.

The central theme of these initiatives is efficient movement of the laboratory's technology into new and improved products and services in small- to medium-sized businesses.

In LANL's proposed spirit of institutional entrepreneurship in a networked, cooperative environment, the laboratory seeks and accepts full responsibility for development and execution of the program. Specifically, the laboratory's approach must provide fair access for small and minority businesses and defensible, clear decision methods when it is necessary to choose one business over another.

In recognition of the important role of small business in the nation's economic health, these initiatives should be supported by DOE programmatic funds. LANL's program would be coordinated with those at other DOE laboratories in both strategy and operation, to help ensure the best value for the federal investment.

To support the above initiatives, the Administration should undertake the following:

- Charge DOE with developing small/minority-owner business technology-transfer programs decentralized to the laboratory level, with *direct funding and reporting.* DOE should maintain oversight responsibility.
- Charge each laboratory with developing implementation programs for these business sectors, stressing *short-term impact.*

The entrepreneurial spirit of small and medium-sized businesses meshes well with the technical entrepreneurship of LANL engineers and scientists. Much of our success with small business preceded the formality of a laboratory "technology transfer" mission. Thus, the current climate can nurture a major initiative. The above efforts could be the nucleus for increased vigor in existing small business programs and for a new era of cooperation across federal, state and private sectors. The resulting success would have enormous impact on U.S. strength in the increasingly global economy.

CONCLUSION

For the United States to regain its international competitiveness, we need to:

1. Provide funding commensurate with the necessary U.S. commitment to achieving economic security. The Clinton-Gore technology policy recommends applying up to 20% of each laboratory's budget to joint ventures with U.S. industry. We recommend the following breakdown of such funding:
   - ° Two percent of the laboratory's budget for direct funding of technical services, outreach, business functions, industrial exchanges, technical information, and pre-CRADA activities necessary to build strong programs with industry.
   - ° Two percent of the laboratory's budget for support of commercialization of the laboratory's technologies and technical assistance and other services to small businesses (economic development and licensing).
   - ° Eight percent of the laboratory's budget for specific collaborative activities stressing market-driven, cost-shared programs with national scope (industrial partnerships).
   - ° Eight percent of the laboratory's budget for participation in industrial national challenges. These programs, funded primarily by the government, would focus on long-term enabling technologies with expected industrial impact in the ten-year range.

2. Develop critical, meaningful performance measurements that reflect the long term nature of benefits from these programs.

If the government continues to invest in the industrial partnering infrastructure of Los Alamos National Laboratory, and if mechanisms can be developed to leverage the laboratory's resources on a broad front, LANL can produce significant and vital contributions well beyond the 21st century.

LANL can gain much from industry. Our mission-directed efforts (e.g., nuclear defense, military weapons systems and large science projects in materials, energy, and energy conversion) can benefit from industry's experience and knowledge in such areas as concurrent engineering, quality assurance, and cost-effective development approaches. In such technology areas, industry's capabilities exceed those in our laboratory. Industrial interactions will allow our laboratories to accelerate the development cycle of their primary missions and reduce the taxpayer's expenses that support these activities.

CURRENT PRACTICES, COMING CHANGES

Norman Peterson

Strategic Planning Group
Argonne National Laboratory
Argonne, IL 60439

INTRODUCTION

There is a consensus being developed in Washington for new government-sponsored efforts to encourage technological innovation as a partial means of increasing the competitiveness of American companies. Toward that end, Congress and industry have begun focusing on the resources of the national laboratory community to help companies use new technologies. It is a foregone conclusion, following the end of the Cold War, that these laboratories need to develop new missions, and that their activities will begin to see a more immediate result in the civilian rather than the military sector.

The laboratories have extensive facilities and equipment, and human resources that can be used by industry either through a single company/laboratory effort, or through multi-company/multi-laboratory cooperative agreements. In the near future, however, a number of obstacles need to be overcome to ensure effective partnering between industry and the laboratory community. For the most part, both sectors are unfamiliar with the capabilities and needs that each has. The laboratories are primarily technology-driven, with a concentration on developing new capabilities and long-term responses to problems. Industry, by contrast, is market-driven. It concentrates on generating technologies to solve specific product development problems in a short time. Another important issue is that industry is very attentive to the costs of developing new products or processes, while the laboratories have traditionally not given attention to this concern.

Even as these issues continue to be addressed, we are witnessing a number of successful partnerships being developed that take full advantage of laboratory resources in meeting and solving current problems. The following three projects illustrate these new efforts.

THE AMTEX PARTNERSHIP: A ROLE MODEL FOR U.S. COMPETITIVENESS

The AMTEX partnership is a new collaborative R&D program developed by the laboratory technology transfer program of the Department of Energy's (DOE) Office of Energy Research, the integrated U.S. textile industry, and DOE's national laboratories. The purpose of this program is to engage the unique technical resources of DOE labs to develop and deploy technologies which will increase the competitiveness of the integrated U.S. textile industry.

From Lab to Market: Commercialization of Public Sector Technology,
Edited by S. K. Kassicieh and H. R. Radosevich, Plenum Press, New York, 1994

This will be accomplished through multiple cooperative research and development agreements (CRADAs) between DOE laboratories and R&D/educational institutions associated with the integrated textile industry.

The AMTEX initiative has two distinguishing features: 1) joint, long-range strategic R&D planning by the DOE labs and an entire integrated industry, and 2) an operational framework producing up-front coordination of proposed work between energy-research and defense-program laboratories. Through this operating mechanism, AMTEX members and DOE labs will submit recommendations for consideration by DOE program offices that are industry-driven, coordinated, and prioritized. The laboratory technology transfer program in the DOE Office of Energy Research is leading this effort. It is coordinating program direction and support with other DOE program offices, including defense programs and conservation and renewable energy. Other program offices, such as environmental restoration and water management, are likely to participate as the program becomes established.

A significant feature of the AMTEX program is that the national laboratories are working collectively with an entire vertically-integrated industry through five industry-supported, non-profit research, education, and technology-transfer organizations including the nation's leading textile research universities. Through this innovative structure, the results of the AMTEX R&D programs will be available to both large and small businesses. The textile industry has long recognized the interdependence of the fiber, textile, apparel and fabricated products sectors. The individual companies as an industry are linked to strong interaction and mutual support in technology development and deployment.

The AMTEX program focuses on the unique capabilities and resources of DOE laboratories on making major advances step-wise in the industry's competitiveness. Manufacturing process advances will be coupled with information technologies to create a customer-driven, integrated network of companies capable of responding rapidly to both collective and individual customer preferences.

The U.S. apparel industry has been severely hurt in recent years by imports from countries with very low labor costs. The entire U.S. industry, including all the interdependent sectors of fibers, textile apparel and fabricated products, and retailing, is responding to this challenge. Its approach is to change the competitive battlefield from low labor costs and mass production to product quality, value and responsiveness to the customer. This strategy will be implemented through new manufacturing and retailing paradigms such as "demand activated manufacturing," direct fiber-to-apparel weaving techniques and semi-automated custom apparel, services and products that only American companies can provide. This strategy will preserve and expand the number of well-paying jobs for American workers.

The "leapfrog" technologies expected to flow from AMTEX research programs will create demands for new types of textile and apparel manufacturing equipment not existing today. The technology needed to develop these new industries will be in the U.S. Hence, the manufacturing job base will be in America. The contract governing performance of AMTEX programs has strong provisions for keeping resulting technologies within the United States.

THE FORUM: A MARKET-DRIVEN PROCESS FOR TECHNOLOGY APPLICATION

As U.S. major manufacturing companies feel increased global and domestic marketplace pressures, those pressures are transferred downward to their small and medium-size subcontractors, translating into demands for greater versatility in production capacity, documented state-of-the-art quality control, and flexible delivery schedules that can meet immediate production needs ("just in time").

These and other demands create imposing challenges to the subcontractors, many of whom are not prepared to meet the new demands.

Past efforts at enhancing competitiveness through technology application have been less than successful, largely because they have focused on a technology-push rather than market-pull process. Federal and state governments have long been concerned about the plight of U.S. small and medium-size manufacturers (SMMs), citing the need for modernization and an enhanced technology base. Federal laboratories, universities, and other research organizations have sought to transfer technology to industry. Business and financial assistance has been available to help SMMs bolster the management aspects of their businesses. Community colleges have long provided the business and technical training needed to support the implementation of new technologies and technical practices. In spite of this wealth of available resources, job retention/creation has not occurred to the extent desired. The reasons for this are complex, but center on several key issues, the most important of which is the need to start with a statement of supplier requirements by major manufacturers.

Argonne National Laboratory (ANL) has created a process, the FORUM, to explore and test how DOE and other federal laboratories, in collaboration with state governments, can help small and medium-size manufacturers meet the increasingly competitive technical and business demands of the domestic and global marketplace. The FORUM process culminates in an exposition where technology applications are presented at conference halls or other centers located at universities, community colleges, and other community facilities convenient to SMMs. The process comprises the following elements and steps.

1. The market

   Step 1: Define the supplier requirements of major manufacturers in a single industry, in a way that is understood by that industry's SMMs.

   Step 2: Identify the impact of those requirements on the needs of the SMMs. Steps 1 and 2 are conducted through focus groups comprised of industry representatives.

2. The technological base

   Step 3: Identify technologies and technical practices in the public and private sectors (multiple federal laboratories universities, engineering firms, equipment manufacturers) that can address the technological needs of the SMMS.

3. Business/financial assistance

   Step 4: Identify business and financial organizations that can provide the business and financial management support that SMMs need to apply new technology/technical practices (Small Business Administration banks, venture capital firms, leasing companies, trade associations, and community organizations, including colleges and local business and political leadership)

4. Resource accessibility

   Step 5: Present the information created in Steps 1–4 at strategic locations at times convenient to SMMs to create an accessible integrated approach to technology application.

DOE's Sandia and Oak Ridge national laboratories have agreed to join Argonne as partners in the FORUM, focusing on the vehicle manufacturing industry (both on- and off-road) serving the states of Illinois, Indiana, Ohio, Michigan, Minnesota, Tennessee, and Wisconsin. The FORUM will be guided, directed, and coordinated by a secretariat comprised of representatives of the vehicle manufacturing industry (large manufacturers and small suppliers), federal and state governments, technology providers, and business and financial assistance providers. The secretariat's first steps involve the formation of two business focus groups, one involving the major vehicle manufacturers (Chrysler, Ford, General Motors, Caterpillar and Deere), who will define their requirements of subcontractors, and an impact of these requirements on the subcontractors' needs. The second group will be comprised of their subcontractors. This information will then be reviewed and analyzed by the technology providers, who will coordinate the presentation of information at the FORUM.

In each participating state, the FORUM will be conducted in several communities selected through a bid process. An operations contractor will be used to handle the logistical elements of each FORUM. Local community, trade, and professional organizations will publicize the event. The shows will be scheduled at times convenient to SMMs.

At the FORUM, the SMMs will work with representatives of the organizations providing technology and business/financial assistance. These discussions are expected to result in the development of consultation agreements between the SMMs and the technology providers. These agreements and the ensuing technology applications will be documented through the technology providers.

Teaming three major DOE laboratories, each with special technological strengths and research facilities and committed to working with small business, is one of the most significant elements of the FORUM approach. These premier research and development organizations command a combined total annual budget of well over $2 billion and employ more than 17,000 scientists and engineers. Argonne is an acknowledged leader in transportation technology and Oak Ridge in advanced materials for manufacturing; Sandia is well known for its highly developed engineering and industrial practices. Combined with other federal laboratory, university, and private-sector capabilities, the FORUM technology base is an unprecedented and powerful tool that may be brought to bear synergistically on the need for modernization and growth by SMMs in a geographically focused, industry-specific, market-driven way.

ARGONNE NATIONAL LABORATORY/BETHEL NEW LIFE PARTNERSHIP: AN EXPERIMENT IN URBAN TECHNOLOGY TRANSFER

Argonne National Laboratory and the Bethel New Life Inc. (a community development corporation) in Chicago's West Garfield Park neighborhood have entered into a unique partnership based on the premise that there may be technical solutions to the problems of this low-income inner-city neighborhood. The two organizations have focused on the development of new industries that can be the basis for jobs in this community, on establishing training programs appropriate to such industries, on revitalization of the neighborhood's industrial and residential infrastructure, and on a program in higher education incorporating working knowledge of city needs with high-quality traditional engineering curricula.

The program presently consists of related efforts based on the community's resources and requirements. To revitalize industry in the area, provide training in an expanding field, and restore the commercial infrastructure, Argonne and Bethel will assess the environmental status of abandoned factories in West Garfield and devise plans for cleanup and reuse. Argonne and Bethel will develop a process for plastic recycling, using steam from a municipal trash incinerator to develop raw materials for new area small businesses. These two resources (steam and recycled plastic) may attract new and existing plastic manufacturing companies to the area.

The partners will also develop, in a two-stage program, new affordable and energy-efficient housing in West Garfield. In the first phase, state-of-the-art techniques will be used in the reclamation of two twelve-unit apartment complexes. Based on lessons learned in this first phase, materials/practices will be developed appropriate to more affordable, efficient, and durable new construction/reclamation. In both phases of this program, emphasis will be placed on training opportunities, possibilities for new business development in the neighborhood, and new product/service development for existing small businesses. Argonne, in this project, will serve as the window to other national laboratories and universities with major efforts in the area of energy conservation/construction technologies.

Finally, Bethel, Argonne, the University of Illinois at Chicago, and the Illinois Institute of Technology, in conjunction with Chicago city colleges and high schools, will develop a program in urban engineering that couples a degree in traditional engineering curricula to job experience in area businesses and work with area community development organizations. This program will establish a cadre of highly trained engineers cognizant of the social, economic, environmental and technical requirements for restoring American cities.

The Argonne/Bethel partnership encompassed in these projects is an experiment in technology transfer. If it results in better jobs, higher living standards, and a more habitable neighborhood, its experience will be transferred to other neighborhoods in Chicago and to other cities.

The three above-mentioned projects can be precursors of other collaboration projects to be developed between the laboratories industry, and communities. Each has brought together a set of unique laboratory capabilities that have been able to respond to a set of industrial and community problems. All three of these efforts required a number of constructive planning meetings addressing the obstacles listed earlier in this paper.

Making these mechanisms work requires an intensive hands-on effort, a familiarity with the participants, and, most importantly, a degree of flexibility that can accommodate change. In the final analysis, these types of partnerships will make their mark in contributing to national competitiveness.

THE ROLE OF THE RESEARCHER

Lee W. Rivers

National Technology Transfer Center
Wheeling, WV 26003

INTRODUCTION

Facilitating the application and commercialization of federal technology is a complex process, but not necessarily a chaotic one. Professionals in technology transfer are trying to understand the diverse factors in the process so that they can rationally control it.

To understand technology transfer, one cannot simply describe the researcher's role, the engineer's role, the executive's, manager's, lawyer's, marketer's, technology-transfer specialist's, or recipient's roles. It is a complex group process that requires a broad approach.

The federal government spends in excess of $70 billion annually on research and development—about half of the R&D done in the United States. "Some" (rather than "a lot") of the federally funded technology has been commercialized. There is a tremendous amount of potentially useful technology in databases and files. There also is a treasure house of scientific knowledge in the minds of the 100,000 federal researchers, which I believe is the primary source of technology for industry.

Congressional acts of the past decade and legislation currently being considered recognize the value to American industry of this mother lode of R&D products, skills, and facilities. The question is, how to get the public sector and the private sector to work well together. The rest of this paper will focus on this question.

The two sectors—federal government and industry—represent different cultures:

Federal: R&D has a specific public purpose, such as a tank cannon, a re-entry shield for a spacecraft, a new road construction technique in a national forest, or a method for measuring the flow of coal slurry. Until recently, not much thought was given to commercial uses for such technologies or for the processes used to manufacture them. And, except when the national security was at issue, most technologies were available on a non-exclusive basis.

Industry: American companies have depended mostly on their own R&D to come up with targeted products and processes. In the competitive commercial world, it is important that they keep their new technologies close to the vest and that they move them to the market as fast as they can.

From Lab to Market: Commercialization of Public Sector Technology,
Edited by S. K. Kassicieh and H. R. Radosevich, Plenum Press, New York, 1994

INCENTIVES TO WORK TOGETHER

Legislation during the past decade, as well as new acts being considered by Congress, have been aimed at bringing government technology developers and industry into a working relationship. Rather than listing the laws of the past twelve years, I will mention some incentives for cooperation and collaboration which they have provided for the public and private sectors. Dick Chapman and Fred Grissom, in "Mining the Nation's Brain Trust," organized these ideas starting with industry incentives:

- Systematic and fair access to federal technology.
- Enablement for cooperative R&D.
- Reduced R&D costs.
- Use of federal expertise, data sources, and unique facilities.
- New licensing opportunities.

Incentives for government organizations to cooperate with industry are:

- Outside support for laboratory and agency R&D programs.
- New external sources of funding for agency R&D.
- Access to advanced industry technology.
- Royalties for researchers and laboratories and agencies.

Industry Incentives

- Access to federal technology is freer and fairer now that legislation encourages the labs to cooperate with the private sector. Uniformity of the access process is spreading across the agencies. One problem that remains is that small and medium-size, and sometimes even large, companies don't know where to start looking for assistance or technologies. NTTC is the hub of a one-stop-access national network that has removed much of the difficulty from the first approach.
- Enablement of collaboration grew out of the Technology Transfer Act of 1986, which gives federal laboratories the authority to enter into cooperative R&D agreements (CRADAs). This permits federal and industry scientists and engineers to work together in laboratories, share the costs, and share the results. The 1986 act requires the labs to give preference to small and medium-size firms, although many CRADAs have been signed with large ones.
- Company R&D costs can be significantly lowered through joint ventures, use of unique facilities, and assistance from federal researchers. Also, the cooperating companies have access to all the scientific results of the ventures while paying only part of the total cost.
- Use of unique federal facilities not only lowers the cost of doing research, but allows companies to do R&D that they otherwise would have no way of doing. And their capabilities may be highly leveraged by working with federal researchers who may have unique expertise.
- The new authority of agencies to grant exclusive licenses to companies is a major incentive to do business with the government. A lead time of a year or two, or even six months, can help keep a firm in the forefront in the competitive, fast-changing world of high technology.

Government Incentives

- Probably the most powerful incentive is the demands of Congress and the administration. With the new emphasis on technology as the fuel for American economic competitiveness, agencies have started viewing technology transfer as a way to impress the executive and legislative leadership with their responsiveness. The more collaboration with industry (including CRADAs) and the more useful assistance they provide to companies, the better they hope to be treated at budget time. Praise from pleased companies in their local areas can influence representatives or senators.
- Additional funding through collaboration is another important incentive. The CRADA authority allows agencies to accept private funding for joint R&D, thereby advancing their own research as well as the companies'.
- The benefits of collaboration work both ways. By interacting with industry, laboratory scientists gain access to company expertise and technology that may contribute to their own work.
- The final federal incentive, sharing of royalties by laboratories and scientists, has not yet fulfilled their hopes. But royalty paybacks can take a long time to grow.

BEST TECHNOLOGY-TRANSFER PRACTICES

Bob Carr, in his *Journal of Technology Transfer* article (Spring/Summer 1992), conducted a survey of technology-transfer practices at federal laboratories, including some operated by research universities. He then produced a menu of what he thinks are the best practices. A selection of these practices can give one an idea of the scope of technology transfer in the federal government:

- *Agency and laboratory attitudes and techniques supporting technology transfer.*
  - Encouragement from the highest to the lowest management levels in such areas as funding, program direction, and staff activities.
  - Formal recognition, such as career paths, better job descriptions, and appropriate personnel evaluation.
  - Creation of a technology-transfer network to help individuals keep up to date in techniques and to keep each other abreast of technologies with commercial potential.
- *Intellectual properties for use by collaborating companies.*
  - Training of technology-transfer professionals, members of the technology-transfer network, and scientists to recognize the commercial value of their intellectual properties.
  - Simplified disclosure and patenting procedures to encourage the inventor.
  - Fee access to research facilities by "technology ferrets" (representatives of an industry or a consortium of firms) to identify technologies of interest to their clients.
  - Use of personnel with relevant technology and marketing experience to make patenting decisions.
  - Collaboration with business schools to do market research.
- *Preparing technologies for commercialization.*
  - Internal and external maturation funds dedicated to bringing technologies closer to commercialization.
  - Use of CARDAs for maturation.
- *Using technology-transfer intermediaries.*
  - Working with industry associations to identify generic technology needs and to find laboratories that can help meet the industry needs.

- ° Use of venture-capital firms as intermediaries to help develop start-up companies.
- ° Use of technology brokers to help small federal R&D labs evaluate and commercialize technologies.

HOW NTTC FACILITATES THE TRANSFER OF FEDERAL TECHNOLOGY

The National Technology Transfer Center's purpose is to strengthen the competitiveness of American industry by making sure that business has rapid and productive access to marketable federal technologies and expertise. To do this, NTTC is organized into three facilitating functions:

- Gateway/clearinghouse providing industry with useful information about and access to the federal laboratories.
- Education and training program.
- Economic-development program.

NTTC has developed alliances with the following organizations: National Institute of Standards and Technology, National Information Service, Federal Laboratory Consortium, Electronic Industries Foundation, PENNTAP, US Navy, Strategic Defense Initiative Office, and Association of Federal Transfer Executives. NTTC also receives thoughtful, knowledgeable help from its three advisory boards representing industry, the federal agencies, and technology mangers.

The Gateway/Clearinghouse

NTTC has compiled a database and indexing system on technology expertise, research in progress, and available unique facilities at the 700 federal laboratories. Industry access to the database is through a toll-free telephone number (1-800-678-NTTC). Calls may have such results as a cooperative R&D agreement, identification of a needed technology, verbal technical assistance, or joint benchwork.

Typically, this is how the NTTC system works:

1. A company scientist or engineer calls with a need or a problem. A worker takes basic information about the caller and the problem, and passes it to the head technology agent.
2. Within two days, a technology-transfer specialist with the appropriate expertise is assigned the task, calls the client, and helps him or her develop a precise definition of the problem.
3. The technology agent or an information specialist enters the database and identifies laboratories and researchers focused on the problem area. The agent contacts the appropriate individuals through each laboratory's office of research and technology applications (ORTA). The agent determines whether a researcher is indeed able and willing to help the client.
4. Within two weeks after calling the 800 number, the client is given the telephone numbers of federal researchers who offer assistance or collaboration.
5. NTTC completes follow-up analysis and evaluation on the effectiveness and impact of the technology transfers resulting from this operation.

Education and Training

NTTC's training mission is to develop and deliver customized technology-transfer courses to government, industry, and universities. The organization is gathering a group of contractors who have expertise in the areas covered and who are highly skilled in the art of training.

NTTC's first major client is the US Navy, for which the agency is preparing a 30-month training program for the Navy's research laboratories. It is aimed at improving the transfer of their technologies to American industry. The participants will be laboratory personnel directly and indirectly involved in technology transfer operations.

Other NTTC projects and plans include:

- Training program at several Environmental Protection Agency locations starting in the spring of 1993.
- Series of short seminars, conferences, and forums to foster innovation and raise technology-transfer awareness among professionals in all sectors.
- Education program, including bachelor's and master's curricula in technology and technology transfer, at Wheeling Jesuit College. The undergraduate curriculum will begin in Fall 1993 and the graduate program will get under way in the following school year.

Economic Development

NTTC is working on various fronts to expand the role of technology and technology transfer in traditional economic-development initiatives:

- Fund for Strategic Partnering, aimed at developing innovative technology-transfer methods. The FSP provides matching funds to encourage universities, industry, non-profit organizations, and state and local economic development groups to team with, federal laboratories for commercializing new technologies.
- Capital-sourcing database for technology-based business start-ups and small and medium size companies.
- Cooperative agreement with the Electronic Industries Foundation, under which a panel of executives from member companies will help the federal laboratories coordinate their research with the needs of the electronics industry.
- Training program for economic development officials from New York, Pennsylvania, Maryland, and West Virginia, aimed at helping them in the areas of technology transfer, management, and program operations.

Another projected program is an award to federal laboratories for excellence in technology transfer and contributions to economic development on local and regional scales. The award procedure will be patterned after the Malcolm Baldrige industry award.

There are other NTTC activities—such as development of measurement techniques and a series of conferences and publications on issues in technology transfer.

THE IMPACT OF FEDERAL TECHNOLOGY TRANSFER ON THE COMMERCIALIZATION PROCESS

Roger A. Lewis

Office of Technology Utilization
U.S. Department of Energy
Washington, DC 20585

Some people find the suggestion that federal technology transfer can impact technology commercialization impossible to accept. Federal technology transfer can, and does, impact the overall technology commercialization process. The U.S. is feeling some of this impact now, and shall experience more in times to come. The author's experience is with technology transfer at the U.S. Department of Energy. Some of the lessons learned about federal technology transfer at DOE may offer ramifications for others in the federal system.

BACKGROUND

Over the past four years, our world has undergone breathtaking changes. The cold war has ended and the former Soviet Union has collapsed. While there is a significant reduction in the number and types of U.S. nuclear weapons that the U.S. will maintain, the potential spread of weapons to conflict-prone "third world" regions continues.

Substantial growth in federal funding for research and development has been seen. Except for the space station, the Department of Energy system plays a dominant role in all the national major research development efforts, such as SSC and the Human Genome Project, while continuing to emphasize full compliance with laws, regulations and accepted standards for all DOE facilities. At the same time, DOE and all other agencies funding research and development are experiencing significant budgetary constraints. These changes open opportunities and create challenges rivaling any we have experienced. They also send a clear signal that the status quo has changed radically. One example from the DOE system that brings this point home is that, for the first time since 1945, the United States is not building any nuclear weapons.

The number one national (and international) priority is U.S. competitiveness in a global economy. The U.S. can no longer rely on domestic markets for the goods it purchases or the services it seeks to sell. The U.S. economy must be revitalized, and it is increasingly evident that technology will play a critical role if it is to succeed. This is the bottom line for federal technology transfer. It is the responsibility of managers, producers, and refiners of technology to do everything they can to make sure that technology plays the critical role it must to keep the United States economically competitive.

From Lab to Market: Commercialization of Public Sector Technology,
Edited by S. K. Kassicieh and H. R. Radosevich, Plenum Press, New York, 1994

Expanding collaboration between federal laboratories and private industry can reduce two key elements: the investment requirements needed to commercialize technology, and the risk that industry must assume to make the commercialization take place. The outcome can be a win/win deal for both industry and government. Industry secures valuable assistance and expertise at minimal risk and cost. The federal research community leverages precious dollars and, at the same time, introduces business expertise and perceptions into government facilities.

LAWS AND REGULATIONS

The laws that govern technology transfer activities and how they are done include:

- The first piece of legislation establishing technology transfer as a legitimate component of the federal R&D process was the 1980 Stevenson-Wydler Technology Innovation Act.
- Since 1980, the Bayh-Dole Act and its amendments have affected intellectual property ownership rights associated with technologies developed by some types of government contractors.
- Since the mid-1980s, legislation has allowed more cooperative research in the commercial world, leading to more cooperative research in the government world.
- In the late 1980s, an executive order from the White House encouraged agencies to step up their efforts to transfer technologies. This was accompanied by a reduction of OMB limitations on working with businesses.
- The 1986 Federal Technology Transfer Act, and the 1989 amendments which implemented it in the DOE system, is the single most important legislative driver in federal technology transfer today.

New legislation expanding the types of things done in technology continues to be passed by Congress. Many of the new laws concentrate on setting up and managing technology transfer processes. The National Defense Authorization Act for fiscal year 1993, for example, authorizes the Secretary of Energy to establish a new program to encourage more cooperative work with small businesses.

Another example is the Small Business Technology Transfer Act of 1992 directing DOE, DOD, HHS, NASA, and NSF to fund cooperative R&D projects involving a small company and a researcher at a university federally-funded research and development center or a nonprofit research institution. Each project must support the mission of the sponsoring agency. The difference between this program (known as STTR) and the Small Business Innovation Research program (known as SBIR) is that STTR will enable researchers at universities, FFRDCs and nonprofit organizations to join forces with a small company to spin off their promising ideas while still employed by the research institution.

Legislation is integrating technology commercialization processes into the mainline research and development. The Energy Policy Act of 1992 extends the use of technology-transfer mechanisms to alternative energy programs, while requiring a minimum cost-sharing commitment of 20% with outside partners for R&D activities and 50% for commercialization efforts. Similar "direct application" technology programs have existed for many years, but now there is more legislative authorization to do so than ever before. Federal agencies other than DOE, will feel the effects of new laws "mainstreaming" technology commercialization as part of their R&D programs in the future.

What is the overall impact of these pieces of legislation establishing federal technology transfer as both a means and an end in the management of federal research and development? I believe there are two major types of impact. The first is immediate and direct—an increased focus on sharing federally developed technologies with external partners as uses are defined. The second is harder to measure, but holds promise for far-reaching impact because it includes concerted efforts to work with industry from the outset to define and carry out technology programs.

DOE TECHNOLOGY TRANSFER

Almost 160,000 people work in DOE sites and programs; half of them are technically trained in science and engineering. All the DOE technologies have been developed with dollars from U.S. taxpayers; replacing the unique sites and equipment managed by DOE on a daily basis would require a capital investment in the hundreds of billions of dollars. Because most of these sites are managed by management and operating contractors, there are additional dimensions to technology transfer that differentiate DOE facilities from other agencies. There are, however, several government-owned, government-operated sites within the DOE system.

The DOE Enhanced Technology Transfer Program was put in place in January 1991 as a systemwide response to the National Competitiveness Technology Transfer Act of 1989. The NCTTA established technology transfer as a component of all DOE programs and R&D missions; it further allows the DOE laboratories to enter into cooperative research and development agreements (CRADAs).

Efforts to institutionalize technology transfer at DOE over the past four years include:

- Guidance and policies promoting collaboration between the DOE system and industry, state and local organizations and educational institutions.
- Formal meetings with industry partners to share information about what is working for them, and what is not.
- Reaching out to contact potential partners in industry, other government agencies, and schools on a regular basis.
- Memoranda of understanding with state agencies as well as other federal agencies to provide channels for cooperative efforts.
- About 2000 invention disclosures within the DOE system each year. From these 2000, about 800 patent applications are filed and almost 200 patents are awarded. An increasing portion of these patents are being licensed.
- Opening the gates (literally) of formerly secret plants for the public to see the available expertise, facilities and equipment—the Y-12 in Oak Ridge in June 1992, the Kansas City plant last fall, and the Panex Plant in Texas in February 1993.

DOE has learned a few lessons from experience. Some of these lessons hold promise for others in the federal technology transfer community. These lessons also reflect individual impacts on the overall technology commercialization process.

LESSONS LEARNED

- Present DOE partners indicate a need to decentralize the processing of technology-transfer agreements. Now, CRADAs are negotiated and approved in the field without going to Washington.
- Industry must be sure that the information provided and developed in concert with us will be confidential and protected (the 1989 Act lets us provide exclusivity for up to five years under CRADAs).

- Model CRADAs can expedite the negotiation process, but one size does not fit all. Instead, DOE has developed special model CRADAs for specific industries (computer, manufacturing) as well as a modular CRADAs with alternative "plug-in" provisions for particular requirements.
- Successive deals get easier. A company's first collaboration with a laboratory often takes a lot of time and effort; after the first round things go more smoothly because both parties have learned from one another and assumptions are understood.
- The advance payment required by law is especially burdensome for small businesses. Alternatives are being developed to make this easier. These include reducing the amount needed up front from new partners, or paying funds into an escrow account and allowing the partners to continue to draw interest.
- An on-line system is in place to make sure that deadlines set in the 1989 law for reviewing and approving CRADAs are met.

And these things are working.

At the end of January 1993, there were more than 350 CRADAs in place across the DOE system. The average approval time for a CRADA over the past year has ranged from four to twelve days (this is down from more than 20 days per CRADA the previous year). More than one quarter of our current CRADAs have small business partners, and 7% include university partners. These CRADAs include a federal contribution of more than $278 million (44%) and participant contribution of more than $358 million (56%).

CRADAs are easy to count and have a special visibility because they are legislatively mandated, but they are by no means the only way to develop working relationships with outside partners. Deciding which mechanism to use depends on individual circumstances. Every case is different in some way. Technical assistance continues to be a major component of federal technology transfer. Consulting, personnel exchange, user facilities, work-for-others and cost-shared procurement are valuable mechanisms for transferring expertise and resources, and expanding the possibilities for commercializing federal technology.

There are still some outstanding issues to resolve. At the top of the list is product liability. Industry does not want to assume full liability for injury or damage occurring from products that result from collaboration with us. Outcomes resulting from emerging federal policies in this area will have a substantial impact on the private commercialization process.

U.S. competitiveness needs to be better defined. While the overriding purpose of our technology transfer activities is to benefit the U.S. economy, it is not easy to define the best ways to make sure that is the case. What constitutes a U.S. company? What types of jobs should be required to remain within the U.S. ? How much impact should federal policies on technology transfer have on U.S. company commercialization practices?

The critical role played by small businesses, particularly in a technology-based economy, is apparent. Small businesses tend to be more flexible, innovative, agile and non-risk-averse than large industry. Small businesses also provide three out of five new jobs in this country. How is the U.S. going to encourage more cooperative efforts with small businesses? How do small businesses commercialize technologies differently from large industry? What are the different types of interactions that small businesses have with technology brokers and other intermediaries, compared to large companies? What kinds of assistance do small businesses need from us compared to that needed by large industry?

Some issues are hard to wrestle with and require participation by others to solve. Laws and regulations focusing on conflict of interest, federal procurement, and antitrust enforcement continue to change as the entire federal system gains more experience in technology transfer. The last issue of this paper is intellectual property rights. Although progress is being made in defining ownership of rights, this continues to be a sticking point in laboratory-industry negotiations and must be handled on a case-by-case basis. The lessons to be drawn from any single instance are limited by the diversity we encounter with each new opportunity. And yet intellectual rights, and the way they are managed, represent one of the most significant factors in defining the impact we can have on the technology commercialization process.

WHAT'S NEXT?

The issues the U.S. must tackle grow more complex. Over the next year or so, time and energy will be devoted to addressing the following questions:

Technology Transfer and Financial Conflict of Interest

By law, federal employees may now receive royalty payments for their inventions. However, if the government does not patent the invention and the employee does, the payments would come directly from a commercial enterprise rather than through the government agency. This may put the employee in a conflict-of-interest position under laboratory COI policy.

Copyright of Software Developed by Federal Employees

Although federal employees may patent their inventions, they have no right to copyright protection for the software they develop. In contrast, employees of federal contractors (including management and operating contractors) can obtain copyrights on software they design. For several years now, there have been bills proposed to enable federal employees who develop software under a CRADA to file for copyright.

Research and Experimentation Tax Credit

A temporary tax credit for research and experimentation, designed to encourage private-sector R&D, expired on June 30 1992. Both the DOE National Energy Strategy and the Clinton/Gore policy statement *Technology: The Engine of Economic Growth*, recommend making the tax credit permanent.

Modernization of Intellectual Property Law

The intellectual property provisions of the Bayh-Dole Act, hailed in 1980 and 1984 as innovations, now present a barrier to involving federally-funded small businesses, universities and not-for-profit organizations in cooperative R&D ventures with private sector organizations. Provisions that made sense when most R&D projects received 100% federal funding no longer are the best way to go when more government-supported R&D is cost-shared with a private sector partner.

Protection of Commercially Relevant Data in Cooperative Research Generated by Technology Transfer Agreements

Current arrangements for protecting commercially relevant, jointly developed, scientific and technical information may not be adequate in today's technology transfer climate. Industry representatives are especially concerned that the five-year protection period for data may not be long enough and that protection from release under Freedom of Information Act requests is uncertain.

Coordination of the Treatment of R&D Agreements and Trade Agreements

At present, there is some disagreement between U.S. trade negotiators and U.S. federal R&D agencies. The issue is whether joint R&D activities between federal and nonfederal organizations should be treated as subsidies under trade agreements. One of the reasons this is a problem is that the trade and R&D communities have not traditionally worked together. This situation is changing to accommodate the expanded role of federal technology transfer.

SUMMARY

The federal technology-transfer community is complying with both the spirit and the letter of the laws and policies that define their operations. They are learning to transfer and commercialize federal technologies faster, better, and with more impact that in the past.

The resolution of many of the issues described in this paper will help decide the ultimate scope of the impact of federal technology transfer on the commercialization process.

THE IMPACT OF THE FEDERAL TECHNOLOGY TRANSFER ON THE COMMERCIALIZATION PROCESS CONFLICT OF INTEREST

Albert Sopp

University of New Mexico
Albuquerque, NM 87131

Like Roger Lewis (see p. 109), this paper focuses primarily on the DOE component of federal technology transfer and, in that area, the issue of conflict of interest.

The bad news about conflict of interest, according to some experts, is that it is vexing. That is, it is usually characterized by vaguely written policies full of generalities and catch-alls. There are few legal guidelines (because so few court cases). Conflict of interest can be easily politicized, is often subjectively applied, is subject to emotional treatment and public relations outbursts, and is burdened with a doctrine like criminal loitering—the mere "appearance" of conflict being sufficient to establish wrongdoing.

The author's perspective of DOE in the technology transfer context is conditioned by views from both inside and outside. The inside part is experience as a DOE patent attorney involved in intellectual property and technology transfer, both in DOE HQs and in the field, for ten years ending in 1986. Since then, he has handled intellectual property, including licensing, for the University of New Mexico, involving many dealings with DOE and the nearby Sandia and Los Alamos national labs.

The term "technology transfer," in the sense of commercialization of federally funded R&D as a programmatic concept, probably originated in 1971 with a gentleman named Metcalf, an employee of the Navy's Weapons Center at China Lake, CA. Metcalf suggested that the Navy and other DOD components ought to look at commercialization of weapons technology as a kind of bonus return on the military R&D investment dollar which could help justify the very substantial Cold War and Viet Nam War military R&D and procurement expenditures.

There were two objections to Metcalf's proposal. One was that attention to commercial uses of military technology would conflict with and dilute researchers' concentration on development of military hardware. Another was that commercialization by government agencies conflicted with private industry—i.e., government should not compete with private industry.

Of course, federally-funded "technology transfer" had been going on since the 19th century, as part of federal contracting. A primary example of 19th century technology transfer was shipbuilding. In this century, the most obvious example of technology transfer has been in the aviation industry. Many airframe companies started and grew through government

From Lab to Market: Commercialization of Public Sector Technology,
Edited by S. K. Kassicieh and H. R. Radosevich, Plenum Press, New York, 1994

contracts, beginning with WWI, and continuing through the twenties and thirties into WWII, and on through the Cold War highlighted by the Korean and Viet Nam conflicts.

During those eight decades there developed a close relationship between the military departments and the airframe manufacturers. This relationship was strengthened by the dual-use nature of aviation technology, where many products and processes involving aeronautics avionics, etc. have dual uses, both commercial and military applications.

Aviation technology was generated largely at taxpayer expense, often involving the use of government-owned facilities, government-generated technical data, and government or military personnel, with the civilian contractor retaining invention rights. As authorized by Congress to encourage the maintenance of contractor expertise in military-related technologies, contractors have received public funding, known as Independent Research and Development funds ("IR&D") and have been permitted to treat inventions and technical data generated under IR&D as proprietary.

It seems pretty clear from these arrangements that use of public funds by federal contractors to develop commercial products was not regarded as a conflict of interest, but rather as a community of interest. Perhaps not coincidentally, for most of those eighty years United States military and commercial aviation enjoyed world-wide preeminence. The United States aviation industry is almost the only U.S. industry that has consistently produced a favorable balance of trade.

In the Armed Services Procurement Regulation (ASPR) first promulgated in 1947–1948, DOD contracts had provisions ensuring identification and reporting of contractor-made inventions and furnishing the government rights in technical data, so that in addition to having future or follow-on procurements protected by patent and data rights, the government could procure items based on technical data competitively at lower prices and/or establish second sources.

Over the years, federal contractors have resisted these ASPR patent and data provisions on the grounds that contractor inventions or technical data were developed totally or partially at private expense (these, of course, included IR&D-funded inventions and technical data). Certainly, the contractors preferred treating such inventions and technical data proprietarily for purposes of both commercial and military (including foreign) sales. Proprietary rights in inventions and technical data were essential to establishing "sole source" status. Disclosure to the government was felt to amount to disclosure to the world, especially in view of the Freedom of Information Act (FOIA) passed in 1966. Arguably, technology transfer through federal contracting led to commercial success and advantage for some contractors. To the author's knowledge, this policy was never successfully challenged on grounds of conflict of interest,i.e. the use of public funds to develop technology which provided commercial advantage to one or more particular contractors.

GAO investigations in the 1960s showed that some contractors tried to treat their technical data (and related inventions) as proprietary, pertaining to items, components, or processes developed at private expense rather than at federal expense. The Boards of Contract Appeals and the courts held that the government had unlimited rights in such data and inventions, even if the government had made only a very slight contribution to their development. This situation led at least one respected private industry observer in 1974 to conclude that many contractors of the highest competence might avoid government contracts because of overreaching federal policy, particularly regarding technical data rights.

In 1973–74, when the Energy Research and Development Administration ("ERDA") was established, there was some sentiment in the Congress, especially by Senators Hart and Long, that the public "should not have to pay twice" for technology including inventions developed with public funds. (Contrast this with the rather generous patent and data policies of DOD.)

Thus, an early draft of ERDA's legislative patent policy would have required compulsory licensing of government-funded inventions by contractors. This was later changed to a rigorous title-with-waiver policy vesting title in such inventions in the government and, if the contractor wanted title, requiring the contractor to request, via an involved and rather rigorous procedure, that the government should waive its title rights.

The "title-with-waiver" policy became law. In practice, the grant of such waivers to large businesses, covering inventions that might be made under contract, has often required substantial cost-sharing. One might argue that this has had the effect of mollifying those advocating that the "public should not have to pay twice" by requiring the contractor (waiver-recipient) to pay for invention rights. In a sense, it might be regarded as a kind of advance recoupment by the government of revenues later received by the contractor through commercialization of such inventions. Perhaps it also can be said that if conflict of interest was perceived to exist by virtue of certain ERDA-DOE contractors being allowed to convert government-generated technology into proprietary, royalty-producing properties, that such conflict was resolved by requiring the waiver-recipient-contractor to pay the government—the taxpayers—for those properties.

In the 1980s, policy-makers in the Reagan Administration embarked on a radical change of federal antitrust and intellectual-property policy. The policies then in place, except for DOD, had been inherited from the trust-busting era culminating in the 1930s, and were: (a) a strong antitrust bias in the Justice Department against patents, (b) a preference for government ownership of inventions and compulsory licensing, and (c) the concept that "the public should not have to pay twice" for government-generated patents and technical data. The message from the Reagan Justice Department was that proprietary rights were the food of a vital economy, and necessary for successful international competition. Rigorous economic analysis showed that the inherited policies were wrong because they gave international competitors a "free ride," i.e., free access to federal contractor-developed technology, thus reducing the economic value of new technology and the incentives for federal contractors and others to bring the technology to the marketplace. This was hurting the U.S. economy. Thus, it would be economically advantageous for essentially all federal agencies to follow DOD policy, under which the government should have "government use only" license rights.

The impact of the new Justice Department approach and other initiatives of the Reagan Administration led to PL. 98-577 (allowing DOD contractors to control the rights in technical data developed at substantial government expense) and other laws and executive orders generally providing greater patent and data rights to government contractors and other types of participants in federal programs.

Of course, this policy change favoring proprietary treatment of government-generated inventions and technical data has enabled federal contractors to establish commercially advantageous proprietary positions more easily, including trade-secret treatment of technical data developed (at least in part) at U.S. taxpayer expense. It is interesting to note, in connection with this marked change in policy, that there was little or no objection made that federal contractors might have a conflict of interest, i.e., might be in a position to steer research efforts and results under federal contracts toward developing commercial products or improving trade secrets rather than concentrating solely on providing the most effective goods and services for the contracting federal agency.

Patent and data policies of almost all federal agencies now allow large business contractors to retain proprietary rights in inventions and technical data generated at substantial government expense and to commercialize them, charge the public royalties. The Energy Policy Act of 1992 (which retains some cost-sharing features of the ERDA title-with-waiver policy for large businesses) intends that technology transfer shall have the same or higher priority than furnishing of goods and services.

Federal employees, as well as contractors, are involved in technology transfer. Under the Federal Technology Transfer Act of 1986 (FTTA) federal-employee inventors at government-owned, government-operated laboratories are statutorily permitted to receive royalties on their inventions, and to consult with contractors or licensees commercializing the inventions, although not for direct payment by them. This function of federal employees is presently (Winter-Spring 1993) under Justice Department review because of alleged conflict of interest due to prohibitions on federal employee conduct existing in previous, unrepealed statutes. This review will probably not change the thrust of the FTTA.

Presumably there will be no further criticism that contractors commercializing government-generated technology are in a conflict of interest situation, even when assisted by government employees whose inventions may dovetail with theirs, because they, as well as the government employees, are able to realize a profit in technology they were allowed to create at public expense under government contracts. Obviously, the effect of the foregoing is that technology transfer, even when directly aided by government employees, has been placed by the Congress at a priority higher than conflict of interest.

Further, if a federal employee invents something which is related, but not "directly related", to the employee's official duties, under Executive Order 10096 the employee may retain title in the invention and if he/she wishes to commercialize it, may do so and still remain "on the job" without a conflict of interest problem. If, under Executive Order 10096, the circumstances surrounding a federal employee's invention require assignment to the government, the employee may obtain an exclusive license through government licensing procedures and commercialize the invention through his/her private company. Any conflict of interest involved would arise from violation of rules in other areas, such as influencing a federal procurement with the inventor's company, self-dealing, etc.

With this background in mind, let us consider DOE's national laboratories, all of which are taxpayer-supported, government-owned facilities operated by contractors under five-year management and operating "M&O" contracts. Until the Bayh-Dole Act in 1981 (P. L.96-517), inventions and technical data generated at these facilities were owned and controlled by the government, and commercialization of these inventions and technology on an exclusive basis by the M&O contractor or any of its employees was generally forbidden unless a waiver was granted by DOE—a rare event.

Beginning with the earliest days of the AEC, through ERDA, and extending to the present day, DOE's M&O contractors have been required to report technology developments to the public before being allowed to commercialize them. This precluded, or at least discouraged commercialization by the M&O's corporate affiliates, including those involved with atomic energy. Of course, Congress, in establishing the AEC, had been concerned that M&O contractors labs should not gain unfair commercial advantage in the atomic energy field.

Part of the rationale for restrictions on M&O contractors was that the M&O contractor acts as an agent or quasi-agent of the government, is expected to maintain objectivity in evaluating other contractor programs for DOE, and theoretically would have an unfair advantage in developing and utilizing technology in a wholly government-funded facility without risk, overhead and direct expenses. In short, the objection seems to be that such a contractor would have an unfair competitive edge over other kinds of contractors and should not be permitted to profit from the operation of these facilities.

Primarily as the result of Bayh-Dole, M&O contractors for DOE's nonweapons laboratories were allowed to commercialize lab-originated inventions and return the royalties therefrom to the government or to the laboratory facility. Since 1981, DOE has permitted commercialization of lab-originated inventions by sponsors who pay all costs, including overhead for the lab research work producing the invention. In addition, such sponsors, under a class waiver granted in 1981, may treat technical data and inventions first produced under such arrangements (called "Work for Others") as proprietary.

Since passage of the National Competitiveness Technology Transfer Act (NCTTA) in 1989 establishing "technology transfer" as a mission for contractor- and government-operated federal laboratories, commercialization of lab-originated inventions and technical data for private gain (other than through "Work for Others") has been permitted through Cooperative Research and Development Agreements ("CRADAs"), which normally involve substantial cost sharing of the research by the CRADA participant.

Recently, with the passage of the yet-to-be-implemented Energy Policy Act of 1992, DOE has received considerably more latitude in granting invention and data rights to contractors. This act has the potential of substantially enhancing DOE's technology-transfer capabilities through direct contracting with small businesses and, with some cost sharing, large businesses.

At present, regarding DOE CRADAs, there is no DOE policy encouraging or expressly allowing CRADA participants to retain as consultants the same technical people in the DOE lab who generated the inventions or technical data to be licensed and retained as proprietary. Policy on consultants appears to be left with the M&O contractors, some of whom do not allow it, probably because of concerns about possible investigations by Congress or DOE which, whether or not well founded, could involve the vexing aspects of conflict of interest mentioned at the beginning of this paper.

Apart from CRADAs, a DOE lab employee who makes an invention and wishes to commercialize it (e.g., by starting a company), or to investigate possibilities of commercialization while working on the same technology at the lab, is generally precluded from doing so for reasons of conflict of interest. Presumably, such an employee would be in a position to steer lab technology, or possibly even subcontracts, to his/her outside business—a situation perceived to involve problems in case of a conflict-of-interest audit by either the agency or Congress. Also, the M&O contractor may not favor a policy that is no doubt rather difficult to administer and would allow its employees to leave, be hired away, or be diverted from other duties for which they were hired.

From a conflict of interest standpoint, what really are the differences between a national laboratory employee commercially using patents and technical data he or she generates as lab employee and a federal contractor commercially using patents and technical data it generates as a federal contractor? Why should the former be a conflict-of-interest situation while the latter is not?

As mentioned above, an M&O contractor may obtain a DOE waiver to commercialize an invention arising at the laboratory. Also, an employee of such a contractor may obtain such a waiver and then perhaps, start a small business to develop/commercialize the invention. As in the case of federal employees, there is nothing prohibiting a lab employee from commercializing the waived invention while remaining as a lab employee.

The distinction between DOE lab employees and federal employees or contractors occurs at this next stage: the federal contractor or employee is permitted to commercialize while at the same time performing work in the same field of technology as the commercialization. The DOE lab employee, on the other hand, is prohibited from commercializing in the same field as his/her laboratory work.

In effect, federal policy conveys this message: it is all right for a federal employee or federal contractor—of either the large- or small-business variety—to be paid by the taxpayers to commercialize the very technology being concurrently developed for the government. In fact, we have seen that this activity by contractors has been encouraged for many decades.

However, it is not acceptable for a DOE lab employee (and starter of a small business) to do the same thing. Instead, the lab employee who wants to start a small business in a field identical or closely related to his/her field of work at the lab is required to change to a different field, or to quit the lab. Such a step obviously removes the employee from his/her place of expertise, cuts off technology flow and prevents leveraging of research effort.

Thus, in the case of the federal employee or contractor, the expertise and benefit gained through federal work may flow immediately and continuously to commercial development. In the case of the DOE lab employee, this flow and benefit is lost. Apparently, the concept driving this policy distinction is conflict of interest.

Somehow, the DOE lab employee situation is different conceptually from the situation of a federal employee or private company-federal contractor, even though all would be in positions making it possible to steer technical information, generated at federal expense and not then available to the public, to a private company/contractor for that company's private benefit.

Traditional conflict of interest and property rights analysis would conclude that the DOE lab employee should not be allowed to convert the employer's property to his/her own use without the employer's permission. However, does or should this rule apply where the employer (in terms of ownership of property rights and beneficial interest) is not the contractor but, in reality, the federal government? Is this not the case for M&O contractors who operate facilities owned by the government? Can the above distinction mean that, by selective application of conflict of interest principles, the government in effect favors transfer of technology through established businesses (small or large) as contractors or federal laboratory employees rather than through individual employees (of established businesses) whose inventions can form the basis of small business start-ups? More simply stated, does not conflict of interest policy militate against formation of small businesses by DOE lab employees?

CONCLUSION

The author recommends that, because the mission of DOE's national labs includes technology transfer, and because it is recognized as an economic reality that small business start-ups substantially contribute to economic well-being, conflict of interest policy should be construed to permit small business start-ups by DOE lab employees even while (if not especially because!) they remain in lab positions directly related to the technology of the small business.

While the conditions recommended to govern this activity are somewhat complex because of other potentially competing technology transfer efforts of DOE labs via CRADAs, etc., the appropriate rules governing business start-ups by DOE lab employees might take the following general form:

a. The lab employee's start-up business grants to the government a nonexclusive, nontransferable (except as provided below) irrevocable, paid-up license to practice or have practiced any invention and technical data generated by the business for government purposes (including disclosing and granting of a nonexclusive license therein to CRADA participants and lab sponsors to whose projects the employee's immediate technical unit is assigned at the lab) for the period during which the lab employee involved in the business start-up is an employee of the lab in technology directly related to that of the business, and for one year thereafter;

b. The lab employee's business shares royalties with the lab or government at a modest level. (Note: This operates as a conceptual substitute for up-front cost-sharing by large businesses, as under DOE waivers and as is indicated under the Energy Policy Act);

c. The employee's business follows government policy on U.S. preference regarding treatment of inventions and technical data generated by the business during the period the employee is employed by the lab as in a) and for one year thereafter, and

d. The employee may request or agree to be reassigned to another component of the lab or leave the lab based on a commitment by the lab to perform work under a CRADA or other arrangement for a third party in a field directly related to the employee's field.

In conclusion, in the 1980s the government had little difficulty in making radical changes in antitrust and intellectual property policy based on a rigorous economic analysis. Can we not now look at another such change in how conflict of interest is viewed based on the same kind of analysis? Is not the perceived conflict of interest regarding DOE lab employees really not conflict of interest but actually community of interest enabling added dimensions of technology transfer and leveraging of research effort? Should we not reexamine priorities as times change, and make the indicated adjustments in order to fully exploit technology transfer opportunities?

II

PARTICIPANT ROLES IN PUBLIC-SECTOR TECHNOLOGY COMMERCIALIZATION

INTRODUCTION

There is no single process description which can encompass all instances of technology transfer and commercialization. Those who transfer technologies from public sector laboratories to the private sector should recognize the many possible mechanisms (or methods) available (use of facilities, work-for-others, cost-shared contracts/cooperative agreements, scientific and technical publications, presentations at technical meetings, workshops and demonstrations, bidding conferences, employee consulting, third-party agents, exhibits, participation in electronic data bases etc.) and make use of the most appropriate combinations, considering the nature of the transfer, the return to the taxpayer, and the benefit to the U.S. economy. For each technology transfer, a comprehensive strategy should be devised that recognizes the efficacy of employing multiple transfer mechanisms over the sequence of activities which comprise the entire transfer cycle. Publication of technical papers in scientific or technical journals may be an appropriate mechanism to create awareness of the commercialization opportunity, followed by a CRADA or personnel exchange as mechanisms most appropriate to stimulate additional technical work to advance a particular application. One very comprehensive model of the process is provided in the article by Eliezer Geisler and Albert Rubenstein, *The role of the firm's technical entrepreneurs in commercializing technology from federal laboratories*, in this section.

Geisler and Rubenstein analyze the issues related to the low historical incidence of technology transfer and commercialization from federal laboratories. Part of the solution they propose is the use by the recipient firm of an internal champion or entrepreneur. However, to overcome the substantial number of barriers identified by the authors, they recommend a significant number of changes in the firm (such as incentives) to support the efforts of the internal entrepreneur.

Academics have studied the process of technology transfer with considerable effort for several decades. The literature is now replete with studies of processes in specific contexts, such as transfer via multinational corporations from developed to less developed countries, or intraorganizational transfer from one functional area to another. Several efforts have been made to describe "generalized" technology transfer models, but most critics of these models have felt that to make the models sufficiently general, they become too simplified to be useful; conversely, if they remain specific enough to provide useful guidance in the study of the process, they are inappropriate for all but a few circumstances similar to the context for which the original model was developed.

Presented below is a very simple model of technology transfer which suffers from the above justifiable criticism of highly generalized models, but provides a simple framework for discussing the format of this section of the book. Many of the articles in this section provide more specific, context-dependent models to amplify the discussion of different processes and roles.

The intent of these models is to build on the existing relevant knowledge base, and to focus particular attention on the context in which specific transfer process dictum has been identified and verified. For example, much work has been done to understand the technology transfer process from developed countries to less developed countries (LDCs) through multinational companies bilateral agreements between governments and multilateral foreign assistance programs. Although many aspects of these foreign transfer models are not relevant to the process being examined here, certain insights can be gained from this specialized body of knowledge; that is, several considerations derived from models and cases of international transfer can be very useful for some of the instances in which one might be more generally interested. Since we are interested in transfer to small or start-up companies from national labs, we can learn from these studies about the impact of cultural differences as barriers to transfer, since the culture of the inventing lab is radically different from that of a small business. Similarly, if a technology is developed by a national lab for application in the defense sector, it may have useful applications in other industries, but the efficient factor mix (labor, capital, technology, etc.) may be as different between industries as it is with international transfer.

Another general situation indirectly related to our interests and for which considerable research has been done, is intraorganizational technology transfer. Although the focus of the federal government technology transfer effort is interorganizational from government agencies and contractors to businesses and universities, much can be learned from phenomena studied in the intraorganizational case. For example, the infamous "not-invented-here" syndrome and functional barriers resulting from different perspectives, attitudes, and knowledge bases in different organizational divisions apply equally to interorganizational situations. In addition, since our interests include technology commercialization as well as transfer, it is important to consider models which provide insights into continued development and interfunctional transfers within recipient firms. As Geisler and Rubenstein contend, most transfers from government labs to private firms are initiated by exchanges involving principally the technical staffs of both entities. Most funds-in agreements to laboratories and CRADAs promote the interaction of basic and applied researchers. Mere transfer of intellectual property rights and understanding of technology from laboratories to private-sector scientists and engineers begs the issue of movement to commercialization within the firm. Roles of interest in this process include not only the intrafirm entrepreneur described by Geisler and Rubenstein, but also possible expanded roles of technology source engineers, some of whom have supported the manufacturing activities of government facilities, especially in the defense sector. Additionally, as Robert Keeley points out in his comments on the Geisler and Rubenstein article, technical personnel in both technology source and recipient can serve as new-venture entrepreneurs to create additional novel innovations from the technology.

In addition to technology transfer models developed for situations differing from our primary interest, there are models of related processes, such as technological innovation and technology diffusion, that can provide insights for our model development. Since technological innovation is frequently defined to include the steps of invention as well as commercialization, the latter stages of the process have considerable congruence with technology transfer; indeed technology transfer can be an important part of the innovation process. Technology diffusion is usually defined as the passive acceptance over time of a given technology into common practice without a purposeful attempt by the source to cause adoption in any particular instance. Since successful technology transfer is frequently found to include both "push" and "pull" mechanisms, it is likely that diffusion models can provide insights into the latter category of mechanisms, some of which may be valuable for technology transfer, especially when the transfer is on a nonexclusive basis.

Thus, while we are not trying to be all-inclusive in our review of existing concepts, models and studies of technology transfer and related processes, it is the goal of this book to include sufficient discussion of general models and examples of specific applications to cover most important issues and provide guidance to practitioners in their diverse situations.

SCIENCE AND TECHNOLOGY DEFINED

Before developing a model of technology transfer, it is necessary to establish the meaning of the term. Before we define and describe the term, we should agree on a meaning for "technology," and for the related concept of "science."

Technology is applied knowledge imbedded in tools, equipment and facilities, in work methods, practices and processes, and in the design of products and services. It is "know-how" in contrast to the "know-why" that characterizes science. Science is the non-application-oriented base of fundamental knowledge and understanding of phenomena from which developments in technology are derived. Science explains why things happen as they do, or why they are structured as they are, by investigating and determining cause and effect relationships and by synthesizing existing knowledge into new concepts, theories, and models. In terms of traditional stages of innovation, science is most closely related to pure, fundamental or basic research, while technology is most closely associated with applied development and engineering.

As an interesting aside, analysis of the early efforts of industry and federal laboratories to collaborate through the mechanism of CRADAs suggests that the vast majority are oriented toward partnerships which advance basic research (science, by our definition). We should also note the dramatic formation of research consortia since the 1984 legislation permitting joint "precompetitive" research. Obviously, the transfer of these results to members of the consortia entails more science transfer than technology transfer.

The common use of the term "technology transfer" does not distinguish science from technology, because of their inherently close relationship. For example, if technology were transferred in the form of a product prototype without also transferring associated science (which explains why the invention works as it does), the user would probably have to develop the prototype further to suit his or market needs through a relatively inefficient process of trial and error. If the associated science is also transferred, theory can guide the redesign in a much less expensive process of experimentation and evaluation. Thus, when we use the term "technology transfer," we will mean the transfer of technology as defined above and also (it is to be hoped) the associated science.

TECHNOLOGY TRANSFER DEFINED

Technology transfer is an extremely complex process with many factors that can influence its conception, operation and result. The process may be partially defined by describing the many mechanisms that can be employed, the kinds of institutions or individuals involved (such as the source, transfer agent and user), the barriers and enhancing forces, and the environmental or contextual factors. In its simplest definition, technology transfer is the process in which technology and its associated science is moved from the source of invention or discovery to a user, with or without adaptation, and with or without an intermediate agent. In graphic form, this simplified process may be represented as shown in Figure 1.

This simple graphic representation identifies five types of entities which influence the majority of transfer and commercialization processes. Three of these entities are directly involved in each episode of transfer—namely, the technology source, the recipient or user, and the agent or intermediary. Two other types of entities are indirectly involved, most commonly by influencing the general environment in which transfer is attempted, rather than directly in

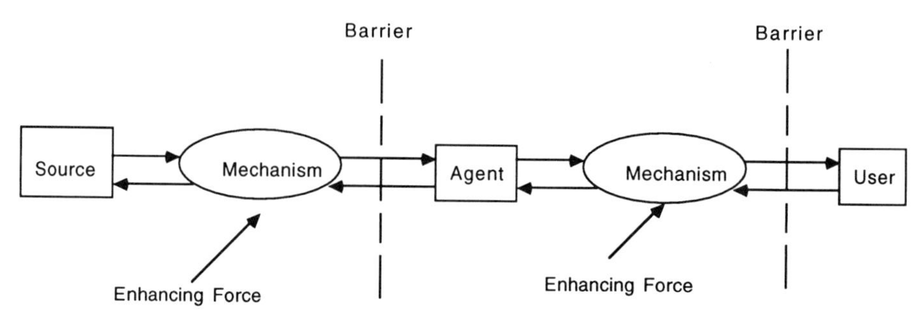

Figure 1. A simple technology transfer process model for role identification.

each episode. One of these indirectly-related entities is that set of institutions and individuals that provides enhancing forces to promote technology transfer and commercialization, in general and in specific instances. Examples are professional societies and associations, such as the Technology Transfer Society or the Federal Laboratory Consortium, whose primary missions are to be enhancing influences for technology transfer. Others, such as Congress, only occasionally concern themselves with technology transfer, but in isolated incidents provide positive influences such as supporting legislation.

The other type of entity indirectly influencing the transfer process is that set which, often unintentionally, places barriers in the transfer path. Examples include academics and government agencies who perform ill-conceived studies using poor measures of success and failure which mislead policy makers. Other examples are managers and policy makers who are driven to avoid the appearance of impropriety or conflict of interest, and who, in the process, provide disincentives to undertaking the risks inherent in innovative activities.

The roles of the three types of directly related entities are discussed in this second section and the processes and mechanisms, including enhancing forces and barriers, are examined in the third section of the book. The roles of each type of entity are also discussed briefly below.

This generic description of technology transfer may include elements which are unnecessary or even dysfunctional in some instances. For example, an agent may not be useful if the technology is "packaged" in an appropriate manner and both the source and the user are sufficiently sophisticated in the technology, the application and the transfer process.

We will now elaborate our definition of technology transfer by examining in greater detail the respective factors identified above: source characteristics, agent characteristics, user characteristics, barriers, enhancing forces and environmental influences. As each factor is elaborated and discussed, we will cite examples so that the correlates of success or failure are understood in terms of these factors. The prescriptive paradigms to be presented in the last section of this book will then represent the most significant "clusters" or combinations of factors which seem to be frequently associated with success and are frequent occurrences in the set of opportunities available to the national laboratories.

ROLE DIFFERENCES IN PROCESS VARIATIONS

The taxonomies of technology transfer have defined a number of dimensions which may serve to differentiate one process from another. For example, technology "push" processes are often compared to technology "pull" processes. Since there are many different forms of processes, the roles of the respective participants will obviously vary from one process to another. In the example of "push" vis-à-vis "pull" processes, the roles of the technology sources and users, as well as the relationships, are quite different. In a push transfer process, the role of the source is highly active while the recipient is relatively passive; this situation is likely to be reversed in pull transfer processes.

The degree of radicalness of the innovation or technology advance is another dimension of the technology commercialization process which significantly affects the roles of the participants. As Betz explains, radical innovations often restructure entire industries [1]. For example, the advent of PET bottles created a major new player in the beverage container business and lost significant market share for users of the old technology. Radical innovations are more likely to evolve from a "push" technology transfer technique than are incremental innovations. Incremental innovations, which enjoy a higher success rate than radical innovations, are more likely to result from a pull transfer method in which user problems or needs provide parameters for design innovation.

In this second section, Tornatzky and Ostrowiecki (*Technology needs: The art and craft of identifying, articulating, and communicating* p. 137) examine the process of identifying needs for technology advances. As they explain, the roles of the technology source and the recipient or user change substantially with increases in the degree of innovativeness required. The more radical the innovation, the less likely it is that the user will be able to define needs in a manner that translates into innovation attributes. With incremental innovations, the technology source and the user are more likely to interact effectively to visualize characteristics of the new technology.

As Tornatzky and Ostrowiecki point out, the transfer and commercialization process has issues at both the organizational level and the level of individual participants. Thus, the process dependency of roles not only pertains to the types of organizations involved (sources, recipients, intermediaries etc.) but also to the organizations themselves. That is, a source organization can have a highly-centralized process with transfer responsibilities primarily vested in a formal office, or, as the current trend seems to be, a decentralized process with technical staff having greater responsibility. In the latter case, role responsibilities at the individual level become increasingly important.

Tornatzky and Ostrowiecki list critical dimensions to consider in selecting these individuals, including personality, experience, background, and preparation and training (including creativity training and industry education). At the organizational level they describe the composition of groups, as well as their roles and functions as the critical dimensions of transfer process design.

Having established that roles are process-dependent, we will now examine roles in terms of the generalized process described above. In addition, several of the articles in this section describe role variations dependent on differences in processes and situations.

THE ROLE OF THE TECHNOLOGY SOURCE

As described above, the technology source assumes primary responsibility for successful technology transfer and commercialization when the push mode of transfer is employed, when radical innovations are involved, and when spinoffs result in source personnel becoming the primary agents of commercialization. Many public-sector technology sources are still groping with the issues involved in defining their roles. From the research concluded to date on technology transfer, the issues to be resolved have become more obvious.

Source Characteristics

The existence of the following characteristics in the technology source has been de strated to increase the probability of successful transfer, although the importance characteristic depends on the type of transfer mechanism employed, the user characteristics, and the type of agent used, if any:

1. Incentives to facilitate technology transfer (e.g. financial rewards to the inventors of the technology such as sharing royalties).
2. Existence of a formal, well-staffed technology transfer function.
3. Systems and procedures within the source which allow technology users access to facilities and staff in the source organization.
4. Willingness by the source to provide continued technical assistance as the recipient adapts the technology and refines its application, especially when transferring technology to small businesses.
5. A technical staff with strong involvement in extra-organizational professional activities which facilitate interaction with industry.
6. A multiplicity of funding sources and sponsors, some of which are industrial.
7. A comprehensive base or inventory of intellectual property and the ability to establish and protect strong proprietary positions in technology, including trade secrets and know-how as well as patents.

The role of technology source scientists or engineers in the identification of applications for technology advances is often assumed to be quite limited, especially if the institutional setting of a university or government laboratory discourages significant communication with industry. For this reason, federal research laboratories were required to set up Offices of Research and Technology Applications (ORTAs). It was assumed that the technology source would staff this technology transfer function with people who have the appropriate skills, personalities, and knowledge to be creative intermediaries between their laboratory's scientists and the potential industrial users of laboratory technology. As Tornatsky and Ostrowiecki point out in their article, few ORTAs have been staffed by using the personnel selection tools available to find the best candidates. They hypothesize that needs-driven technology transfer can be best effected by laboratory intermediaries who are creative, smart, interdisciplinary, empathic communicators. While most laboratory researchers and ORTA personnel are obviously smart, some of the other traits are not ordinarily found in researchers.

THE ROLE OF TRANSFER INTERMEDIARIES

Although the concept of transfer intermediaries usually implies third-party entities, laboratory ORTAs can be seen as intermediaries between the technical staff of a lab and potential industrial partners and technology recipients. Several laboratories are moving toward decentralization of this "internal" intermediary function. For example, Sandia National Laboratories has defined "product managers" in the technology transfer office, each assigned to a specific lab division. Other laboratories are experimenting with roles assigning formal responsibilities to "bound[illegible]anners" or "gatekeepers" in technical areas to represent their areas to [illegible]

[illegible]son explains in his article, *Technology transfer from federal labs—* [illegible], there are important roles to be performed beyond such "internal" [illegible]chnology sources, and the private-sector technology acquisition or [illegible]icle Anderson builds two comparative models of technology transfer [illegible]del and a "control system" model, in order to define the range of [illegible]an play in the process. In the conduit model, the intermediary is interposed directly between the source and the recipients, often taking ownership or a license to the technology rights, and serving as the primary communication channel between source and recipient. The control system model implies a much more modest role for the intermediary, who serves primarily to introduce the source to the user and to control information flows to ensure a successful transfer.

After describing the range of technology transfer processes in which agents can play significant roles, Anderson examines the various functions provided by agents and presents a taxonomy of types of intermediaries, from both public and private sectors. His conclusion is that most federal laboratories and universities can make more effective use of intermediaries if they carefully assess the roles which agents can perform.

The commonest conception of technology transfer intermediaries is based on functions such as agents, brokers, market researchers, creativity enhancers, or technology application specialists. However, as Tornatzky and Ostrowiecki explain, there are many aids and tools beyond human organizations that can facilitate the process. These include such instruments as computer mediation and expert systems, visual imagery enhancers, and facilitating protocols or manuals. Some of these tools are examined in more depth in the last section of this book.

The ultimate agent, of course, is the entrepreneur who secures the rights to laboratory or university technology (perhaps as the inventor) and proceeds to commercialize it. Robert Keeley, in his article *Transferring technology to start-up companies*, takes the lessons learned from the Geisler and Rubenstein article and extends their arguments to the external entrepreneur as intermediary. He describes the three parts of the process of entrepreneurship and explains how the traditional methods of federal laboratories must be modified to accommodate the entrepreneurial means of technology transfer. He effectively argues that such a method may result in more novel applications of the technology relative to the more usual ways of transfer to an established company.

Glenn Bacon, in his article, *Guerrilla technology transfer—The role of the researcher*, also argues strongly that the laboratory should recognize the historical effectiveness of researchers in the entrepreneurial act. He describes instances in which researchers carried their technology from the source organization through the commercialization stage, and, in the process stimulated the substantial economic development of regions like Silicon Valley. The "ladder-based" process of innovation described by Bacon is most likely to result in the novel (vis-à-vis routine) innovations represented by Keeley as more effectively commercialized through entrepreneurship. Bacon also describes the process of "cyclic" development of technology which commonly leads to the routine innovations portrayed by Keeley. Bacon explains why the U.S. system of technology development is superior in ladder-based, sequential, novel innovations but inferior to the Japanese and other systems which excel at cyclic, routine innovations and provide improved applications.

Bacon also explains why American companies are more successful at generating radical (novel) innovations in basic technologies, while the Japanese provide the product applications and incremental improvements which often capture large market shares in technologies invented elsewhere. He provides prescriptions for the federal laboratories to play a more significant role in the cyclic development processes of U.S. firms in order to gain and control market share.

In addition to the traditional concept of transfer intermediary, other kinds of agents have been shown to contribute to successful technology transfer, although the value of each type of agent and the services performed depends on the mechanisms used and the characteristics of the source and user. Although licensing agents are more commonly used by universities than by national laboratories, some of them represent considerable expertise and experience relative to particular industries or technologies. Their role with respect to federal laboratories may be related to introducing source and recipient rather than negotiating terms which effect the relationship.

Since the entrepreneur has been described by several authors as an effective intermediary, there are entities related to the entrepreneurial role which can facilitate the transfer of technology to an entrepreneur and assist in the commercialization process. Research parks have played effective roles in the transfer of university technology, and several federal- and state-supported laboratories are experimenting with these relationships. Research parks which include business incubators may offer special attractions to laboratories that are encouraging local economic development through spinoff entrepreneurship. Government agencies, including state universities, have joined federal laboratories to provide support services to local companies which are trying to transfer and use government technology.

THE ROLE OF TECHNOLOGY USERS

While the responsibility for technology transfer in the source organization is usually focused in a formal office, the user or recipient organization almost always does not have an identifiable responsible group. There are many different potential technology users in any sizable industrial concern. Process technologies may be imported by manufacturing engineers, product technologies may be acquired by designers or service departments, and R&D departments may infuse their projects with technologies of external origin. Although some U.S. firms have now moved toward greater centralization of technology planning and acquisition processes through the establishment of a chief technology officer, the identification of a specific potential technology user in a large recipient organization is a complex and difficult process.

User Characteristics

The following characteristics of technology recipients or users have been found important for a successful technology transfer event.

1. Scientific or technical capacity sufficient to understand the technology, to communicate effectively with the agent or source, and to modify the technology for new applications, operating environments, cost structures, manufacturing processes, etc., which facilitate commercialization.
2. Sufficient capital, management acumen, and experience competitive presence in the anticipated market and other elements of "business" capacity to ensure successful commercialization.
3. Existence of an internal champion for the technology whose enthusiasm for it is sufficient to offset any "not-invented-here" resistance.
4. Management commitment to long-term internal development of new business opportunities.

If the technology to be transferred and commercialized represents only an incremental step, the user organization is more likely to play a dominant role in specifying the desired functionality or other parameters of the application. If the advance is more radical in nature, the user organization often depends upon the source to identify opportunities for improvement and to test the feasibility of these advances. The user organization can then help formulate the final configuration. Tornatzky and Ostrowiecki argue in their article that better management of the relationship can include formal tools to specify future states and effect creative processes for finding technical solutions.

Often, the ultimate user of the innovation is not the recipient of the technology. The recipient may be a manufacturing firm which interprets the needs of the marketplace. Trade associations, third-party servicers and test facilities, and financing sources (e.g., leasing companies for capital equipment innovations) may all have a role to play in defining the innovation attributes which will invoke their enthusiastic endorsement and will facilitate their introduction and diffusion.

INDIRECT SUPPORT ROLES

Although not as directly involved in specific instances of public-sector technology commercialization as the sources intermediaries and recipients discussed above, there are important support roles to be performed. The most obvious is Congress and various policy makers, who create both a regulatory environment and a commitment of resources, including sources of funding and personnel. Legislative efforts since 1980 have made a dramatic impact on federal laboratories and contractors, providing both incentives and a more permissive milieu. Several articles in this book trace federal legislation and its impact on the process of public sector technology commercialization.

Since the 1984 act which permits precompetitive research to be jointly performed by private firms, the advent of research consortia has promoted cooperative work between firms, just as the use of CRADAs has fostered collaboration between the federal laboratories and industry. In the last section of this book, the article by Gibson and Rogers (*The evolution of technology transfer at MCC*) describes the operation and performance of one such consortium. Although research consortia are relatively new, it appears that they will be a useful mechanism to move research from the public to the private sector, once experience produces improvements in the process. The expectation of many observers of the development of research consortia is that they may prove to be especially effective in advancing the technology position of the United States in international commercial competition, at least in select industries where other comparative advantages also exist.

Other types of organizations have also had significant effects on the technology transfer process. Professional societies and associations have developed new functions, varying from training members in intellectual property issues or technology transfer mechanisms to serving as clearinghouses for technology searches or licensing opportunities. State and local governments, as well as nonprofit economic development organizations, have implemented a wide variety of support programs to facilitate technology commercialization, especially in the area of start-up businesses and small firms. Many of these local support organizations have evolved because the taxpaying public has become more informed and supportive of the technology transfer and defense conversion efforts of the federal government.

One area of support which seems to remain at an inadequate level is involvement of the risk capital community. Private investment in early-stage, public-sector technology is insufficient to commercialize the radical innovations which create new industries or recapture world markets. Many venture capitalists bemoan the "lack of good deals" but feel little obligation to commit early seed capital or effort to packaging advanced technology into commercial deals. And Congress has not constructed tax or other incentives to lengthen investors' horizons or broaden their scope to technology-based deals. Incentives by the government to induce activity in special industries, such as alternative energy, proved highly unprofitable to those entrepreneurs and their backers who were unfortunate enough to be drawn in. Their memories are unlikely to fade without concerted effort to build more functional relationships between government laboratories and private capital.

Entities Which Inhibit Technology Transfer

Whether intended or not, barriers to technology transfer and commercialization are inherent in the bureaucratic systems of the government and large firms. The adoption of new methods of interaction is slowed by the fear of change and its unknown impact on organizations and jobs. Bureaucratic processes and structures in the federal agencies, as well as in large universities and private firms, have inhibited the implementation of supportive policy. For example, from the passage of the 1989 National Competitiveness Technology Transfer Act until the implementation of the CRADA process authorized by the Act in the DOE Defense Programs laboratories, more than 18 months elapsed.

However, some of the reluctance of these large organizations seems to be reduced as a result of studies of practices in other countries and companies. For example, the not-invented-here (NIH) syndrome seems to be weakening in those progressive firms that have designated chief technology officers (CTOs) whose responsibilities include finding and acquiring interesting external sources of technology.

Some organizations fail to contribute their potential to the process of technology transfer because of an inability to deal with the conflicting attitudes and values between inventors (scientists and engineers) and persons responsible for commercialization (business-oriented people). The failure of these entities has been cited as a stimulus for entrepreneurial spin-offs in the private sector, but, in contrast, federal laboratories and most universities are not likely to have many inventors who would choose this mode to circumvent their institution-based frustrations.

Attitudes toward wealth generation and accumulation by federal employee-inventor (or employees of federal laboratory operators) have slowed the development of incentives that can motivate inventors to take a more active part in the commercialization of their technology. For example, until very recently Sandia National Laboratories management refused to let employees consult for remuneration or share in royalties derived from their inventions, in spite of the 1986 Technology Transfer Act which required a minimum of fifteen percent of royalties to be paid to employees. Similar attitudes about "conflict of interest" pervade the thinking of many administrators in state universities. Most regulations regarding conflict of interest require persons with an interest or potential interest in the outcome of the organization's decision process to report the existence and circumstances of the interest to a superior who will judge the possibility of conflict of interest. The regulations do not intend persons with interests to be prohibited from employment or actions which involve their interests, but rather, that they make superiors aware of those interests so the situation can be monitored for possible conflicts. The conservative stance of some administrators has discouraged many inventors from being involved in technology transfer.

CONCLUSIONS

The roles of individuals and organizations involved in technology transfer and commercialization are being reconsidered as new emphasis and resources are committed to the processes. As will be discussed in Section III of this book, the mechanisms and processes of technology transfer are continuing to evolve so that new roles and role modifications are required. We have presented a simple model of technology transfer so that we could define several types of roles—those of technology sources, recipients and intermediaries between the two. The papers presented in the following section elaborate on these generalized roles and discuss role adaptations to fit special circumstances and contexts for technology transfer and commercialization.

REFERENCE

Betz, F. (1987). *Managing technology: Competing through new ventures, innovation and corporate research.* Englewood Cliffs NJ: Prentice-Hall, 7.

TECHNOLOGY NEEDS

The Art and Craft of Identifying, Articulating, and Communicating

Louis G. Tornatzky[1] and Beverly Ostrowiecki[2]

[1]Southern Technology Council
Research Triangle Park, NC 27709

[2]Industrial Technology Institute
Ann Arbor, MI 48106

DEFINING THE PROBLEM

"Technology transfer" is one of those unfortunate terms that seem dense with meaning, but which on closer inspection are sources of definitional confusion. Let us define it as a set of activities engaged in by groups of people (researchers, product developers, marketers), who usually happen to work in different organizations, but who are more or less dedicated to accelerating the pace at which knowledge-embedded artifacts find themselves in the hands of another class of people, called "end-users." Looked at this way, technology transfer is really a special case of what might be called "R&D management," "engineering management" or "product realization" if it occurred under the organizational and legal roof of a single company. Since it often doesn't occur under one organizational roof, we have evolved a set of concepts, practices, and issues that are presumably specific to the special case of technology transfer as practiced across different organizations with presumably noncoincident missions and goals. (In fact, we can learn from the larger body of literature and practice dealing with product development, as will be seen.)

One of the recurrent issues of technology transfer derives from the observation that this kind of brokering operation, in which disparate people and organizations are presumably knit together into a "process" [1], would probably work better if all the role incumbents therein shared a common vision of the goals served by the process. Since the goals presumably involve satisfying the needs of end-users, there is great credence put into a *needs-driven* process of technology transfer. For many years, based on careful retrospective analyses of cases of successful innovation [2], it has been legitimate dogma. However, what does it mean operationally to create and implement a needs-driven process, and what are the features of human nature and organizational life that conspire against the attainment thereof?

From Lab to Market: Commercialization of Public Sector Technology,
Edited by S. K. Kassicieh and H. R. Radosevich, Plenum Press, New York, 1994

DEFINING NEEDS OF END-USERS

Defining end-user needs sounds like a fairly straightforward activity. Unfortunately, since we are trying to use needs to "drive" a process by which previously nonexistant knowledge is expressed in previously nonexistant tools and artifacts, the task is a bit daunting. In effect, we are asking people to think about the future in radical or nonincremental ways. In addition, we would like this process of envisioning needs to be somehow "valid" or true, so that when we arrive at that future time we will have planned and acted correctly—in effect, be creative and visionary, but feasible.

Defining future technology needs is also different from doing "technology assessment" or discerning broad trends in technical areas for planning purposes [3,4]. There are various measures of assessment including empirical analysis of patent data [5], systematic identification and evaluation of precursor events [6], or getting input from focus groups. On the other hand, we are wondering about the processes by which people envision fairly specific "futures."

Unfortunately, one of the facts about cognitive functioning and cultural socialization is that we tend to view the world through a limited set of conceptual and perceptual categories. Most of us avoid uncertainties in that conceptual-perceptual structure most of the time, and even more avoid using our energies to create more uncertainties. Moreover, the way in which we "vision" technology is very much a function of cultural context and socialization [7]. This state of affairs creates interesting problems for defining the user needs which might be met by next-generation technology. Let us describe a case.

Within the last year we have been involved in two interrelated projects [8,9] trying to define next-generation needs in processing or manufacturing technologies in a group of industrial companies. Both projects involved an interview/survey approach, although they differed in the extent of human contact involved (e.g., mailed surveys versus telephone interviews). The general strategy was to ask people to think about recurring functions or activities in their companies (e.g., design, fabrication, assembly) and the technologies and procedures pertaining thereto, and to identify for us "unmet needs" or needed new technologies, ideally with some sense of relative importance.

In the "set" we conveyed to our respondents, we explicitly invited them to dream and to be creatively vague. For example, we encouraged them to think about a needed functionality ("something that would do X") with no regard to what that something might look like.

Without going into detail, we can report several conclusions:

1. Given any alternative, most people will avoid thinking about future technology needs. This is particularly true when one asks them to attach words or descriptions to some future state. On the other hand, if one can provide end-users with prepackaged future scenarios they are slightly more willing to mull them over. People will comment on Rorschach cards, but they won't design them.

2. For most people, most of the time, the future is incremental. It is unusual and difficult to get people to express, visualize, or ponder *replacement* technologies or to talk about needs that are radically different from current practice. There is a strong thread in the innovation literature about the difficulties of adopting/implementing radical or complex versus incremental or simple technologies, and there seems to be an analogy in visualizing future technologies [10,11]. That is, people seem to have a hard time coping with the use of new technologies when those technologies "push the envelop" of their norms, values or ways of looking at the world. What could be more perturbing than thinking about what doesn't exist? The danger, of course, in attending primarily to incremental improvements is that we will become very good at hearing the train coming (by having our ear close to the track), just before it hits us [12].

3. It is difficult to get people to make comparative judgements, or establish priorities, between future need states or technologies. This is particularly true when the options represent radical departures from present reality. One can get people to rank options, but only with great interpersonal hand holding.

So much for a sleek and easy needs-defining activity from the perspective of end-users; the customer or end-user may not be terribly good at perceiving the future [13]. While we will continue to ask and poll end-users about their future research and technology needs [14,15], experience suggests that we will also need to explore novel ways and contexts to make that inquiry, and to involve other categories of informants. One important potential category of informants is those who are presumably at the cutting edge of the innovation-creation process. What about the apperceptive powers of researchers and developers?

WHAT DO RESEARCHERS AND DEVELOPERS DO ALL DAY?

The Richard Scarry question is an important one. If researchers and developers have special powers to anticipate the needs of end-users and/or to communicate about evolving technologies, the problems just described become less disabling. After all, since R&D people supposedly have been given time and jobs to develop more attention to thinking, speculating, and envisioning, perhaps they have unique capacities to divine the need-state of the world. Unfortunately, the following are either true or reasonable hypotheses:

1. R&D people, scientists and engineers, are somewhat more willing than the rest of us to tolerate experiential, conceptual, or perceptual ambiguity. That is, they are a bit more creative, but there are also major differences within these categories of people, and thus great overlap with the rest of us [16].

2. Scientists and engineers are probably further disabled by their training and professional socialization. For one thing, they are rewarded for looking at the world in terms that are discipline-based in physical or biological processes, and through theoretical concepts that pertain thereto [17]. Much of the current generation of scientists has been socialized and rewarded for *not* considering mundane end-user needs or applications. Moreover, the most fervent end-user needs may be inarticulate expressions of business concepts or philosophical values, and thus outside the cognitive categories of researchers.

3. Research people are probably among the more poignant examples of the ghettoizing of American professional life. Scientists and engineers tend to talk to scientists and engineers, and talk only to engineers and scientists in comparable-sized institutions. That is—large; there is little contact between technical people in small companies and cohorts in research institutions [18]. One of the ingredients in understanding the needs of others is the ability to appreciate the experience of those others. There are major institutional impediments to being able to do this, and, in many contexts, communication between scientists and end-users is fairly unproductive [19].

4. Surprisingly, incrementalism is a much more powerful strain in the culture of research than some might surmise. Kuhn's [20] classic distinctions between normal and paradigm-shifting sciences are still valid. Most academic science is still little more than "coprolitic accretion."

Even if end-users were better at articulating needs, or research people were better at being mindful thereof, there are nonetheless some other major obstacles to getting better at all of this.

COMMUNICATING ABOUT THE FUTURE OR ABOUT ANYTHING

Humans have developed communication systems to a high level. Unfortunately, not all of what has developed is applied to the problem of technology transfer. For example, a common medium of exchange among scientists and technologists is the technical paper and/or its accompaniment, the technical briefing replete with overheads. This mode of communication is best, and preferred, when dealing with content that is relatively well known and/or well operationalized [21]—the "as-is", rather than the "to-be".

One weakness of the written technical paper is the limited power of the printed word to convey nuances and ambiguous speculation and hints. Moreover, few scientists and engineers are given training in using language systems for creative expression. Some of the current efforts in "creativity training" [22] involve playing with a variety of alternative communication media: pictures, paints, drums, toys, and so on.

It is noteworthy that some of the more technically advanced organizations, those on the cutting edge of technology practice and achievement, are making greater use of nonprint media. These include video recording, CD/ROM, various approaches to computer-mediated group processes, animation, and so on [23]. Some organizations and companies are experimenting with various versions of "virtual reality" or "virtual proximity" as a way of allowing people to communicate about the future.

Another characteristic of traditional modes of communication is their "dyadic bias"—that is, they are often designed for, or imply, a relationship between two communicators. There is little ability to engage parallel and simultaneous channels of communication among general communicators. Again, cutting edge communication systems permit quite different kinds of interaction [24]. Too often, we limit ourselves to networks of text, with marginal opportunities for interaction. As bandwidths permit, and our timidity declines, technology will permit highly interactive, multimedia national town meetings in which needs and potentials can come together.

It is not only the medium which we use to communicate about future technology needs, but also the context or how we bound, or better yet not bound, the problem. The "law of the lens" has a great impact on what discoveries we will make [25]. Researchers and practitioners bring a plethora of assumptions and biases to their work; these make up the "lens" through which they see the world, and it is often subconscious. We see some things and not others. Becoming aware of these frames of reference frees scientists and engineers to explore alternative ways to frame a problem. Different perspectives can bring new solutions or turn constraints into strengths.

Tools and techniques that can help enrich our communication about the future include brainstorming, synethics, word association [26] attribute listing, morphological analysis [27], concept mapping [28] breakthrough thinking [29,30], analogy and anomaly [31], biassociation or fusion [32], visual excursion [33], various computer tools [34] and exploration beyond traditional market assumptions [35]. A common thread running through these techniques is the encouragement of new associations. Psychologists have long thought this is the key to creativity. Most of the techniques have been formally or informally shown to increase the quantity of ideas. A criticism is that it is unclear whether they produce creative and useful ideas. This criticism may stem from the difficulty in judging creativity. An increase in the quantity of ideas, nonetheless, probably increases our chances of discovering novel ideas. For example, the excursion technique was used by NASA in solving a closure problem. NASA engineers used word association to find a word with high potential for visual imagery. The next step was to fantasize. They selected "rain forest" and imagined "someone running through the forest and having thorns stick to his or her clothes." This gave them the idea for Velcro [36].

WHITHER THE VISIONARY?

There is, of course, a technology transfer analogy to the old saw about leading horses to water and only some will drink; that is, some people by dint of personal characteristics will be better at a needs-driven process than others.

This is not a new theme in innovation literature and in other allied fields. Over 20 years ago Ev Rogers [37,38] summarized the research on what distinguishes innovators from noninnovators; they tend to be smarter, have broader experience, be more cosmopolitan, and so on. There is a parallel literature in psychology defining the distinctive characteristics of "creative" people [39]; for example, they are likely to engage in more intuitive behavior [40], to have certain kinds of cognitive styles such as a proclivity for divergent thinking, to be more internally-motivated or controlled, and to have more visual imagery [41]. They also tend to score differently on standard or "objective" tests of creativity, or components thereof, although this tends to be only weakly related to tests of intelligence [42]. Several cautionary points are worth noting in this context. One is that the "validity" of creativity measures or tests seems to vary widely. That is, their relationship to objective indicators of performance or behavior tends to be quite variable, depending on the nature of the test and/or the behavioral outcome. Second is the "domain specificity" of creativity and its measurement. Creativity in science is probably poorly related to creativity in art, and so on. There are other fields that have relevance here as well; from market research literature, we know that "key" or "lead" users tend to be different from those who follow.

Unfortunately, these somewhat disparate veins of literature can lead to quite different implications in taking action, depending on how one comes down on the nature v. nurture or make v. buy choice points.

For example, one might conclude that the secret to an effective needs-driven technology transfer presence is to staff it with the right kind of people (those who are creative, smart, interdisciplinary, empathic communicators and so on). This leads one down a path of creating and deploying selection tools to pick those right people. In all honesty, we have probably not used this approach enough. We have assumed anybody can be an ORTA, or a technology sourcing specialist, or that all scientists who are technically competent can relate to industrialists. There is, of course, a large literature in the personnel selection field [43] that indicates that an investment in selection is worth much more than money spent on training.

Unfortunately, we in technology transfer have also not spent the money for training. There is a large, often underground, set of practices intended to train people to be more creative, perceptive, or empathic. Private companies, on the other hand report a sharp increase in investments in programs and participation therein [44]. For example the Center for Creative Leadership, in North Carolina, has built a multimillion dollar business in this area, and across the country there is a wide variety of training offerings. These range from programs which are fairly straightforward classroom experiences, to others which are reminiscent of the 1960s human potential experiences [45,46]. Unfortunately, the validity of any single approach—or combination thereof—is at best weakly documented, and the practitioners of these approaches tend to be epistemologically at odds with any quantitative data-gathering about effects. There are also no documented, validated training programs that are *specific to technology transfer*. Again, there are approaches, programs and tools—but no data indicating the magic bullet.

ORGANIZATIONAL SETTINGS

Earlier, our problem definition suggested that one of the knottiest issues of technology transfer is how to link different stakeholders in different organizations, via a shared and communicated vision of needs (and solutions). Even if we solve the questions of who and how (above), there is still the question of the *organization context* in which a needs-driven process of technology transfer occurs. We know that, in general, organizational composition, size, and the like affect the creativity and problem-solving of groups [47], but what about specific settings of interorganizational technology transfer?

Given the multiple organizations involved in a typical technology transfer engagement, there are three subsets of organization context questions: the organizational context of user organizations, of technology "source" organizations, and of boundary-spanning structures.

User Organizations

It must be realized that the "sourcing" of external technologies is not a standard function or organizational subgroup or job description. There are several aspects of this nonroutinism. One is that there are many different subcategories of users within a company, which reflect different discipline/technology domains, process v. product, and different stages of the life cycle (early development v. downstream implementation). The question is, who or how can one person speak for these multi-variable needs? Is the "gatekeeper" role a person or an organized function, and are there other complementary functions [48]? How can the babble of needs-sayers be orchestrated to speak with validity and clarity?

Technology Source Organizations

If research organizations are to *anticipate* end-user needs, how does this happen? Is this a market-research function, or one which involves technological "visioning" on the part of presumably industry-hip researchers? One long-standing issue is where one places the research organization, organizationally speaking—whether it is better to have it as a central R&D facility and free it to think great thoughts or to have it physically located with and responsive to an operating division [49]. These interorganizational linkages are essential to the "ideation" process at the onset of development [50].

Technology Transfer Brokering Organizations

Should such an organization be technically visionary, industry-smart and knowledgeable about the transfer process? Alternatively, should it be merely a facilitating or non-value-adding broker? For example, ITI has tended to be a *player* (in terms of both value-adding technology development and understanding industry needs). Other organizations have seen themselves as more content-free.

Nonetheless, one of the major trends in U.S. R&D has been the growth of various kinds of boundary-spanning organizations. This has included more deliberate transfer and needs-sensing organizations in universities [51] as well as industrially-led consortia [52,53]. One of the more interesting operational aspects of consortia is their recent attempts to do "concurrent R&D," particularly at the early agenda-setting (and needs-defining) part of the process [54]; that is, involve all the stakeholders in the transfer process—researchers, developers, and users working together.

Given these various options, what is to be done?

THE MISSING MEGA-PROJECT

For those who are familiar with the history of experimental design and multivariate statistics, this problem (of needs-driven technology transfer) has many of the aspects of agricultural research in the 1940s (e.g., the development of high yield corn). That is, it is a multivariate phenomenon, longitudinal in nature, with the potential for highly complex interactions among contributory factors.

Perhaps what we need, to understand or describe a valid needs-defining process, is a large field experiment [55], conducted over a five-to-ten-year time frame, possibly in several technical domains. Based on a quick survey of the literature, it appears that we have independent variable "clusters" to work with. It seems to make no sense to manipulate individual variables precisely; rather, we should try to determine first order effects of "chunky" variable clusters, and sort things out later. For example, we need to know if training per se has an impact on valid need-setting, and then ascertain what kinds of training.

Nature of Participants

There are several dimensions of "who" is involved in needs determination that are worth examining:

- *Personality.* Assuming an appropriate battery of tests could be assembled (Myers-Briggs, internal-external control, visual imaging, divergent thinking, intelligence), it would be useful to sort participants on some composite dimension.
- *Experience and background.* There are a variety of Rogers'-type indicators (e.g., cosmopolitanism) that could be aggregated. Findings on the situational-specificity of creativity may mean that participants need some baseline level of experience and knowledge of a field in order to be "validly creative".

Preparation and Training

For even the "right" kind of participants, there may be benefits from training:

- *Creativity training.* As mentioned above, a variety of approaches have been taken to sensitize people to nonobvious solutions.
- *Industry education.* Some type of briefing/education on industry trends, policy shifts, business developments, and current technologies can give people uptodate or broad gauge understanding of an industry.

Social-Organizational Context

There are several aspects of organizational context to examine:

- *Groups v. individuals v. composites.* Are technology needs best articulated by appropriately selected and prepared individuals and then somehow aggregated, or is this an example of "group work"? Alternatively, are these different approaches to combining group and individual contexts?
- *Stakeholder organizations.* Are future needs best articulated by individuals in end-user organizations, or is there an "anticipation" function played by researchers and developers, or perhaps third party organizations?
- *Roles and functions.* Even assuming that we reach conceptual/empirical coherence on what kinds of organizations are at the needs-setting table, it is still unclear what organizational roles or functions within these organizations are represented. Is it the research and development function? Product group? Marketing or market research? Resident gurus or "core competency" leaders? Manufacturing engineering or operations? Deciding can lead to identifying different types of people.

Aids and Tools

As usually practiced, the craft of needs articulation is a mix of the intellectual and the intuitive "vision thing," and a variety of tools/technologies and aids have been developed and deployed to aid the process. To our knowledge, however, there are no methodologically-respectable comparative evaluations of major categories of tools. Some of these tools might include:

- *Computer-mediation.* "Group work" systems [56], more individually-oriented tools [57], or various kinds of expert systems.
- *Visualization aids.* A variety of technologies, some quite high-tech (computer-mediated virtual reality) and others more mundane.
- *"Manual" facilitation aids.* Questionnaires, rating scales, surveys or facilitating protocols (for consultants or other kinds of third party help-givers).

WHAT NEEDS TO BE DONE FIRST?

The above mega-project will probably never happen as a single research effort. Nevertheless, we can begin to do some tangible things:

1. Document in more explicit detail various approaches to articulating, uncovering, and communicating next-generation technology needs. It might be useful to have a conference dedicated exclusively to research and practice therein.
2. Define what we mean by an effective needs-driven process of technology transfer.
3. Gather empirical, quantitative data comparing different approaches to needs defining.

In summary, we need to become more analytical and disciplined in understanding future technology needs. This will be another step in creating the ultimate tool kit for technology transfer [58], one which focuses on the high leverage, leading wedge of the process. As technology policy becomes more politically acceptable (including picking technology winners and losers), building more valid lists of "critical technologies" will become more important. Judging by how that process is currently conducted [59], some of the issues discussed here could bear closer attention.

REFERENCES

1. Tornatzky, L. G. & Fleischer, M. (1990). *The processes of technological innovation.* Lexington MA: Lexington Books.
2. Rothwell, R. (1977). The characteristics of successful innovations and technically progressive firms - With some comments on innovation research. *R&D Management, 7,* (3), 191–206.
3. Brownlie, D. T. (1992). The role of technology forecasting and planning: Formulating business strategy. *Industrial Management and Data Systems* (2), 3–16.
4. Coates, J. F. (1977). The role of formal models in technology assessment. *Technological Forecasting and Social Change, 9,* (1,2) 139–140.
5. Mogee, M. E. (1991, July-August). Using patent data for technology analysis and planning. *Research and Technology Management, 34* (4) 43–49.
6. Dutton, W. H. (1991). *Visions of communications and society: A comparative perspective.* Unpublished paper, Annenberg School for Communications, University of Southern California.
7. Martino, J. P. (1983). *Technological forecasting for decision making* (2nd ed.). New York: North-Holland, 139–144.

8. Tornatzky, L. G., & Hochgreve, D. (1993, January). Assessing future needs of metalforming. *MetalForming* 31–39.
9. Tornatzky, L. G. & Ostrowiecki, B. (1992). *Technology needs in four industrial sectors.* Unpublished paper, Industrial Technology Institute, Ann Arbor MI.
10. Nord, W. R. & Tucker, S. (1987). *Implementing routine and radical innovations.* Lexington MA: Lexington Books.
11. Tornatzky, L. G. & Klein, K. J. (1982). Innovation characteristics and innovation adoption-implementation: A meta-analysis of findings. *IEEE Transactions on Engineering Management, EM-29* 28–45.
12. Bennett, R. C. & Cooper, R. G. (1982, July-August). Managing our way to economic decline. *Harvard Business Review.*
13. Bennett, R. C. & Cooper, R. G. (1982, July-August). Managing our way to economic decline. *Harvard Business Review.*
14. Swanson, D. (1984). Research needs of industry. *Journal of Technology Transfer, 9* (1), 39.
15. Lann, R. & Pouffier, C. (1991). *1991 Technology needs survey.* Unpublished paper, Georgia Institute of Technology, Economic Development Laboratory Research Institute.
16. Barron, F. & Harrington, D. M. (1981). Creativity, intelligence, and personality. *Annual Review of Psychology, 31* 439–476.
17. Shapley, D. & Ray, R. (1985). *Last of the frontier.* Philadelphia: I. S. I. Press.
18. Pelz, D. C. & Hart, S. L. (1986). *Business firm study: Basic data or relations with universities.* Unpublished paper, CRUSK, Institute for Social Research, University of Michigan.
19. Dearing, J. W., Meyer, G., & Kazmierczak, J. (1992). *Portraying the new: Environmental technology innovations and potential users.* Unpublished paper, Department of Communication, Michigan State University.
20. Kuhn, T. (1970). *The structure of scientific revolution* (2nd ed.). Chicago: University of Chicago Press.
21. Chappanis, A. (1971, November). Prelude to 2001: Explorations in human communication. *American Psychologist, 26* (11), 949–961.
22. Heguet, M. (1992, February). Creativity training gets real. *Training* 41–46.
23. Council of Consortia (1992, December). *Electronic communication media.* Unpublished inter-organizational memorandum.
24. Schrage, M. (1990). *Shared minds: The new technologies of collaboration.* New York: Random House.
25. Barabba, V. P. & Zaltman, G. (1991). *Hearing the voice of the market: Competitive advantage through creative use of market information.* Harvard Business School Press.
26. Watson, D. L. (1989). Enhancing creative productivity with the Fisher Association lists. *Journal of Creative Behavior, 23* (1), 51–58.
27. Kotler, P. (1991). *Marketing management* (7th ed.). Englewood Cliffs NJ: Prentice Hall, 318–325.
28. Truchim, W. M. (1989). An introduction to concept mapping for planning and evaluation. *Evaluation and Program Planning, 12*, 1–16.
29. Nadler, G. (1991). *Concepts of purposes for liberating creativity.* Paper presented at Design West/ 1991 Conference.
30. Nadler, G. & Bond, A. (1991). *Breakthrough thinking: A challenge to conventional approaches to planning, design and problem solving.* Paper presented at International Strategic Management Conference, Toronto.
31. Grudin, R. (1990). *The grace of great things.* New York: Ticknor & Fields, 24–33.
32. Kodama, F. (1992, July-August). Technology fusion and the new R&D. *Harvard Business Review*, 70–78.
33. Fernald, L. W. (1989). A new trend: Creative and innovative corporate environments. *Journal of Creative Behavior, 23* (3), 208–213.
34. Proctor, T. (1991). Brain: A computer program to aid creative thinking. *The Journal of Creative Behavior, 25* (1), 61–68.
35. Hamel, G. & Prahalad, C. K. (1991, July-August). Corporate imagination and expeditionary marketing. *Harvard Business Review* 81–92.
36. Fernald, L. W. (1989). A new trend: Creative and innovative corporate environments. *Journal of Creative Behavior, 23* (3), 208–213.

37. Rogers, E. M. (1962). *The diffusion of innovation.* New York: Free Press.

38. Rogers, E. & Shoemaker, F. (1971). *The communication of innovation.* New York: Free Press.

39. Barron, F. & Harrington, D. M. (1981). Creativity, intelligence, and personality. *Annual Review of Psychology, 31* 439–476.

40. Agor, W. H. (1991). How intuition can be used to enhance creativity in organizations. *The Journal of Creative Behavior, 25* (1), 11–19.

41. Wheatley, W. J., Anthony, W. P. & Maddox, E. N. (1991). Selecting and training strategic planners with imagination and creativity. *The Journal of Creative Behavior, 25* (1), 52–60.

42. Barron, F. & Harrington, D. M. (1981). Creativity, intelligence, and personality. *Annual Review of Psychology, 31* 439–476.

43. Schmidt, T. L., Hunter, J. L., McKenzie, R. C. & Muldrow, T. W. (1979). Impact of valid selection procedures on work-force productivity. *Journal of Applied Psychology, 64* 609–626.

44. Heguet, M. (1992, February). Creativity training gets real. *Training* 41–46.

45. Feldhausen, J. F. & Clinkenbeard, P. R. (1986). Creativity instructional materials: A review of research. *The Journal of Creative Behavior, 20* (3), 153–182.

46. Wheatley, W. J., Anthony, W. P. & Maddox, E. N. (1991). Selecting and training strategic planners with imagination and creativity. *The Journal of Creative Behavior, 25* (1), 52–60.

47. Thornburg, T. H. (1991). Group size and member diversity influence on creative performance. *The Journal of Creative Behavior, 25* (4), 324–333.

48. Chakraarti, A. & Hauschildt, J. (1989). The division of labor in innovation management. *R&D Management, 19* (2), 161–171.

49. Bean, A. S. (1990). *Does 'Closer to the Customer' mean 'Further from the Truth'?* Paper presented at Conference Board/IRI Conference on "Winning in the Global Technology Game", New York, March 6–7.

50. Rubenstein, A. H. (1992). *Ideation and entrepreneurship.* Unpublished paper, Northwestern University.

51. Mitchell, W. (1991, June). Using academic technology: Transfer methods and licensing incidence in the commercialization of American diagnostic imaging equipment research, 1954–1988. *Research Policy, 20* (3), 203–216.

52. Smilor, R. W. & Gibson, D. V. (1991, February). Technology transfer in multi-organizational environments: The case of R&D consortia. *IEEE Transactions on Engineering Management, 38* (1), 3–13.

53. Rhea, J. (1991, September/October). New directions for industrial R&D consortia. *Research-Technology Management, 34* (5), 16–26.

54. Link, A. W. (1990). Perspectives on cooperative research: Learning from U.S. experiences. *International Journal of Technology Management, 5* (6), 731–738.

55. Fairweather, G. & Tornatzky, L. (1977). *Experimental methods for social policy research.* London/New York: Pergamon Press.

56. Olson, G. M. (1990). *Collaborative work as distributed cognition.* Unpublished paper, University of Michigan.

57. Proctor, T. (1991). Brain: A computer program to aid creative thinking. *The Journal of Creative Behavior, 25* (1), 61–68.

58. Tornatzky, L. G. (1992, Spring/Summer). Whatdyaneed, buddy? *Journal of Technology Transfer, 17* (2/3), 5–7.

59. Mogee, M. E. (1992). Technology policy and critical technologies: A summary of recent reports. Discussion Paper Number 3. *The Manufacturing Forum* 37.

TECHNOLOGY NEEDS

Commercializing It "Backwards"

Lanny Herron

University of Baltimore
Baltimore, MD 21201

Lou Tornatzky and Beverly Ostrowiecki [*Technology needs: The art and craft of identifying, articulating, and communicating, p.* 137] have suggested that we need to become more analytical in understanding future technology needs in the market place, in order that both the research and development (R&D) and the technology transfer (TT) processes can be rationalized. They have based this prescription on their view of TT as consisting of R&D across, rather than within organizations. Their call for a more market-driven process echoes that which has been ubiquitous throughout both R&D and TT circles for many years, yet adoption of this prescription seems as illusory now as it has ever been.

Unfortunately, the transfer of public-sector technology to the private sector is unlikely ever to become successfully market-driven for several basic reasons. In the first place, public-sector R&D is not a creature of the market place, but rather of the political process. Both federal laboratory R&D and federally contracted R&D are ultimately driven by public policy as defined by Congress and the Administration. Whether it be directed toward defense, public health or some other area, such policy is driven through the political process by perceptions of the public good, rather than arising through economic market mechanisms. This central fact has heretofore precluded the public R&D process from accomplishing a market-connected rationalization. Furthermore, even though public R&D policy is now being redefined to include the goal of market-driven rationalization, this goal is largely doomed to failure, for two reasons. One, the political process which funds public R&D will inevitably divert this R&D from a market focus; this is the very nature and definition of a political as opposed to a market process. Two, the internal incentive structures of the public R&D process are not aligned with those of the economic markets and can never be as long as the R&D is public. The more public R&D incentive structures become identical with those of economic markets, the more the R&D will cease to be public. Thus, transfer of public-sector technology to the private sector is something very different from R&D within organizations. Not only does it, per se, extend across organizations, but also the organizations which it extends across are driven by very different mechanisms, purposes, and incentives. These circumstances make market-driven TT extremely problematic.

This problematic nature of the market-driven commercialization of public-sector technology is exacerbated by the very nature of both R&D and TT, even if they be wholly public or wholly private. R&D, particularly basic R&D, tends to be technology-driven rather than

From Lab to Market: Commercialization of Public Sector Technology,
Edited by S. K. Kassicieh and H. R. Radosevich, Plenum Press, New York, 1994

market-driven. This is partly because technology itself defines the limits of the possible and thus the parameters of problem definition, and partly because technological innovation requires very different skills than does either economic innovation or public implementation. Such skill differences commonly lead to divisions of labor within organizations (both public and private), which in turn lead to intraorganizational TT problems.

The conclusion to be drawn from the above discussion is not that transfer of public-sector technology to the private sector is doomed to failure, but rather that the solution to the problem does *not* lie in organizing the public R&D process around market information. In fact, the public R&D process has always generated, and will always generate, multitudinous technologies which are not market-driven, and much the same can be said for private R&D as well. Therefore, the solution to commercialization of public-sector technology is not to be found in designing market-driven R&D processes, but rather in designing technology-driven processes. Such a contrarian or "backwards" process view runs counter to conventional wisdom, which fact has put us in a position where the design of technology-driven processes is at present fairly embryonic.

The challenge of designing a technology-driven commercialization process comes in matching the subject technology with actual or potential needs in the marketplace, and, as Tornatzky and Ostrowiecki rightly point out, needs are not usually well-defined by users. But as they also point out, needs have not historically been defined with much success by researchers or R&D managers. In fact, the function of defining or discovering marketable needs has historically been performed by entrepreneurs. Thus, in constructing a model for matching technologies with needs in an economically marketable manner, the practice of entrepreneurship would appear to contain many applicable parallels.

The economist Kirzner [1] points out that the entrepreneurial discovery of marketable opportunities rests on the state of awareness of the incipient entrepreneur and that this state of awareness is tightly linked to economic incentives. In other words, opportunities are discovered in a positive relation to potential rewards offered. As an example of this phenomenon, Kirzner notes that money lying on a sidewalk will easily be found even by the most unobservant. Thus, incentives are a key to a technology matching process, but, as has been pointed out above, such incentives are problematic in a public R&D situation. This is exactly why conflict of interest problems arise so often in public organizations. Everyone wants the economy to prosper through the auspices of the government, but as soon as any *one* prospers, a hue and cry immediately arises. Such a milieu will never do for a successful commercialization process, so it appears that these processes must be driven from outside the public sector. This is not to say that incentives for public R&D personnel are not helpful (or even necessary) to the TT process, but that an incentive system able to drive a healthy commercialization process must be implemented largely from outside the public R&D domain.

The economist Simon [2, 3] points out that significant discoveries (in either a scientific or a business sense) take place within a background of prior knowledge. In other words, the value of a discovery will tend to bear a positive relationship to the expertise of the discoverer. Thus it would appear that not only are incentives necessary to successful commercialization, but also that knowledge and experience in the business world are necessary as well. This is not to say that R&D workers can never discover valuable commercializations on their own, in fact quite the contrary, because we *all* have experience in the business world, at least as consumers and readers. The point is that increased business experience will increase entrepreneurial discovery, particularly if the business experience is in a field, industry, or area addressed by the subject technology.

A third ingredient necessary to commercialization of public-sector technology is access. If incentives and knowledge for commercialization must exist largely outside the public R&D process, individuals and organizations from the private sector must have easy access to public-sector technologies. Thus, mechanisms such as CRADAs, technology extension services, and various other functions performed by public-sector technology transfer offices are a step in the right direction, and more such mechanisms will abet any commercialization process.

The final ingredient necessary to the commercialization of public-sector technology is time or person-hours. In the past, when technologies were simpler, the functions of invention and economic innovation were commonly carried out by the same person, for whom technology transfer was largely a mental process. Today, whole teams of people are often needed, not only to invent and develop a technology, but also to plan, produce, and market it as well. This is because of the increasing amount of knowledge necessary for each of these processes. Technology itself is the means to an end, while what we call technology transfer commonly refers not to transfer of the technology per se, but to the transfer of necessary knowledge about the technology. Thus, TT person-hour needs increase as knowledge increases. Those who enter the TT process with specialized knowledge must not only disseminate that knowledge to others, but must also absorb the knowledge of other specialists in order to participate in the process. This fact makes TT and technology commercialization highly labor-intensive, with the necessary labor being highly skilled as well.

Thus, the four major ingredients required for a successful technology-driven public-sector commercialization effort are incentives, expertise, access, and labor (skilled person-hours).

Many incentives currently exist in the marketplace, as does much business expertise. The major problems at present are access and labor, and until these problems are reduced in relation to the size of the incentives for TT, public-sector commercialization efforts will not live up to their expected potential.

There are at least several short-term solutions to these problems, two of which are computerization and utilization of college students. Computerization consists of building large-scale data-bases to match technologies to markets. Many of the techniques which would be applied to these data bases have yet to be developed, but the need is obvious and the work has begun. On the other hand, the use of students in public-sector commercialization has barely begun. Students have, in their course work or as student assistants, ample incentive to help match technologies to markets. The experiential learning is invaluable, the possibility of market rewards alluring, and the opportunity costs are low. Probably nowhere in our economy is so much talented labor being underutilized as in the learning process in applied fields. Well-supervised technology transfer activities in matching technologies to markets could provide improved quality of learning in business and engineering, as well as generate market opportunities for the students. Such programs addressing the "backwards" methods of matching existing technologies to new or existing markets could create tremendously productive synergies.

REFERENCES

1. Kirzner, I. M. (1985). *Discovery and the capitalist process.* Chicago: The University of Chicago Press.
2. Simon, H. A. (1986). How managers express their creativity. *Across the Board, 23* (3), 11–16.
3. Simon, H. A. (1985). What we know about the creative process. In R. L. Kuhn (Ed.). *Frontiers in creative and innovative management.* Cambridge, MA: Balinger. 3–22.

COMMENTS ON, AND ENHANCEMENTS TO, *Technology Needs: The Art And Craft of Identifying, Articulating, and Communicating* (Louis G. Tornatzkay and Beverly Ostrosiecki)

Barry J. Lerner

Concurrent Technologies Corporation
Johnstown, PA 15904

INTRODUCTION

This paper provides some general comments on, and enhancements to, the Tornatzky and Ostrowiecki (T&O) paper entitled, *Technology needs: The art and craft of identifying, articulating, and communicating.* (p. 137)

The purpose of this document is to provide further insights into the identification and communication of commercial technology applications, based on my experiences at Concurrent Technologies Corporation (CTC). The company operates three national centers of excellence (COEs): the National Center for Excellence in Metalworking Technology (NCEMT); the National Defense Center for Environmental Excellence (NDCEE); and the CALS Shared Resource Center (CSRC). I would classify myself as a practitioner and facilitator of technology transfer (T^2) at CTC.

CTC is a six-year-old firm dedicated to supporting the development and transfer of federal and university technologies to the domestic industrial base. In many respects, the firm is a new model for a technology transfer (T^2) organization; its mission statement includes the following phrase: "Improvement of the Nation's industrial competitiveness through the development and transfer of leading edge technologies."

This paper is divided into three sections. The first section "Divergence Among Needs Drivers," describes a key element I perceive to be missing from the discussion—the recognition that the needs of the researcher or funding agency often differ from those of commercial industry. To resolve the conflict, the second section "Centers of Excellence as Identifiers and Communicators," describes the role of COEs as agents for T^2. These organizations are charged with identifying end-user needs, identifying research efforts to solve their requirements, developing the research into viable commercial products and processes, and placing these in the hands of industrial end-users. The third section provides further insight into maximizing communication between researcher and end-user, and is entitled "Techniques for Enriching Communications."

From Lab to Market: Commercialization of Public Sector Technology,
Edited by S. K. Kassicieh and H. R. Radosevich, Plenum Press, New York, 1994

DIVERGENCE AMONG NEEDS DRIVERS

Tornatzkay and Ostrosiecki specified that, "since the goals (of T^2 efforts) presumably involve satisfying the needs of categories of people called end-users, there is great credence put into a *needs-driven* process of technology transfer." I certainly agree that a needs-driven approach is the best practice for the rapid transition and application of a newly sponsored technology. I also agree that defining needs is not an easy task; T&O's efforts clearly demonstrate the divergence between researchers' and end-users' perceptions of new technology requirements.

As a practitioner of T^2, I believe a more fundamental divergence exists between the research and/or sponsoring organization and the potential end-user. The sponsor of the research—DARPA, DOE, DoD, NASA, NIH or other agency—often has a particular need or requirement in mind that may (and often does not) have commercial viability. An example is the recent headlines about multi-million dollar toilets for the space station. As a result, the problem is greater that the lack of good communication tools for T^2 practitioners; one must identify end-user needs through the rigorous processes mentioned by T&O and cross-reference these requirements against the needs of the user. Presumably, the proliferation of firms like the Industrial Technology Institute, CTC, and others which identify technologies, add value as developers, and understand industry needs, will make the T^2 process more efficient over time.

CENTERS OF EXCELLENCE AS IDENTIFIERS AND COMMUNICATORS

As stated above, value-adding "players" have a key role in the continuing improvement of the T^2 process. To that end, many such organizations have developed; one organizational model is the center of excellence. These centers play a key role in the successful transition of new technologies to industry.

Too often, T^2 professionals concern themselves with maximizing the transfer of technology from the laboratory to industry, and finding end-users for sponsored technologies; T^2 also includes identifying end-users' problems and searching the research base for sponsored technologies.

TECHNIQUES FOR ENRICHING COMMUNICATIONS

No single solution exists for communicating new technologies to the end-user, or gathering new technologies from a source. Tools, such as "written technical papers" (for R&D personnel), or alternative communications, such as "video recording, CD/ROM, various approaches to computer-mediated group processes, animation, and so on" (for end-users) are important, but can never eliminate the need for one-on-one researcher-to-developer communication. Of course, several other communication techniques exist that COEs can use to maximize the T^2 process. CTC uses a variety of these; several are listed in Table 1.

Table 1.

Type	Description	Methodologies for Insertion
In-plant implementation	Direct manufacturing product and process solutions, which *implicitly include the requisite hands-on education and training* for continued implementation of advanced technologies and methodologies.	• Technology application • Product/process improvement • Cooperative agreements/teaming • Rapid response • Direct solution by CTC
Education enterprise	Development and application of resources for educating individuals and organizations in the use of advanced manufacturing techniques.	• Teaching factory • Classroom training • Training resources • Workshops and seminars • Curricula development
Government, industrial and academic outreach	Large scale dissemination of technologies	• Development and implementation of military and industrial specifications • Conferences/exhibits • Project demonstrations/open houses • Trade journal articles/proceedings • Databases/networks/shared library resources • National & regional trade organizations/consortia participation/local economic development organizations • Publications and videos

THE ROLE OF THE FIRM'S INTERNAL TECHNICAL ENTREPRENEURS IN COMMERCIALIZING TECHNOLOGY FROM FEDERAL LABORATORIES

Eliezer Geisler[1] and Albert H. Rubenstein[2]

[1]University of Wisconsin/Whitewater
Whitewater, WI

[2]Center for Information and Telecommunication Technology
Northwestern University
Evanston, IL 60208

1. INTRODUCTION AND PRIOR RESEARCH

This paper describes the role of the corporate internal technical entrepreneur in commercializing technology from federal laboratories. The study of the process of technology transfer from federal labs has been well documented in the literature. Chakrabarti and Rubenstein [4], for example, studied factors which significantly affected the adoption of 45 NASA-disseminated technologies in 68 organizations. Whiteley and Postma [21] studied several federal laboratories and their past and current users and collaborators. They suggested that the main mechanism for interaction between the federal laboratory and industry is at the individual level, between researchers in the interacting organizations. Wolek [22] studied the technology transfer process at the USDA's Agricultural Research Service (ARS). He surveyed 54 cases of transfer in over 50 ARS laboratories, and concluded that "the polices, procedures and resources of ARS are not designed to support effective transfer to industry" (p. ii). Allen et al. [1] studied three large national DOE (Department of Energy) laboratories. They concluded that "those inventors who become entrepreneurs in order to commercialize their inventions have different attitudes ex post from those inventors who do not participate in the commercialization of their inventions through the act of entrepreneurship" (p.12).

These and other studies have shown the complexity of the transfer process between federal laboratories and industry. Although, as described below, the literature identifies several categories of factors affecting successful transfer, there is little empirical evidence to establish clearly how such technology transfer can best be successfully accomplished.

Paramount among the factors which have not been well investigated is the role technical entrepreneurs play in the corporate environment. In a major study of factors influencing technology transfer from DOD (Department of Defense) laboratories to law enforcement organizations at the state and local levels, Hetzner [5] concluded that, although the DOD laboratories were successful in making information available about their programs and

From Lab to Market: Commercialization of Public Sector Technology,
Edited by S. K. Kassicieh and H. R. Radosevich, Plenum Press, New York, 1994

technologies, they nevertheless have not been "nearly so successful in getting this information to a decision-maker in the potential user organization" (pp.224–245). Such factors are also found in Rubenstein [8], Chakrabarti and Rubenstein [4], and Rubenstein and Geisler [15]. In studies of the ARS laboratories (which are some of the more successful federal laboratories in transferring technology to industry), Rubenstein and Geisler [15], Rubenstein Geisler and Souder [17] and Rubenstein and Geisler [13] have found that successful transfer requires intensive interpersonal interaction between researchers in the laboratory and the industrial firm. This is also corroborated by Wolek [22].

In addition, prior research has also identified two aspects of the federal lab: 1) the "double-bind" in which the potential technology transfer agent in the federal lab finds himself/herself, and 2) the "critical gap" between exploratory and applied research in a federal lab. These phenomena are part of the environment in which the corporate technical entrepreneur must function.

We view the corporate "internal technical entrepreneur" as a critical player in the industrial commercialization of technology from federal labs. This role is described in terms of its effect on initial technology transfer and the overall commercialization process. It is strongly influenced by the framework within the firm of incentives and facilitators to commercialization.

The importance of the issue of commercializing technology from federal labs has recently gained impetus in various reports and pronouncements, particularly with regard to the role federal labs are called on to play in the nation's quest for global industrial competitiveness [23]. The potential for such a role is evident in the size of the federal R&D network. There are currently 726 federal laboratories, with a total annual budget of about $20 billion. Although there are at present almost 850 cooperative research and development agreements (CRADAs) between federal labs and industrial companies, there is consensus among laboratory directors, corporate officers and academics that "very little technology that industry can actually incorporate into commercial products has been forthcoming" [23, p.4].

In this context, it is important to improve our understanding of internal corporate processes which may lead to or inhibit successful transfer of federal technology. The role of the technical entrepreneur in the large firm is a critical aspect of the transfer process.

2. THE ROLE OF THE INTERNAL TECHNICAL ENTREPRENEUR IN THE FIRM

2.1. Some Aspects of the Transfer and Commercialization Processes

The basic flow of technology transfer from the federal lab to the company is shown in Figure 1. The figure shows the intermediaries in the flow, the corporate or divisional cooperative R&D office, and the technology transfer office of the federal lab. The last was established pursuant to the 1980 and 1986 Federal Technology Transfer Acts [19, 20]. Although the main linkage (in technical terms) is between the scientists or engineers on both sides of the interacting organizations, the commercialization process inside the firm is influenced by a variety of factors, as shown in Figure 1.

Chakrabarti and Rubenstein [4], in their study of 73 cases of NASA innovations transferred to industry, concluded that, in the 28 *process* cases, the success of adoption by the company was directly associated with: 1) general degree of connection with the firm's existing operations, 2) specificity of relationship between the technology and some existing and recognized problems, 3) quality of information from source of innovation, and 4) availability of personnel to implement the technology. In the 45 cases of *product* innovations, the success of adoption was directly associated with: 1) availability of personnel, 2) ability to obtain

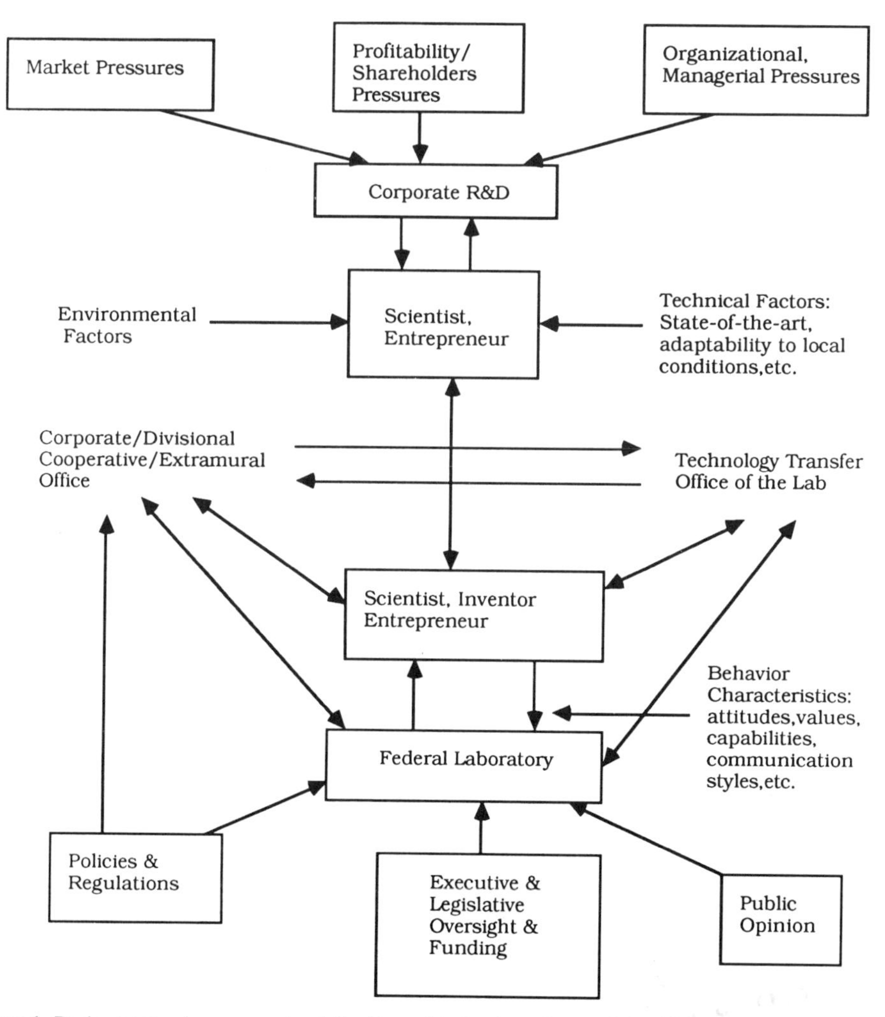

Figure 1. Basic structural components of the flow of technology from a federal laboratory to an industrial company. *Source:* This figure is a composite of related flow diagrams suggested by Rubenstein (1972), ORS/TT Model (document 72/64). Chakrabarti and Rubenstein (1976), Arnow (1983), Hetzner (1973).

financial resources for implementation, 3) top management support for the technology, and 4) emphasis on meeting functioning standards.

Hetzner [5], in his study of DOD technologies and their transfer to law enforcement agencies (state and local), concluded that lack of adoption was because of: 1) need to modify the technology for use by state/local agency, 2) lack of agency funds to pay for such modifications/adaption, and 3) lack of adequate linkages between local and state agencies, and industry.

In these and other studies [17,15,14,13,11,24,6,3,22], two importa
were identified—the "double bind", and the "critical gap."

2.2. The "Double Bind" of the Federal Lab's Technology Transf

Most federal labs have created, by law, offices for the transfer of
publics, particularly to industry and universities. Although most con

ducted between scientists and engineers in the labs, in industry and in academia the tech-transfer officer's job is to promote and facilitate the transfer of technology from the laboratory. This officer usually finds himself/herself in a double bind of apparently conflicting pressures. On one hand, this officer is pressured to transfer and move technology which is essentially basic or exploratory in nature or specifically mission-focused. Usually it is of little use to the corporate technological unit in its initial form. This is the "push" aspect of the process, in which the transfer officer is pressured to "push" the technology out the door. Yet, concurrently, this officer is also under pressures from the potential user firm to make sure the transfer is successful and the technology is viable, relevant, implementable and acceptable to the potential user.

However, the "double bind" situation goes beyond the role of the official or appointed technology transfer agent. There is general agreement among many participants in and observers of the TT process that leaving it solely in the hands of TT agents is doomed to failure. We believe that successful technology transfer requires the direct involvement of the technology source, the scientists and engineers on both sides of the interface between industry and the federal lab. Their double bind is even more severe than that of the formal TT agent. Typically, the technology source is a technical person assigned to projects within the mission of the federal lab who, especially under downsizing conditions and the era of "lean and mean," is hard pressed to keep up with assigned projects and tasks. Even if he/she is strongly motivated to help in the transfer process and to see that the technology is used by industry, his/her supervisor may not share that interest. That is, the mission of the lab may include the tech transfer activities mandated by law and, typically, endorsed by the upper lab management. However, when the need to devote time and other resources at the bench or project level encounters the "main mission" demands for those same resources (time, facilities, energy commitment), the TT activities often take a distant second place. A highly-motivated government scientist or engineer may want to participate in or even lead the "technology push" activities of the lab, or go out and "prospect for" opportunities to "sell" the technology. However, internal lab pressure and career considerations may present significant personal barriers to this kind of involvement.

Chakrabarati and Rubenstein [4], quoting a study by the Denver Research Institute, concluded that "the adoption of NASA technology is directly related to the linkage between the technology and some existing problem which the potential user had at the time of evaluating the technology" (p.30). Studies of other federal labs also emphasized the roles of the scientist and engineer in these labs in the TT and commercialization process. NIH (National Institutes of Health) laboratories encountered situations similar to those described as "double bind" [3]. In the case of DOD, Berkowitz [2] proposes some incentives for improved commercialization of the $39 billion spent on defense R&D in 1992. Although he is skeptical that defense research can revive US industry, he nevertheless emphasizes the difficult position in which DOD scientists, engineers and R&D managers find themselves, where they must balance their traditional mission as contributors to their constituencies (weapons development and acquisition), with the difficult (as well as largely unappreciated) effort to commercialize such technology.

In a study of US Army laboratories, Rubenstein and Geisler [12] identified several issues and difficulties for the transfer of DOD technology to private industry. They studied DOD/Army laboratories engaged in R&D on nonweapon-material equipment for the individual soldier. Some of the technologies developed by these labs had potentially wide applications for commercial firms. Yet the transfer process was hindered by excessive regulations and DOD bureaucracy lack of resources for TT, lack of personal rewards for the scientists and engineers involved in potential TT, different standards of production and usage, adherence to mission-riented targets, and deadlines which frequently excluded other activities not directly mission-ted.

Similarly, Wolek [22] in a study of ARS labs, also concluded that successful technology transfer depended on 1) "clearly perceived *needs* of targeted user" (p. 2), 2) initiation by *industrial sources* and 3) *"interaction* with industry to assure that the researchers consider practical constraints and conditions of use" (p.2). This moves tech transfer into a "pull" mode which can make it difficult for lab personnel to be effective.

The double-bind phenomenon can also affect the effectiveness of the internal corporate entrepreneur, much as it influences the behavior of the tech-transfer agent and professionals in the federal lab. The lab TT agent (as shown in Figure 1) typically interacts with the corporate or divisional office of extramural research, as well as with the corporate technical entrepreneur and professionals in the firm.

The double bind phenomenon may be further aggravated by the second phenomenon, the "critical gap."

2.3. The "Critical Gap" Phenomenon

The "critical gap" phenomenon is illustrated in Figure 2. The technologies developed by the federal labs are either specifically mission-oriented (e.g., military or space applications) or enabling technologies, more basic research than applied, with a longer-term view of problems and multidisciplinary in nature. Argonne National Laboratory, for example, is working on technologies geared toward high-risk, long-term solutions for toxic waste, acid rain, and advanced nuclear reactors. Some of these technologies are not directly transferrable to industry in their "raw" form. Most of them are at the precommercialization stage, and most industrial firms cannot use them directly. There is a need for investment to modify and adapt these technologies to industrial application. This is one aspect of the critical gap. The federal labs lack the resources as well as the charter for such adaptation. Most federal labs are not currently chartered to perform industrial-type R&D, and are, indeed, cautioned to avoid competition with private industry.

For many years, the critical gap has been almost a matter of ideology or government policy. The argument goes that federal labs, despite the congressional mandate to transfer technology to industry, must not enter "commercial" areas in potential competition with industry. The theory is that taxpayers should not pay for developing technology that is of direct benefit to industrial firms or that duplicates what those firms are, presumably, doing in their own labs. It has also been clear that the time horizon, depth, and scope of industrial R&D portfolios have been shrinking in recent years—witness the cutbacks and closing down of many corporate research labs (CRLs). Therefore, in many instances and fields of technology, the gap between where federal labs leave off and where industrial labs are willing and able to pick up has been increasing.

As a counter-trend, many joint ventures, consortia and "special arrangements" have been made on an ad hoc basis in particular fields to encourage faster and more effective commercialization. The mixed signals that many federal labs and their professionals have been receiving have exacerbated both the double bind perception and the critical gap. Much clearer signals and sincere support for tech-transfer activities are needed, along with significant incentives (and reduction of penalties) to address both issues.

On the other side of the transfer process, private industry normally refrains from making investments in adapting such technologies. Rubenstein and Geisler [13, 14], and Rubenstein et al. [18], in a study of agricultural federal labs, have found many industrial companies in the food and chemical sectors which purchase rights to USDA/ARS patents apparently to "own" them without any specific plans to implement or use them.

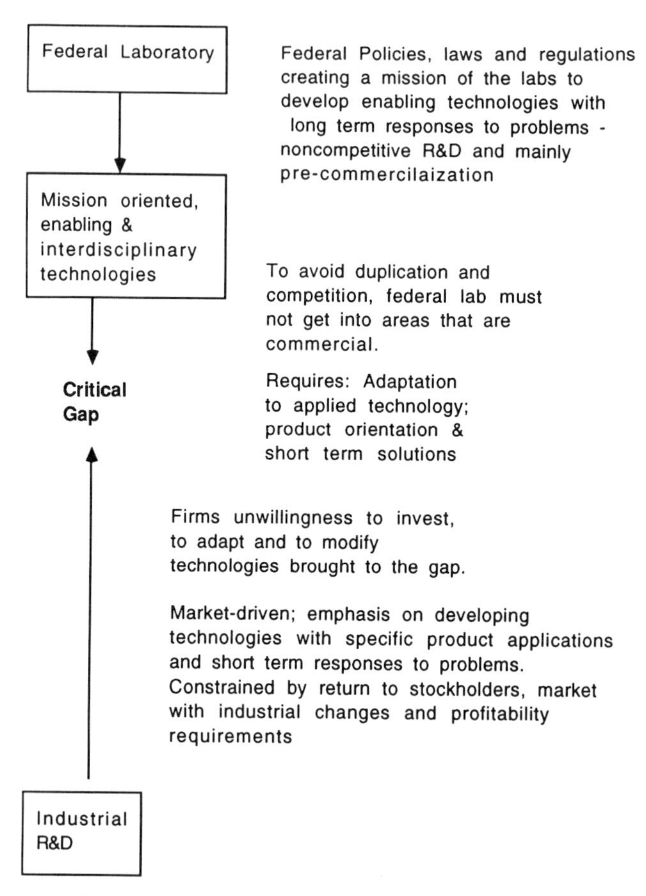

Figure 2. Graphic display of the "critical gap" phenomenon.

Several researchers have described the circumstances of the critical gap. Wolek (1984) suggested that when technology developed by the ARS reaches the field test and development stage, all efforts by ARS cease, whereas no effort has been started by potential users (industrial firms). Wolek also concluded that, "In several of the cases we studied, the work of ARS was failing despite the quality of ARS science, because private sources lacked the capability needed for market success" (p. 67). Earlier, Hetzner (1973) concluded that: "there are few indications that DOD laboratories are willing and able to put forward the…effort that is needed to actively transfer technologies to state and local agencies" (p. 247).

The critical gap creates a discontinuity in the flow of technology from federal laboratories to industrial firms. In addition, there is evidence in previous studies of federal laboratories that only a relatively small number of companies have continuous or routine working relations with federal labs. Although convincing empirical data is unavailable, the conventional wisdom is that most CRADAs (cooperative research and development agreements) are legal cooperative agreements, with little active and continuous technical exchange. In our own experience, in the mid-1980s, we helped the ARS prepare a scientific conference on biological control of insects. Very few chemical companies were initially invited, although biological control of insects would have a critical and long-lasting effect on the chemical industry. Lack of ongoing working relations with industry was the reason given by the organizers for such oversight.

We suggest that the company technical entrepreneur may be able to play a role in overcoming the barriers imposed by the critical gap.

2.4. The Role of the Corporate Internal Technical Entrepreneur

The company's internal technical entrepreneur performs the following tasks in his/her role as facilitator transfer of technology: (summarized in Table 1.)

1. The entrepreneur may help reassign and channel resources to "fill in" the critical gap, by creating projects or mini-projects to modify and/or adopt the technology received from the federal labs.
2. The entrepreneur provides technical expertise, technical assistance, and technical consulting to R&D managers as well as to middle and senior operating managers of the corporation or division.
3. The entrepreneur maintains and tries to increase interaction with parties in the federal labs, including the technology-transfer agents scientists and engineers. Thus, the entrepreneur intensifies the quality of interaction and creates and maintains on going and perhaps lasting linkages.
4. The entrepreneur takes the lead, perhaps in a team approach, in overseeing all aspects of the transfer. These include the transfer agreement, project management, resource allocation and obtaining organizational and managerial support.
5. The entrepreneur provides the enthusiasm and aggressive spirit needed for successful technology transfer. (This is particularly important in those cases where companies are downsizing and eliminating corporate R&D centers). Technical entrepreneurs direct the attention of the company to external technologies which are available for adoption, albeit with a need for additional work for adaptation.
6. The internal technical entrepreneurs provide technical expertise in showing linkage between existing R&D portfolios/projects and technical problems in the division or company, as well as the technology choices available at the federal labs. This linkage makes the federal lab's technology a needed or desirable solution to an existing problem, normally oriented towards a product or process with constituencies in the business side of the company.

Table 1. Tasks of the Firm's Internal Technical Entrepreneur in Comercializing Technology from Federal Laboratories

- Reassignment and channeling resources to fill in the "critical gap."
- Provision of technical expertise, technical assistance, and technical consulting to management (R&D and corporate/divisional).
- Maintenance and increased interaction with relevant parties in the federal labs.
- Overseeing issues of transfer, such as technology agreement, project management, resources allocation, and obtaining organizational and managerial support.
- Provision of enthusiasm and aggressiveness, directing the attention of company to available technologies from federal labs.
- Establishment of linkage between technologies from federal labs and existing problems in the company, product or process-oriented, with constituencies in the business side of the company.

Source: Rubenstein, Geisler, Rawlins and Elliott, [18]; Rubenstein, [8]; Zimke and Lasse, [24]; Rubenstein, Geisler and Lai, [16].

Thus, the internal technical entrepreneur assumes the *active* role of a transfer agent. The entrepreneur links decision-makers and technical people in the two organizations while serving as an advocate for the technology originating from the federal lab. Typically, the internal technical entrepreneur in a firm has a clear project or mission of his/her own, to develop a new or improved product, process, or service and "get it out the door into the market." Where this is the case, and he/she perceives a need or opportunity associated with some technology in a federal lab, the transfer may be "made in heaven." The energy and commitment of the technical entrepreneur can help make sure that the federal technology will be sought out and, if applicable, used in his/her project. If, on the other hand, the technology is peripheral to his/her immediate interest, it may be ignored or set aside. The entrepreneur is primarily concerned with his/her own project and with the benefits this project may gain from the federal lab's technology. The entrepreneur has a vested interest in his/her project, thus having a better, albeit possibly a biased, understanding of the potential benefits of the technologies available from federal labs. He/she may serve as a *catalyst* for the transfer, and may appreciate the fact that technology from federal labs can be initially almost free of charge although it generally requires resources for adaptation. Furthermore, when such technology is linked to the entrepreneur's own project transfer is desirable, and it may be possible that the cost of adaptation (thus leaping over the critical gap) may be lower than the cost of producing these technologies *de novo* in-house.

3. NECESSARY CONDITIONS FOR TECHNOLOGY TRANSFER

Apart from the specific issue of transfer from federal labs to industrial firms, there is a general set of issues related to the TT process, the venture/entrepreneurial (V/E) process, and the role of the individual technical entrepreneur, whatever the source of the technology. At best, successful technology transfer, even between operating units of the same firm, is a low probability event. On this subject, we have studied and consulted with many large firms and not-so-large firms [10, pp.131–160, and 343–404]. A key element in successful transfer seems to be the role of the internal technical entrepreneur, who is committed to his/her own project or program. A recent analysis (by the second author) of over a dozen consulting engagements in this field with medium and large companies suggests that the probability of success for successful V/E projects is inherently low and that there may be a set of "necessary" conditions for the success of *internal* transfers, let alone technology transfers from *outside* (e.g., from government labs). Research on and experience with this phenomenon have not yet yielded a set of "sufficient" conditions for success. We suspect there are some individual factors, such as massive top management support and/or a clear and present or desperate need for a new product/process, that may overcome many of the barriers that ordinarily exist and "ensure" success. Here are some of the potentially "necessary" conditions that have emerged from this analysis:

1. Clear need: *specific*, e.g., crisis-driven; or *general*, e.g. building or maintaining a technology base supporting current products processes and services.
2. Sponsor or champion(s): time protector, funding source or influence, access for tech transfer, career path provider interference runner.
3. TT structure/process: template or model.
4. Reward/recognition/incentives/motivation.
5. Monitoring and evaluation process, measures, indicators.
6. Ability to deal with the organizational, political, and economic environment.

7. Ability to move people across organizational boundaries—temporarily or permanently.
8. Tieing the technology to core competencies, technology policy/strategy, and business/marketing/technology strategy.
9. Clear identification of payer, beneficiary, technology vouchers cash payments.

4. INCENTIVES AND BARRIERS TO SUCCESSFUL TECHNOLOGY TRANSFER

4.1. Incentives to Transfer within the Federal Lab

The inventor/scientist/entrepreneur within the federal laboratory is a potential transfer agent. His/her activities, facilitated by the services of the technology transfer office (see Figure 1), contribute to the transfer of technology from the laboratory to the industrial firm. However, the environment within the laboratory may be conducive to the transfer process, or it may be unfriendly and even hostile. Factors which may make a significant difference as incentives or facilitators to the transfer process include (See Table 2):

1. Availability of "umbrella agreement". An overall agreement standardized and simplified, would facilitate the commercialization process. Industry is weary of government bureaucracy. To have such an agreement available would reduce the uncertainty of dealing with federal institutions and their legalistic structure.

Table 2. Illustrative Incentives to Transfer within Federal Laboratory

- Availability of an "Umbrella Agreement"
- Consensus among scientists and managers at the lab regarding the lab's mission (or one of its roles) to commercialize technology
- Availability and effectiveness of mechanisms within the lab to promote and facilitate commercialization and technology transfer
- Support from top laboratory management for the transfer process and the idea of commercializing the lab's technology
- Increased flexibility in government bureaucracy to allow the laboratory to channel and redistribute resources towards commercialization
- Provision of organizational support for lab's scientists, such as training for technology transfer, joint seminars and conferences with industry, etc.
- Targeting projects from the start towards potential transfer to industry (e.g., "dual targeting")
- Provision of internal personnel policies which diminish or even eliminate laboratory scientists' fear that commercialization activities would threaten their professional careers by exerting pressures on their time and energy that prevent them from publishing and professional/scientific development
- Use of existing linkages between laboratory scientists and the private sector, such as informal interactions with industry scientists and private sector associations; memberships in networks and professional associations; and role of laboratory scientists as experts and consultants to private industry.
- Provision of organizational mechanisms to "open the door" of the laboratory to private industry, "put out the word" on the laboratory's technology portfolio through newsletters, publications, and other means of "spreading the word" and advertizing the laboratory's capabilities.

Source: Rubenstein, Geisler, Rawlins and Elliott [18]; Allen et al. [1]; Rubenstein, Geisler and Lai [16]; Rubenstein and Geisler [15]; Rubenstein [8].

2. *Consensus on lab's mission.* As described in the previous sections, the phenomenon of the double bind has implications for the perceived (as well as declared) lab mission. Although there are pressures on federal labs to help resolve problems of industrial competitiveness, there are conflicting pressures on scientists in federal labs to avoid competition with the private sector. They are told to concentrate on mission-oriented, long term research and to avoid commercially "relevant" projects. Thus, consensus among scientists on the mission of the lab would help reduce uncertainty and feelings of undue risks in commercialization.

3. *Availability of mechanisms for transfer.* Available mechanisms for transfer in the lab are important facilitators for commercialization. In general, federal labs are not structured for commercialization, particularly in light of the policies imposed on them since the early 1980s, by which they were instructed to avoid competition with the private sector. However, some mechanisms may exit. For example, individual scientists in the lab have established ongoing contacts with industry, and there are conferences, seminars and other avenues for cooperation.

4. *Support from top management.* Because, in part, of the double bind phenomenon, commercialization efforts in federal labs would require open (declared) support from top lab management and behavior consistent with that declaration. Scientists must know that their senior managers openly support activities leading to the transfer of technology to private industry.

5. *Flexibility to redistribute resources.* As federal organizations, the labs have more difficulty redistributing their budgetary and human resources for commercialization that do private firms. Therefore, increased flexibility to redistribute resources is a powerful mechanism and incentive for potential commercialization. In a discussion of defense R&D, Berkowitz [2] recommended that DOD allocate funds "to pay the salaries of personnel dedicated to commercialization" (p. 75). Citing the Technology Transfer Act of 1986, and amendments in 1989 [19, 20], Berkowitz calls for increased flexibility through the elevation of technology transfer to a higher grade of priority in DOD attitudes towards its R&D laboratories.

6. *Provision of organizational support.* Although there are several provisions in the technology transfer legislation for funding commercialization activities, there is a need for *organizational* support in the form of, for example, specialized training for commercialization.

7. *Targeting projects.* Many federal labs are dedicated to research which satisfies the mission of their constituencies. DOD laboratories, for example, are geared toward the development and measurement of weapons systems and other material of the armed forces. From their inception, DOD R&D projects are not designed for future transfer to commercial entities.

8. *Provision of internal personnel policies.* Perhaps one of the more powerful incentives, this mechanism influences the key motivator of scientists' efforts in the federal R&D lab. Radosevich [7] Rubenstein, Geisler and Lai [16], and Rubenstein et al. [18] have emphasized the issue of providing incentives and rewards to scientists and their managers engaged in commercialization. Berkowitz [2] offered a comparison between DOD and DOE laboratories in their technology transfer activities. He concludes that DOE has been more successful than DOD because, among other factors, of the incentive structure imposed in its labs. This structure includes commercialization success as "a criterion in evaluating lab and department directors. DOD does not include these criteria in its evaluations" (p. 78). Incentive personnel policies are more powerful that simple removal of the barriers or punishments to behavior leading to commercialization. That is, it is not enough simply to allow scientists in federal labs to engage in and promote commercialization. Success can be achieved only by an active and dynamic effort anchored in a structure of incentives and rewards to commercialization.

9. Utilization of existing linkages. This mechanism supports existing linkages between federal scientists and their colleagues in private industry as a pivotal point for commercialization activities.

10. "Spreading the word". Finally, a strong incentive is a public relations effort to make constituencies of the federal laboratory and private industry aware of the laboratory's increased efforts towards commercialization.

4.2 Barriers to Internal Technical Entrepreneurship and the Transfer Process within the Firm

Rubenstein [8, 9] and Rubenstein, Geisler and Lai [16] described some barriers to TT within the firm. They also listed propositions which describe the TT phenomenon within the firm. Among the barriers mentioned by these authors are:

1. Top management factors. Negative attitudes towards the TT effort can be considering it a low priority activity, lack of overt and declared support, and behavior which demonstrates the low importance of the TT activity in the eyes of the top managers in the firm.

2. Characteristics, capabilities of V/E team. These include weak V/E leaders, lack of reputation of top V/E people, pace of hiring of V/E personnel, V/E people displaying "amateurish" behavior or capabilities, and the fact that V/E people are often "new" in the firm.

3. Relation of V/E group to rest of the organization. Poor relations result from rivalry with other functional or professional groups in the organization, the relation of V/E to existing businesses, the perceived level of commitment of V/E people, and attitudes of "old boys" (established managers and professionals).

4. Resources and constraints on V/E group. The V/E group can suffer from lack of initial and sufficient funds, lack of autonomy, lack of cooperation from divisions, and/or lack of a declared and prestigious champion for V/E activities and goals.

5. Content of V/E projects and programs. Problems include the newness of product and technology, lack of clear strategy as to what the V/E effort will achieve and how, and limited size and variety of portfolios.

6. External Factors. These include actions of competitors and customers, (such as customer behavior which does not encourage the V/E effort on risky new technology), reputation of the firm (when low reputation of firm affects the acceptance of the technology by outside organizations), pricing uncertainty, and the effects of an outside advisory group or panel.

REFERENCES

1. Allen, N., Kassicieh, S., Radosevich, R., & Soderstrom, J. (1991). *Attitudinal and situational differences between national laboratory inventors and inventor-entrepreneurs.* Working paper, University of New Mexico.
2. Berkowitz, B. D. (1993). Can defense research revive U.S. industry? *Issues in Science and Technology, 9* (2), Winter, 73–81.
3. Booth, W. (1982, January 6). NIH scientists agonize over technology transfer. *Science, 243* pp. 19–21.
4. Chakrabarti, A., & Rubenstein, A. H. (1976, February). Interorganizational transfer of technology: A study of adoption of NASA innovations. *IEEE Transactions on Engineering Management, 23,* (1) 20–34.
5. Hetzner, W. A., (1973). *An analysis of factors influencing the transfer of technology from DoD laboratories to state and local agencies.* Unpublished doctoral dissertation, Northwestern University Evanston, IL.

6. Louis, K., Blumenthal, D., Gluck, M., & Starto, M. (1989). Entrepreneurs in academe: An exploration of behaviors among life scientists. *Administrative Science Quarterly, 34* 110–131.

7. Radosevich, R. (1993, Summer). A mixed strategy model and case example of federal technology transfer. *International Journal of Technology Management,* forthcoming.

8. Rubenstein, A. H. (1992). Research opportunities in the study of technical entrepreneurship in the large firm: A propositional inventory. *The Journal of High Technology Management Research, 3* 83–110.

9. Rubenstein, A. H. (1990) *Barriers to successful technical entrepreneurship in the large firm: Some case studies.* Paper presented at the meeting of the Institute of Management Sciences, Philadelphia.

10. Rubenstein, A. H. (1989) *Managing technology in the decentralized firm.* Wiley Series on Management of Science and Technology. New York: J. Wiley and Sons.

11. Rubenstein, A. H., & Geisler, E. (1988). The use of indicators and measures of the R&D process in evaluating science and technology programs. In D. Roessner (Ed.), *Government innovation policy.* St. Martin's Press, pp. 185–204.

12. Rubenstein, A. H., & Geisler, E. (1985) *First technical report on a study of indicators of the productivity/effectiveness of Army R&D laboratories.* Report submitted to the U.S. Department of the Army. IASTA, Inc.

13. Rubenstein, A. H., & Geisler, E. (1984) Multidisciplinary research in a large federal agency. *Proceedings of the Third International Conference on Interdisciplinary Research* Seattle.

14. Rubenstein, A. H., & Geisler, E. (1983) *Development of output and impact indicators flow diagrams for the objectives of ARS program plan.* Technical Memo, No. 8.

15. .Rubenstein, A. H., & Geisler, E. (1980) A methodology for monitoring and evaluating the outputs of a federal research laboratory. *Proceedings of the 9th DoD/Federal Acquisition Institute Acquisition Research Symposium* Annapolis MD.

16. Rubenstein, A. H., Geisler, E., & Lai, P. (1989). Barriers and facilitators to technical entrepreneurship in the large firm. *Second International Conference on Engineering Management* Toronto.

17. Rubenstein, A. H., Geisler, E., & Souder, W. (1979). An organizational design approach to project management in government R&D organization. *Proceedings of the IEEE Engineering Management Conference.*

18. Rubenstein, A. H., Geisler, E., Rawlins, & Elliott, L. (1985). Issues in transition from single disciplines to multi-disciplinary research in a large federal agency. In: G. Marr, W. Newell & B. Saxberg (Eds.), *Managing High Technology* Elsevier Publishers, North-Holland.

19. U.S. Department of Commerce (1989). *The Federal Technology Transfer Act of 1986: The first two years.* Washington: U.S. Government Printing Office.

20. U.S. General Accounting Office (1991). *Diffusing innovations: Implementing the Technology Transfer Act of 1986.* Washington.

21. Whiteley, R., & Postma, H. (1982, November). How national laboratories can supplement industry's in-house R&D facilities. *Research Management* 31–42.

22. Wolek, F. (1984). *Technology Transfer & ARS.* Wharton ARC.

23. Yates, R. (1992, October 26). Refitting Cold War science; New role for national labs: Make U.S. competitive. *Chicago Tribune.*

24. Zimke, M. D., & Lasse, J. (1988). "Affirmative technology," spinoff 1988: Bringing federal and university research to market. *Proceedings of the 13th Annual Meeting* Technology Transfer Society, Indianapolis.

COMMENT ON *The Role of the Firm's Internal Technical Entrepreneurs in Commercializing Technology From The Federal Laboratories*

Robert H. Keeley

Colorado Institute for Technology Transfer and Implementation and
University of Colorado at Colorado Springs
Colorado Springs, CO 80933

Professors Geisler and Rubenstein [p. 155] give us a sobering view of technology transfer. Based on their survey of existing research, in much of which they have been personally involved, and on their personal experiences with corporations and federal laboratories, they voice concern about "the complexity of the transfer process between federal laboratories and industry." Later they say "At best, successful tech transfer in the firm, even between operating units of the same firm, is a low probability event." The reasons for low probability are found throughout their paper, and add up to an imposing set of impediments within firms and laboratories alike. They observe that a corporate entrepreneur is present in most instances of successful transfer, and may well be a "necessary" condition for success, though probably not a "sufficient" condition. Their paper explores the reasons, and derives a set of prescriptions for labs and firms that hope to succeed in technology transfer.

These comments begin by summarizing the authors' analysis—with a few changes in emphasis that helped clarify my own understanding of their work. The comments will extend the Geisler-Rubenstein subject to encompass technology transfer from federal labs to start-up companies.

GEISLER AND RUBENSTEIN'S DIAGNOSIS AND PRESCRIPTIONS

In section 1 of the paper, the authors note that laboratories often are "successful in making available information about their programs and technologies." The labs broadcast their message, but in most cases no one is listening—at least no one with decision-making power.

Section 2 assumes someone in the private sector has received the message. Then the process of investigating the technology and transferring it occurs within the complex of forces that govern the lab and the private organization. The authors summarize these forces in Figure 1. As their note on Figure 1 indicates, the network of interactions has been understood since at least the early 1980's. The most important players in Figure 1 are found near its center; they are the technical personnel of both organizations. Unfortunately, most of the forces surrounding them seem to impede their successful interaction.

From Lab to Market: Commercialization of Public Sector Technology,
Edited by S. K. Kassicieh and H. R. Radosevich, Plenum Press, New York, 1994

A "double bind" of conflicting pressures, described in section 2.2, snares tech transfer officers and lab scientists alike. Essentially, their organizations push them to transfer technologies in ways that minimize the disturbance to the organization's main mission. That means an acquiring organization is expected simply to pick up the technology without assistance from the lab. However, most transfers require support. Not surprisingly, transferees apply pressure on tech transfer officers and lab scientists to obtain that support, aggravating the opposing pressure from lab management to stick to the mission.

In section 2.3 the authors note that the "double bind" is usually exacerbated by a "critical gap" between the state of the lab's technology and the state in which industry needs it to be. The labs find themselves enjoined by public policy from carrying technologies as far along as private firms would like. The "critical gap" may well be widening in the 1990's, driven by industry cutbacks in applied research and by lab efforts to become "lean and mean."

For the daunting array of problems they have identified, Geisler and Rubenstein propose (in section 2.4) the internal entrepreneur as a solution. The internal technical entrepreneur facilitates the transfer of technology by "scaling the barriers": obtaining resources needed to adapt the technology, coordinating the activities of managers in the firm, communicating with key lay personnel, and in general providing leadership for the entire effort.

At first reading, I was reminded of the old television series "Mission Impossible," in which the heroes (here the entrepreneur) were given orders to conduct a clearly impossible espionage mission. On television, the heroes always succeeded; in the real world entrepreneurs will very seldom succeed when powerful institutions have imposed the sorts of impediments described by the authors.

In fact, their entrepreneur is only part of the authors' prescription. The lab and the private firm must both help the process by creating conditions conducive to the success of the entrepreneur. The implicit solution spans section 2.4 (the entrepreneur), section 3 (necessary conditions for technology transfer), section 4.1 (incentives to transfer within the federal lab), and section 4.2 (barriers to internal technical entrepreneurship). These four sections identify 25 conditions—19 that may lead to successful technology transfer, and 6 that prevent it. The conditions have different levels of urgency. Those from section 3 are "necessary conditions", suggesting that the failure of any one will doom a proposed transfer. Those from section 4.1 are "incentives or...facilitators." Those from section 4.2 are "barriers." The clear implication is that chances of success rise as more "facilitators" are in place, and fall as more "barriers" exist.

The authors stop short of suggesting how to judge whether enough facilitators are in place, and whether the barriers are low enough. They note early in the paper that "there is little empirical evidence to clearly establish how best such technology transfer can successfully be accomplished." But their proposal goes well beyond simply dropping an internal entrepreneur into an otherwise hostile situation and expecting good things to happen.

The combination of existing studies and personal experience cited by the authors persuades me that something close to their set of conditions must be met if we hope to improve on the unimpressive record of the past.

However, is their prescription realistic? Will organizations make the needed changes? The authors end their paper without predicting. From the record of past shortcomings and from studies of organizational change in the private sector, it is hard to be optimistic. Private organizations usually have trouble changing until their performance suffers enough to overcome organizational inertia. Then they often consolidate around their core activities, closing off peripheral experiments. The peripheral activities are simply too small to save the organization during a time of real threat, and they create disproportionate distractions, distractions that are abandoned in times of real stress. The closing by Digital Equipment of its corporate ventures in the last few months is an example.

We can only guess, but the uncertainties posed by the winding down of the Cold War and the continuing movement by corporations to become "lean and mean" does not seem to be a propitious environment in which to expand the role of technology transfer. Nonetheless, if one hopes to have any chance of success, Geisler and Rubenstein's proposals should be taken very seriously.

On the assumption that someone has listened to Professors Geisler and Rubenstein, these comments will now address a related issue: Should national labs deal with start-up entrepreneurs, not just with internal entrepreneurs in large, established firms?

TRANSFERRING TECHNOLOGY TO START-UP COMPANIES

Entrepreneurship has three parts: concept generation/discovery, venture formation, and venture management. The first, concept generation/discovery, deals with finding a sufficiently attractive idea for a new business to be built. Studies of the ideas behind entrepreneurial ventures [4,2,1] find that systematic search is a rarity. Instead, the discovery of an idea often involves an "encounter" between the entrepreneur and the idea. The entrepreneur must have sufficient grasp of his/her field to appreciate the idea's potential and to assess its feasibility. As Vesper [4] notes, once the encounter occurs, the entrepreneur quickly goes through a sophisticated "flow chart" to assess its merit.

In contrast to the novel business idea, systematic search may uncover the technology needed for routine problems. For example, disk memory companies, in order to maintain their rate of improvement in storage density and in cost/performance, searched for processes that would allow them to shrink the magnetic head—eventually adapting thin film processes from the semiconductor industry.

The distinction between routine and novel business concepts is not always clear cut. The novel idea may be simply to recognize the merit of an approach rejected by others as technologically inferior or infeasible. For example, the traditional suppliers of oxide disks to the makers of disk drives believed that metallic disks were not cost effective using technologies of the mid-1980's. Entrepreneurs gambled that they could solve the problems. They established the new ventures that subsequently became the principal suppliers of magnetic disks.

Geisler and Rubenstein's proposals address both routine and novel innovations, but seem better suited for routine innovations. For example, their first "necessary condition" is the perception of a clear need by the private organization. When the need is known, an organization will search for solutions, and a federal lab can attract attention through Geisler and Rubenstein's incentives number 9 "Utilization of existing linkages," and number 10, "Spreading the word." Incentives 7, "Targeting projects," and 8, "Provision of internal personnel policies," should also help. But these prescriptions assume that the private company knows it has a problem, a situation common with routine innovations and not with novel ventures.

To promote the creation of ventures around novel ideas, labs must take a more aggressive stance. Just "spreading the word" will not lead to the exchange of enough information for an "encounter" with a novel idea. The scientists developing the technology must be in contact with a wide set of prospective entrepreneurs. They should be attending conferences frequented by private companies, publishing in applied journals, and initiating contacts with potential users. That is, they should try to find the prospective entrepreneurs and speak with them personally, not just through a database. In many cases, such efforts may lead them away from working with large companies.

In section 4.2, Geisler and Rubenstein identify forces that hinder entrepreneurship within a large company. They refer to a formal venture/entrepreneurial (V/E) group. Thus, they seem to be concerned with non-routine innovation because a separate organization, the V/E group, is unnecessary for routine innovation. Normal development channels will suffice. The more innovative, "encounterlike" innovations lie farther from the mission of the development organization and are often assigned to separate V/E groups. But V/E groups have not often shown much success, for the reasons identified by the authors.

The obvious answer from the lab's perspective is not to limit itself to working with large companies. Its activities for disseminating information should be aimed at individuals rather than institutions. Many of the individuals reached will of course be located in large companies; they will be managers, engineers or scientists charged with developing new technology. However, should the new idea fall outside the interests of their current organizations, they may choose to form an independent venture.

An emphasis on individuals and not on their organizations creates potential problems for federal laboratories beyond those discussed by Geisler and Rubenstein. The problems relate to the second stage of entrepreneurship—forming the enterprise. Venture formation is usually an iterative process as the entrepreneur works at assembling the resources needed for the venture: the people, the rights to technology, the funds, the customers and the physical plant. If the venture is being formed within a V/E group or other unit of a large organization, its membership in a larger organization creates an appearance of strength. If the venture fails to materialize, a national lab will be relatively insulated from criticism. Dealing with individuals exposes the fragility of the venture, a fragility that exists anyway but is hidden from view in V/E units.

Aside from the appearance credibility that comes from favoring large companies over individuals, technology ownership issues arise when people leave companies. Case law relating to intellectual property has advanced considerably during the last 15 years, and the labs can probably leave it to the entrepreneurs and their former employers to sort out the issues. Geisler and Rubenstein note a useful caveat, - the labs should make agreements that require private firms to implement the technologies being acquired, not just "own them." Such agreements will facilitate the transfer of the rights to individuals possibly affiliated with the company initially acquiring the technology, should a company choose not to actively pursue the technology. A study by Keeley & Tabrizi [3] finds that start-up ventures often spring from projects abandoned by existing firms, but seldom spring from projects being actively pursued.

A federal lab's ability to deal with individuals, and therefore to create novel (as opposed to routine) ventures, depends on Geisler and Rubenstein's incentives numbered 1 through 5. The specific measures adopted in each of those "incentives" may lead to considerable differences in the amount and nature of tech transfer from a lab. The authors suggest that, if the past is a guide, there will be disappointingly few innovations transferred. We simply have not found the right recipe, if indeed there is one. One way to find that recipe is to generate various processes, monitor them, and select the one(s) that work(s) best. That is, the labs should experiment with processes for reaching large institutions, small institutions, and individuals.

One such "experiment" is to use outside expertise in constructing nonroutine ventures. Specifically, labs might hire venture capitalists and intellectual property lawyers to help address issues dealing with individuals and start-up companies. Such people have accumulated specialized knowledge for assessing the creditability of a would-be entrepreneur, for appraising the technology, and for developing realistic plans for exploiting the technology.

CONCLUDING COMMENTS

The prescriptions of Geisler and Rubenstein deserve the most serious attention of lab managers and of technology oriented companies. As the authors document, past processes have been ineffective. Their paper identifies the reasons, and offers a sensible package of proposals. Readers should keep in mind that the proposals are a "package." Piecemeal implementation will not lead to success.

As one who is particularly interested in entrepreneurship, I hope the labs will focus more on credible individuals as entrepreneurs, with less concern about the organizations to which they belong. The performance of start-up companies in our fastest moving technological industries shows the incredible impact they have, if given a chance.

REFERENCES

1. Davidsson, P. (1989). *Continued entrepreneurship and small firm growth.* Stockholm: Economic Research Institute.
2. Keeley, R. H. & Roure, J. B. (1993, forthcoming). The management team: A key element in technological start-ups and entrepreneurial ventures. *High technology management research series, 8:* New York: JAI.
3. Keeley, R. H. & Tabrizi, B. (1991). Start-ups and spin-outs: Competitive strategies and effects on former employers. *Proceedings of Portland International Conference on Management of Engineering and Technology.* Piscataway, NJ:IEEE
4. Vesper, K. H. (1991). Venture idea discovery mental sequences. *Frontiers of Entrepreneurship Research 1991.* Babson Park, MA: Babson College, pp148–158.

THE OTHER ROLES OF THE INVENTOR IN MIT'S TECHNOLOGY TRANSFER PROCESS

Christina Jansen

Technology Licensing Office
Massachusetts Institute of Technology
Cambridge, MA 02142

INTRODUCTION: MIT'S TECHNOLOGY TRANSFER PHILOSOPHY PERFORMANCE AND PROCESS

This paper focuses specifically on the role of the inventor in MIT's tech transfer process. MIT's office philosophy follows the mandate of Congress as expressed in the Technology Transfer Act; we seek to develop, for the public good, the inventions which arose from the use of federal funds. To do this, the office proactively seeks companies to develop as many MIT inventions as possible. Only by getting them out into the commercial sector can the inventions be tried and tested, and developed into commercial products to benefit our economy and the public.

MIT's technology licensing office is one of the most active university licensing offices in the country. It has over 1000 U.S. patents and many foreign patents in its portfolio. In each of the last three years, MIT has had over 100 U.S. patents issued, with approximately half of these patents optioned or licensed by the time of issue. And in each of the last five years, the MIT office has executed 50–75 options or license agreements.

Annual income consists of reimbursements for patenting cost, option and licensing fees, some running royalties and, in the past few years, substantial sums from the sale of equity obtained in the licensing of start-up companies.

The licensing process, originally modeled on Stanford's, is a proactive marketing process staffed by technically educated, industrially experienced people. Five full-time technology licensing officers, with technical degrees in the primary fields in which they license and direct high-tech business experience, select the inventions they will seek to license. Each licensing officer is then responsible for deciding whether or not to patent the invention (about 40% of the inventions are patented), and selecting an outside patent attorney to draft and prosecute the patent. When a patent application is filed, MIT's office actively contacts companies until a qualified licensee is found. The office tries to find a qualified licensee, not necessarily the optimum licensee. This process has been referred to by Bob Carr as a "first contact marketing process." When the licensing officer contacts an interested company, a technical dialogue is initiated between the inventor and the technical people at the potential licensing company. Usually, the technical people visit the inventor to see the experiments in progress in the lab. At this point, if the company is seriously interested in the invention, it may take an option on

From Lab to Market: Commercialization of Public Sector Technology,
Edited by S. K. Kassicieh and H. R. Radosevich, Plenum Press, New York, 1994

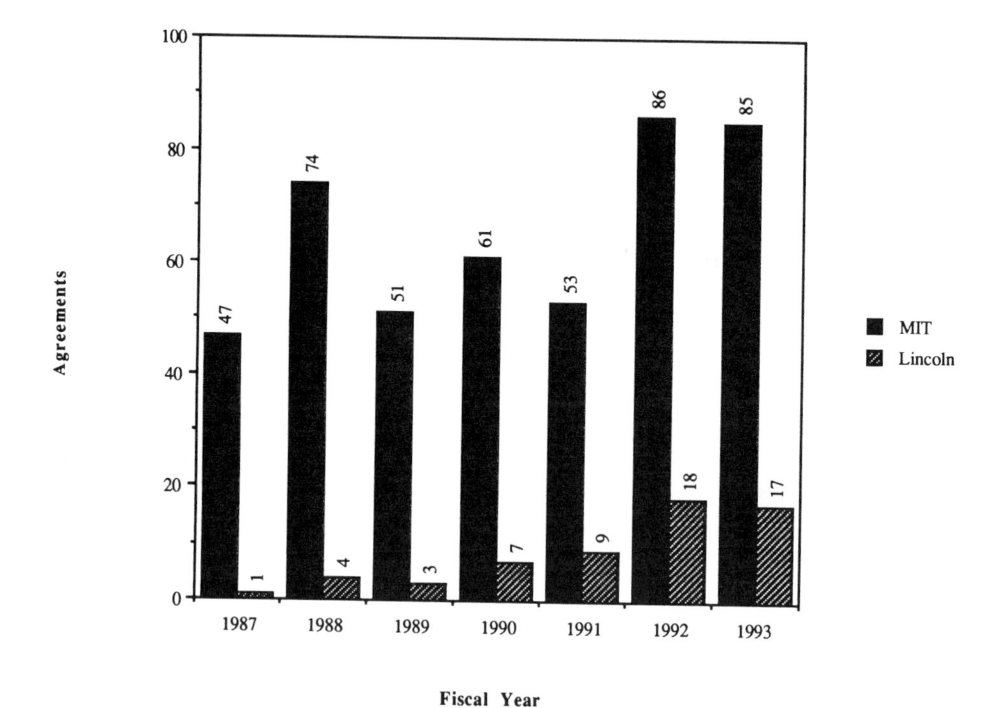

Figure 1. Agreements (licenses and options) by fiscal year.

the invention. The option will allow time for a more complete technical evaluation, possibly some sponsored research, and a business or marketing study. If, after the evaluation, the company decides to proceed, the license is negotiated by the licensing officer. Finally, if all goes well, a product reaches the market, and the inventor collects a share of the royalties (1/3 of net income).

THE MAJOR ROLES PLAYED BY THE INVENTOR

The inventor plays two major roles in the MIT technology transfer process, in addition to the fact that first, and foremost, he or she is the "Inventor" by having recognized a new concept or a new application of technology. A major role played by inventors which may surprise some people is that the inventor also acts as the licensing officers' single best source of licensing leads. A recent study shows that 54% of MIT leads for licenses or options in the past two and a half years originated with the inventor(s). The most important role of the inventor is as the primary person to transfer the technology to the receiving company. We have found this is an essential role. Without the inventor's active participation in this stage of the process, the deal usually cannot be completed. An interesting side note is that a few of our inventors have a well-developed entrepreneurial side and take an active role in starting up companies.

BIT PARTS

There are many minor roles played by the inventors. The inventor must explain the invention to the patent attorney and subsequently read and critique the draft patent application.

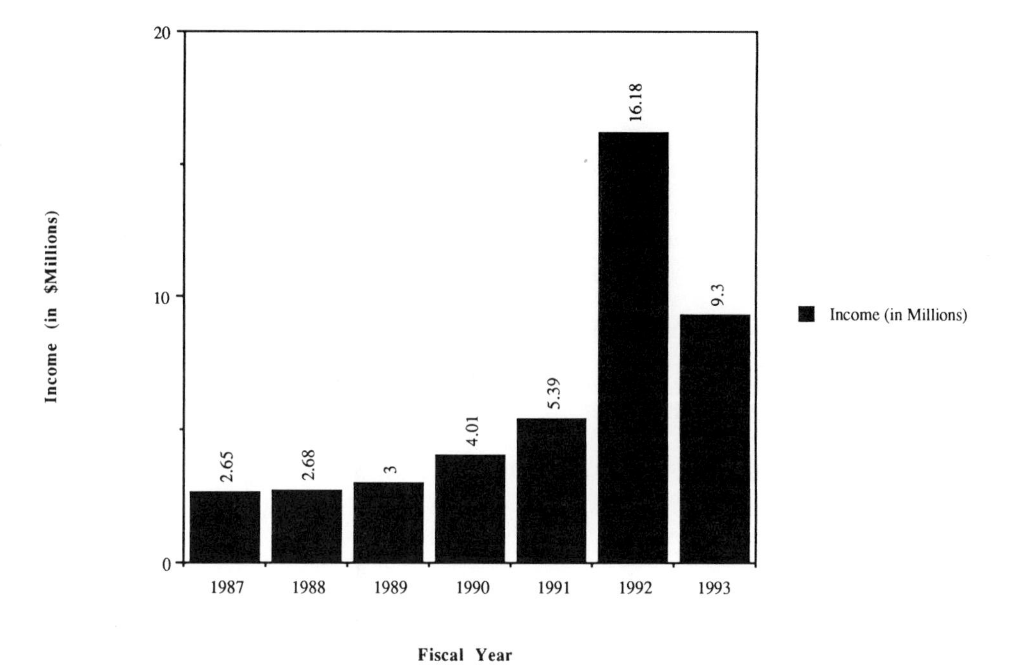

Figure 2. Licensing income by fiscal year.

After the application has been submitted and the patent office responds, the inventor must help the attorney make the arguments distinguishing the invention from others' previous work. During the search for licensees, the inventor is often expected to explain his invention on the phone to potential licensees, to host a string of visitors who may be interested in the invention and to provide laboratory demonstrations. Sometimes a company wants to test lab specimens. Occasionally, it is necessary to carry out several months of experiments to find if the invention is applicable to the company's needs. For example, an invention producing fine metal powders was tested to see if it could be adapted to make fine metal coatings on rods. This was done as a sponsored research project. Sometimes inventors are called on to help with infringements. They may be asked to read descriptions of potentially infringing products, or even test infringing products to see if there is real infringement.

WHO ARE OUR INVENTORS?

From the approximately $800 million in sponsored research funding this year at the MIT campus, Lincoln Laboratory, and the Whitehead Institute, our prolific inventors will have submitted nearly 350 invention disclosures. While most of the inventions come from faculty and staff, a surprising number of students are co-inventors. Two years ago, an MIT graduate student was named one of the National Student Inventors of the Year for his contribution of an instrument to detect and measure toxic and hazardous gases generated in burning coal for fuel. Many inventions have co-inventors from other nearby institutions, such as Harvard and the Massachusetts General Hospital. As the amount of university-industry research collaboration grows, we expect an increase in the number of inventions with industrial co-inventors.

INVENTORS—THE MAJOR SOURCES OF LICENSING LEADS

According to MIT's office lore, inventors are a primary source of direct leads to licensees. That hypothesis has been tested by identifying the source of the lead for each license and option executed by the office in the last two and a half years. In this case, the office lore proved to be fact. The majority of the licensing leads, 54%, came directly from the inventors. Through technical conferences, many inventors have developed extensive technical contacts in the industrial sector. Graduate students who have completed their degrees and taken positions in industry are another major component of the inventors' network of industrial contacts. Companies sometimes seek out professors who publish actively in related fields of interest. Consequently, it is not surprising that inventors are a rich source of leads.

OTHER SOURCES OF LEADS

In this study, the other sources of leads to likely licensees were the technology licensing officer, the licensing company itself, and the company that sponsored the research leading to the invention. In 23% of the agreements, the licensing officer identified the potential licensing company directly. Since all the licensing officers have industrial experience, there are sometimes inventions that fall within the officers' personal areas of technical expertise. When that happens, the officer involved tends to know all the companies likely to be interested and often can call people whom he/she already knows. Some inventions are in very narrow fields; an example of this is an improved material for false teeth. It is a simple matter to identify the half-dozen companies in this niche. Many companies visit the licensing office to make their needs known. By developing this kind of relationship, the company makes sure that the licensing officer will call as soon as a suitable invention comes in to the office. The company itself initiates the licensing process 15% of the time. Although less than 10% of research is industrially sponsored, there have been 25 options or licenses to sponsors in the last two and a half years. The details of the study follow.

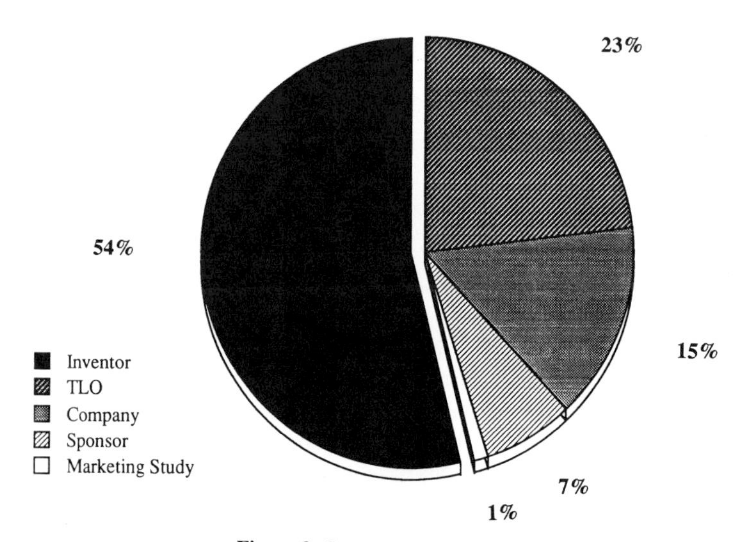

Figure 3. Source of license leads.

LEAD SOURCE ANALYSIS AND RESULTS

Lead sources for licenses or options were analyzed as follows. All options and licenses executed between July 1, 1990 and February 1993 were listed (N = 175) (end-use software licenses were excluded). Each of seven different licensing officers was interviewed and asked to specify whether the agreement lead came about through the inventor, the licensing officer, the licensing company or the industrial research sponsor. In addition, one licensing officer reported that a graduate student had done a marketing study which led to a license. The results are: inventor 54% (N = 95), licensing officer 23% (N = 41), licensing company 15% (N = 25), industrial research sponsor 7% (N = 12), and marketing study 1% (N = 1). These results are presented graphically in Figure 3.

THE INVENTOR—KEY PLAYER IN TRANSFERRING TECHNOLOGY

The participation of the inventor in transferring technology to the receiving company is essential. I cannot think of a case where MIT's licensed technology has been successfully transferred to a receiving company without the direct participation of an inventor or of a student of the inventor who has been involved in the laboratory work. While, in theory, an issued patent is intended to be sufficient teaching of the invention to transfer the technology, in fact the complexity of the kinds of technology we are dealing with requires the personal involvement of the inventor for efficient technology transfer. This is most frequently accomplished by having the faculty or staff member act as a consultant to the company or by hiring a former or present graduate student of the inventor. In addition to consulting, licensing often leads to sponsored research in the inventor's lab. In this way, the inventor has a continuing role in the company, since the research may lead to improvements or even to a potential new product.

THE ENTREPRENEURIAL INVENTOR—A SPECIAL CASE

Some of MIT's more entrepreneurial inventors are interested in starting up businesses with their inventions. Every year several professors elect to become active participants in start-up companies. One example is former professor Greg Yurek, who took a leave of absence from MIT in 1986 to start up a company based on his invention, a method of making a super-conducting wire. Yurek is now the president of American Superconductor, a company employing 75 people. Over the past seven years, MIT has had an average of eight start-ups a year. Since MIT permits professors to take equity in start-up companies, some professors choose to do so. However, to minimize conflict of interest, MIT policy requires that professors who take such equity agree not to do sponsored research that is directly related to the work of the start-up company in their MIT laboratories. Faculty members may consult with companies, (one day a week) and have seats on boards of directors, but may not take a position in line management in a company.

THE INVENTORS ARE ALWAYS WITH US

Inventors play roles in the tech transfer process. Sometimes they feel the need to be involved in license negotiations. Sometimes they decide the chosen licensee is not appropriate. Some inventors say they want to license the technology themselves, but fail to follow through. Some inventors submit half a dozen ideas a year and reduce none of them to practice. Some inventors have inflated ideas about the value of their inventions. Some inventors are perfectionists, and no company is good enough to develop their invention. Some inventors cannot give up control of their invention. Some inventors never return phone calls. Some inventors visit their licensing officer every day.

CONCLUSION—SATISFACTION, COMMERCIAL PRODUCTS ROYALTIES

The majority of MIT's inventors are professors or academic staff members who think that one of their ideas may have a commercial application. At the beginning of the process, most have limited experience with the business world and need coaching from the technology licensing officer in how to participate in the licensing process to help make that idea into a commercial success. Most inventors are very willing partners in the licensing process, in spite of other very heavy demands on their time. Of course, there are a few professors who say, "Here is my idea, you handle it from here." This attitude makes it almost impossible to license and have a successful technology transfer. Professors who are willing to work with the licensing office by giving leads, talking to potential licensees, making samples, and participating in the technology transfer are the ones whose ideas get licensed. If successful, these inventors are eventually rewarded by seeing their own products on the market and by collecting royalties.

GUERRILLA TECHNOLOGY TRANSFER

The Role of the Researcher

Glenn C. Bacon

Technology Management Consultant
Sante Fe, NM 87505

While it is necessary to have a set of rational rules and mechanisms to facilitate technology transfer from the federal laboratories to the commercial sector, these are not sufficient to achieve the full potential of the technologies that have been created. A broad look at the history of technology development shows that most technology flow was not the result of a rational, planned process; rather, it was the result of the "irrational" and unplanned movement of researchers and engineers, often against the wishes of their home institutions. We will explore the conditions which facilitate this very common but somewhat unfashionable mode of transfer.

The history of Silicon Valley shows a migration of key people from Shockley and Fairchild, augmented by a steady stream of new people from Stanford and Berkeley, to form and staff the host of new companies which became the core of the semiconductor industry. Likewise, hard-disk technology developed at IBM was propagated throughout the Valley and around Boulder (Colorado) by key engineers and researchers who once worked at IBM. This also created an industry comparable in size to semiconductors. Arguably, these two streams of technology transfer created more economic value in a shorter period of time than any other example. Yet they were largely driven by the initiative of individual technologists with little help, and often resistance, from the technology-source institutions. This flow of people seems a natural consequence of a healthy and expanding technology interacting with market forces.

While technology transfer officers, and others dependent on transaction fees in public-private technology flow, may find it hard to endorse this mechanism, the taxpayer should favor it. Public investment in technology development is an effort to give an overall boost to the economy, not just an extraction of licensing fees. Obviously identified intellectual-property assets and contractual agreements must be respected, but they can capture only a small part of the value and know-how that has been created. Market forces motivating the career choices of individual researchers are the most powerful force in technology flow. The federal labs must expect and, if possible, encourage this "guerrilla" stream of technology transfer if they are to fulfill their role as major technology-source institutions.

In order to understand more fully the critical role people-movement plays in technology transfer, we must take a deeper look at the flow of technology in value creation. Dr. Ralph Gomory, president of the Sloan Foundation, observed that there are two types of technology flow. Our usual view is the "ladder" process, in which research discoveries are refined through successive stages of prototypes and testing until a manufacturable and marketable product is

From Lab to Market: Commercialization of Public Sector Technology,
Edited by S. K. Kassicieh and H. R. Radosevich, Plenum Press, New York, 1994

achieved. This model assumes that the main technology and product ideas are discovered in a research laboratory. The first transistor resulting from Bell Labs research and relational database software from IBM research are examples. Many science-based biotechnology products come from universities through this process.

Dr. Gomory calls the other process "cyclic." This describes the technology flow when there is already a related product and technology in the field. Here, a company takes what is learned from the development, manufacture and marketing of the previous product and adds new research knowledge to design an improved, next-generation product. Obviously, a successful ladder technology transfer will be followed by a stream of products created through the cyclic process.

The vast majority of new products and technology advances result from a cyclic process, and most technology-based wealth creation results from it. Each new Intel microprocessor or Microsoft operating system has been an incremental improvement on its predecessor. The advance from 16 meg DRAMs to 64 meg DRAMs is a major cyclic process now under way. The rate of this advance is critical to competitive advantage for Korea, Japan U.S. and Europe. To help the U.S., Sematech has focused on aiding companies in the cyclic advance of the semiconductor manufacturing process. Any successful national technology initiative must focus on the cyclic process.

It is also important to observe that product areas in which the U.S. loses to international competition almost always involve the cyclic rather than the ladder process. For example, the U.S. successfully implemented the first ladder transfer of video-tape recording and DRAM technology but lost comparative momentum when these products entered the cyclic path. We lead in ladder-based biotechnology products but are behind in cyclic autos.

The role of the researcher is distinctly different in these two modes of technology flow. The researcher and his/her home institution can play a much stronger role in the ladder process. They possess a larger fraction of the expertise required to find the design-point for the first product, and contractual arrangements such as CRADAs are better suited to the more exploratory and shared-leadership approach allowed by the ladder. These contracts are also better suited to the one-time transfer of the ladder than the continuing and open-ended relationship required by the cyclic process.

Cyclic development requires much tighter and focused management in the product company. Market conditions and manufacturing-process investment dominate the selection of the new product design-point. The research component of the product must fit into a large number of practical constraints and be managed in a tightly coordinated way. The easiest way to achieve this knowledge infusion is to have the researchers on the product development team committed to its priorities and schedules. Given the fast pace and complex trade-offs of this process, it is far more difficult to acquire technology through contractual arrangements with outside research institutions. Even if the technology is acquired outside, the company will probably still need to hire experts in the area, since, with its incorporation into the product, that technology now must be continuously improved.

In order to be more effective in helping the competitiveness of U.S. companies in the cyclic process, the federal labs could move closer to the model of the research universities. These universities have long made a double contribution to the cyclic process through their research and training of graduate students. Most fast-growing technology companies do not have formal research laboratories, and count on a stream of new graduates to bring research advances into each product cycle. These companies are far more interested in the graduate students than in the particular university research projects which trained them. As a result, the great universities are surrounded by a community of successful cyclic-product-development companies staffed by their graduates. This complex of companies can rapidly exceed the size of the university itself. If the federal labs are to undertake a long-term role in commercial technology, they must expect to be compared with these universities. A key indicator of success for a federal lab will be the total revenue of companies that have located nearby in order to have continuing access to its skill base.

Beyond the effectiveness of technology transfer itself, another factor motivating people-transfer as the principal way to move technology is the long-term funding environment of the federal labs. It seems that the trend will be no-growth or downsizing rather than continual growth. If staffing must decline, it is likely that little new blood can be accommodated and the average age of the staff will rise substantially, a very unhealthy situation for a research laboratory. The best way to avoid this is to have a high volume of voluntary resignations allowing a steady input of new PhDs. A positive vision is that the lab becomes like the research universities, a magnet attracting high-quality people to itself and subsequently to the industry which surrounds it. This particularly aids local small business, which does not have broad recruiting scope.

Certainly, a policy encouraging technology transfer through people movement will create new problems for the labs. Managers will argue that productive teams will be disrupted when someone (usually a strong member) leaves. Successful companies will occasionally "raid" a lab for more good people. Increased vigilance will be required in the management of intellectual property. There will undoubtedly be some violations and Congressional inquisitors and bureaucrats will try to create new layers of creativity-suppressing and productivity-inhibiting rules. While these problems are very real and sometimes vexing, it must be pointed out that commercial laboratories and research universities constantly encounter them and manage to succeed. If the federal government wants to be active in the world of commercial innovation at more than a symbolic level, it must be prepared to accept its intrinsically messy environment. Technology transfer thrives within the culture of technologists—not that of lawyers and politicians.

Given the importance of the cyclic development process and the necessity of people movement for most effective technology transfer, some recommendations can be made:

1. Silicon Valley, Route 128 and the L. A. aerospace complex show us that people move technology between institutions and advance it most rapidly when they do not have to relocate their families. Every effort should be made to develop a community of technology companies around each of the federal labs. Capital gains and enterprise zone incentives would accelerate this and provide a net long-term payoff for the taxpayer. The labs should manage their technical strategy in view of the needs of this emerging community.

2. Laboratory managers should be rewarded for their risk in running projects in an environment of higher staff turnover. The manager's performance evaluation should recognize that staff leaving for technically-related commercial ventures is de facto endorsement of the technical importance of his/her project. Further, the manager should be allowed to replace the departed researcher at more than a one-to-one ratio. Such a "Darwinian" staffing approach will expand the productive technical areas and shrink the irrelevant ones through attrition.

3. Lifetime tenure is becoming the exception in the commercial world and should also be the exception in healthy federal labs. If the scientific prestige of the labs is comparable to the research universities, the labs can attract a large cadre of post-doctoral researchers on two-to-three year contracts. Some of these might achieve regular status in a lab, while others are likely to move to local industry. Wherever they go, they are vehicles of technology transfer. It should also be easy for regular employees to move to the commercial world through early vesting of federal retirement benefits,

extended leaves of absence, and flexible consulting arrangements. The existence of a complex of local, technically-related companies will facilitate this.

The most valuable technical asset of the federal laboratories is the knowledge and energy of its people. Some of this can be shared through joint efforts with industry and intellectual property sales. However, a much more substantial and long-lasting transfer from the public to the commercial sector takes place if an experienced researcher, trained in a commercially valuable technology, commits himself to the advancement of the cyclic product development process. This movement also allows the labs to hire the new researchers who will keep them constantly at the leading edge of new and relevant technology creation.

TECHNOLOGY TRANSFER FROM FEDERAL LABS

The Role of Intermediaries

Lawrence K. Anderson

Colorado Institute for Technology Transfer and Implementation
Colorado Springs, CO 80933

INTRODUCTION

For purposes of this paper, I define technology transfer as: "The sum total of the communication processes whereby new technology in the form of research ideas is translated into innovative products and services in the market place." By focusing on technology transfer as a *communication* process with multiple feedback loops, as shown schematically in Figure 1, we can make the role of the intermediary in this process more evident.

The role of the technology transfer intermediary or broker can also be clarified by examining the roles brokers play in other walks of life such as real estate brokers or stock brokers. The role of these brokers is to "make" an orderly market in a given commodity by bringing willing buyers and sellers together—basically a *communication* function. The common practice of buying or selling a home through a real estate broker is a case in point. While it is quite possible to buy and sell real estate without the services of a broker, most people choose to use one; they decide that the value added by the broker is worth the not-inconsiderable fee. This is because the broker, when acting on behalf of the *seller*, by knowing the market far better than the homeowner, can bring the home and its virtues to the attention of a broad set of qualified potential buyers. From the other point of view, when acting on behalf of the *buyer*, the broker can bring to the buyer a broader range of potentially acceptable homes than the buyer is likely to discover on his own.

The same arguments apply to the case of transfer of technology from the federal laboratories to the private sector. Most federal laboratories are not skilled in marketing their technology broadly and, perhaps even more so, most companies, especially small ones, have no idea of the technology available in the federal labs or how to extract [illegible] effectively. Note that in the case of the technology transfer intermediary, just as i[illegible] case of the real estate broker there is a significant issue of whether the broke[illegible] agent of the *seller* or *buyer*, (or attempts to be both) and the related issue of [illegible] agent's commission (fee) comes from. We will examine these issues later, [illegible] analogies for guidance.

From Lab to Market: Commercialization of Public Sector Technology,
Edited by S. K. Kassicieh and H. R. Radosevich, Plenum Press, New York, 1994

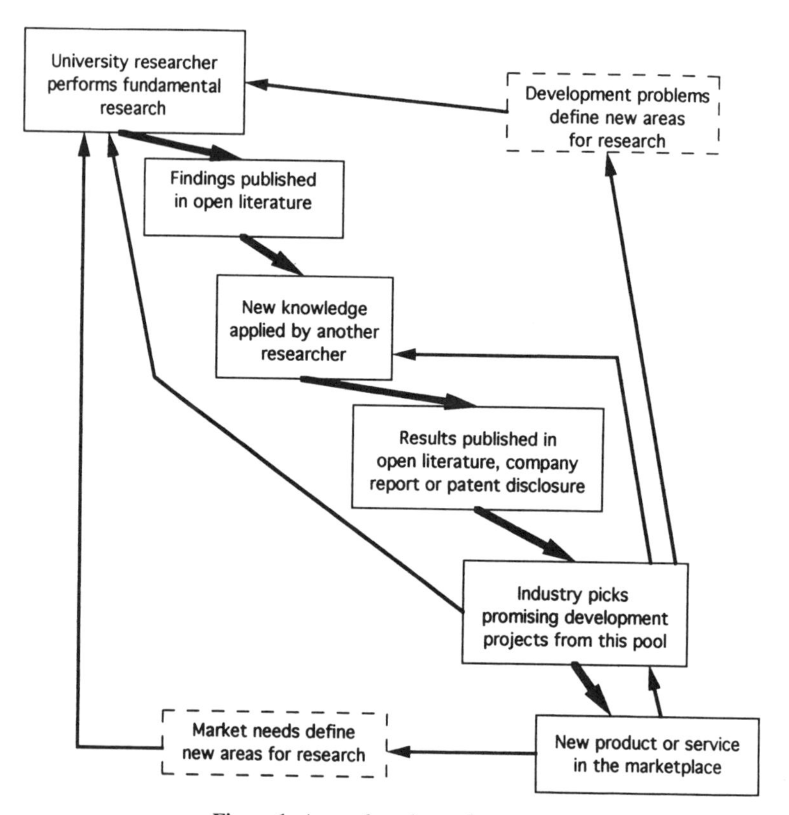

Figure 1. A sample tech transfer process.

MODELS FOR TECHNOLOGY TRANSFER INTERMEDIARIES

In a simplistic way, one can distinguish two extreme models for how a technology transfer intermediary works. These are the "Conduit" model and the "Control System" model, as shown schematically in Figure 2.

In the conduit model, as the name implies, the technology transfer intermediary is directly in the path of flow. All the technology passes through the intermediary on its way from seller (lab) to buyer (customer), as does all the feedback from the customer to the lab. Operating at this extreme is almost never a good thing, for fairly obvious reasons—a universally accepted "best practice" in technology transfer is the direct contact between technical "champions" in the laboratory and the customer company, an element missing in this model. (1) Another
... help us identify when this mode is appropriate. This analogy involves the concept
... channels" for marketing and selling products and services in the commercial
... uit model, the technology transfer intermediary is, in effect, a wholesale
... manufacturer's representative (concepts which are discussed in depth
... s mode of delivering technology to the market place would be most
... inappropriate) when the technology to be transferred is mature (a
... packaged, and when the potential market is large, varied and
... se, it might even be appropriate for the intermediary to take
... n its own name, and then market it selectively by segment in

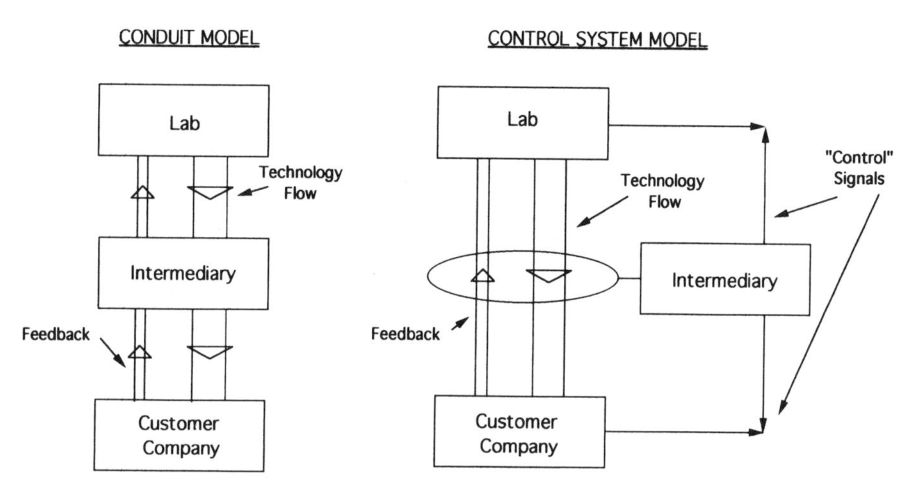

Figure 2. Tech transfer intermediaries—two models.

The control systems model is the opposite extreme. In this model, all the technical interaction occurs directly between the lab (seller) and customer (buyer). The role of the intermediary is to set up the (two-way) communication channels and then monitor them to be sure that all goes well. The analogy is that of a modern telecommunication system in which there is a separate bidirectional signalling path, through a control node, connecting the two parties involved. Another analogy for this model, favored by some technology transfer intermediaries with delusions of grandeur, comes from the world of marionettes. Here the technology transfer intermediary fancies himself an offstage marionette-player manipulating the marionettes (buyer and seller) to achieve the desired result. (It is this sort of behavior that has gotten the technology transfer intermediary a bad name in some circles.)

The concept of technology transfer as a "contact sport" and the best practice of buyer/seller "champions" obviously fits well into the control system model. It is best suited to situations where there are minimum cultural and technical barriers between the lab and customer ("they speak the same language"), and where the deal being brokered can be characterized more as a joint development than as a unilateral sale.

These simplistic, external models have limited utility. However, one can use them to build up more complex, hybrid models which do apply in the real world. For example, suppose one could concoct a hybrid model explicitly for technology transfer between a large federal laboratory and a small customer company. To be credible, this model must contain more detail, as shown in Figure 3. In particular, we must recognize that the federal lab has a technology transfer organization ("tech transfer office"), as well as a technical organization ("bench scientists"). Similarly, the technology transfer intermediary organization may involve technology specialists, interaction specialists and market specialists, or, at the very least, be able to provide all these functions in one or more individuals. In the small company, on the other hand, it is quite likely that research development, engineering, manufacturing, marketing and sales will be tightly integrated (a secret of small-business success). There will be major cultural differences between the lab culture and that of the small company, with the intermediary's technology and interaction specialists doing a lot of "impedance transformations" and the market specialist interpreting market needs to both the lab's development scientists and the customer company. While a lot of the technology, especially at the beginning, is likely to flow through the intermediary in its role as a translator, the goal, even here, is to increase the two-way flow as quickly and comprehensively as possible.

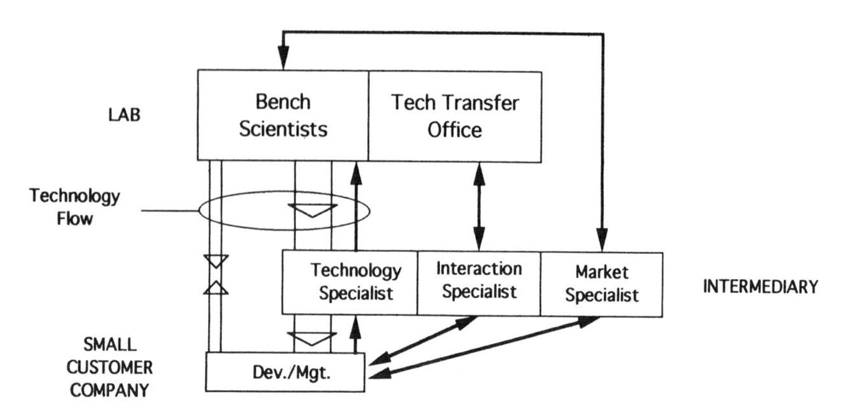

Figure 3. Tech transfer—small company model.

The control system model of Figure 2 probably applies most directly to the case of transfer between a large federal laboratory and a large customer company, as shown in Figure 4. Here, the customer company is likely to have a separate technology acquisition office, as well as a broadly competent research, development and engineering staff. The role of the technology transfer intermediary may be no more than to alert the lab's technology transfer office and the company's technology acquisition office that there appears to be an overlap between the technological capabilities of the lab (seller) and the technology needs of the customer company (buyer). There would be no need for a technology specialist in the intermediary, although, in general, there might well be a role for a market specialist in the intermediary to modulate the two-way flow of technology between the lab/company partners so that what emerges best meets the needs of the market place.

ROLES FOR TECHNOLOGY TRANSFER INTERMEDIARIES

Let us examine now, in somewhat more detail, the various roles that technology transfer intermediaries can play. I have chosen to discuss them in terms of seven functional activities.

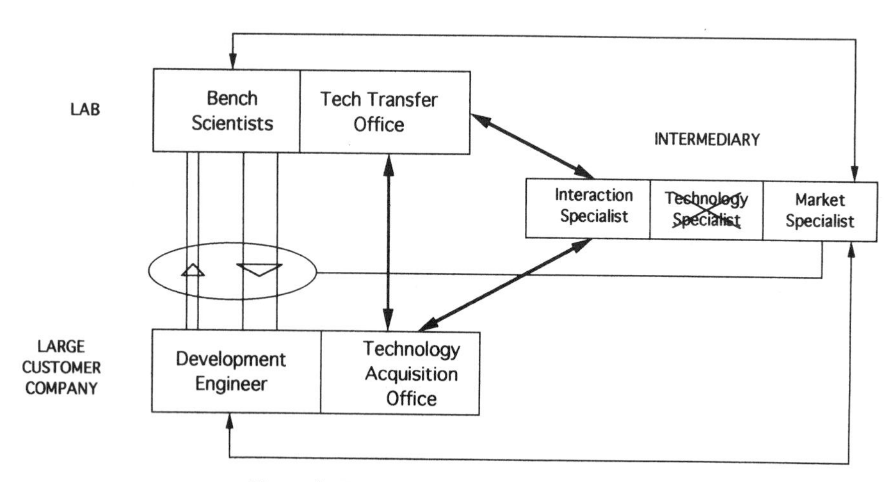

Figure 4. Tech transfer—large company model.

1. Expert business/marketing consulting
2. Brokering small company/large company strategic alliances for technology transfer
3. Networking facilitator
4. Translation between cultures
5. Special project management
6. Education and training
7. Technology distribution channel

1. Expert Business/Marketing Consulting

Serving as a *sellers agent*, the technology transfer intermediary can supplement the federal lab staff's assessment of what existing technology or technological capabilities it has that may have external value, and of the external market for those technologies, and can finally help drive home the actual technology sale. Help with market assessment can often be provided very successfully by university business schools. In a now classic example, students in the Anderson School of Management at the University of New Mexico explored the market potential for a novel explosives initiator originally developed by Sandia for nuclear weapons applications. (2) What emerged eventually, was a start-up business manufacturing initiators for automotive airbag applications.

In terms of surveying broad regional opportunities and needs for technology, public-sector technology transfer intermediaries can often provide a credible third-party mechanism for determining customer needs. For example, the University of Colorado Business Advancement Center (CU-BAC) is currently conducting an in-depth survey of all Colorado high-technology companies to determine their areas of greatest technology need and their awareness of and attitude toward the federal laboratories as a source of technology to meet those needs. (3)

Serving as a *buyer's agent*, the technology transfer intermediary can help a client company identify what technologies are available at which federal labs and "pull" a specific technology out of a particular federal laboratory. There are over 700 federal laboratories and at least half a dozen non-interconnected electronic databases of federal technologies. Knowing where and how to search these requires a professional. While access to all federal technology is supposedly available through a single phone call (to, e.g., the National Technology Transfer Center in Wheeling, West Virginia), the reality tends to be different. There is no substitute for a specialized technology transfer intermediary with specific technical expertise and a network of individual contacts in various federal laboratories.

2. Brokering Small Company/Large Company Strategic Alliances for Technology Transfer

In their book *Winning Combinations*, Botkins and Matthews make a compelling argument for the improved global competitiveness of strategic partnerships between large and small companies. (4) This same formula can be applied to the corresponding three-body problem in technology transfer—a strategic alliance of a federal laboratory, a large company and a small company. Technology transfer intermediaries are often in an excellent position to help establish such partnerships. The Alliance for Photonic Technology (APT) at the University of New Mexico was involved in brokering one such deal. In this case, the problem (read "opportunity") being addressed was the poor yield in fabrication of composite structures; the technology was fiber-optic sensing of chemical constituents, available from a large national lab, and the industrial partners were a large airframe manufacturer and a small maker of fiber-optic sensing systems. (5) In this case, the large airframe manufacturer supplied a market-driven need and relatively deep technological pockets. And the small systems house provided a nimble, entrepreneurial setting in which problem and solution could come together in the form of a marketable diagnostic product. Ironically, this attempt at technology transfer, which seemed to have everything going for it, failed. It did so, not because of any intrinsic deficiencies in the technical or financial arrangements, but because of a lack of flexibility in the DOE CRADA process which was the mechanism chosen for structuring the deal. In this case it was the "substantial U.S. manufacture" clause which derailed the negotiations, a clause which has since been liberalized.

3. Networking Facilitator

Most technology transfer specialists understand the importance of person-to-person "networking" in facilitating technology transfer, and they have their own professional or pseudo-professional societies to promote this—the Federal Laboratory Consortium (FLC), the Association of Federal Technology Transfer Executives (AFTTE), the Association of University Technology Managers (AUTM), and the Licensing Executives Society (LES), for example. While such organizations are useful in bringing together people of similar persuasions to exchange experiences, they are relatively ineffective in bringing together technology buyers and sellers. To do the latter, particularly in the context of encouraging small business start-ups, one needs to bring together university, federal lab and industrial scientists and engineers along with the infrastructure people (lawyers, accountants, venture capitalists, entrepreneurs, etc.) needed to form new, viable high-technology business enterprises. This can be done by forums and roundtables having relatively narrow technical and geographical focus. Technology transfer intermediaries can be an effective driving force in organizing such forums. For example, the Colorado Institute for Technology Transfer and Implementation (CITTI) was instrumental in forming the Southern Colorado Biotechnology/Biomedical Forum, in Colorado Springs, modelled after a similar venture in Denver, and is in the process of forming a Software Roundtable in the region, as well. The MIT Enterprise Forum is an example of this kind of activity coordinated on a national level. These forums/roundtables can be successful to the extent that they focus on the *business* of technology and don't get wrapped up in technology for the sake of technology, on the one hand, or degenerate into a social fraternity on the other.

4. Translation between Cultures

Much has been written about the cultural gulf that exists, even within a single company, between the technology-driven research scientist and the market-driven development engineer. This chasm is even larger between a university or federal lab scientist and a (small) company product engineer. Technology transfer intermediaries can play the role of "cultural translators" here. In some cases this can be as straightforward as helping a laboratory director show a corporate CEO how the laboratory's technology can contribute to the CEO's ROA. Or it could be as daunting as an attempt to shift an entire federal laboratory's focus from technology (for the sake of technology) to the benefits that this technology provides to customers. (6)

In carrying out this cultural translation, the technology transfer intermediary must not lose sight of the fact that what motivates scientists (usually more support for their research) differs from what motivates people at a customer company (usually advantage in the market place.) Deals must be designed to be win-win for all parties. This means not only picking the deals carefully, but also structuring them so that the rewards are appropriate. For example, a research institution may well value getting customer support for sponsored research above getting a royalty fee from a technology license, which may never materialize or, if it does, may not filter down to the demonstrable benefit of the researcher.

5. Special Project Management

Technology transfer intermediaries, especially those with broadly based capabilities, may be in a unique position to manage complex multidisciplinary projects. An example is the current pressing need for defense conversion. Rapid downsizing in some parts of the defense industry is having a profound negative impact on the people, companies and locales involved. Not only is there the obvious loss of revenues and jobs, but also the company and community can suffer a permanent loss in valuable high-tech resources as the people affected move out of the region. One experimental initiative aimed at ameliorating this problem is being conducted by a group from the Colorado Technology Action Consortium (COTAC), a loose federation of Colorado technology transfer intermediaries. The experiment is targeted at a large aerospace defense contractor in the Denver area. The COTAC plan is to identify a few entrepreneurial engineers who are in the process of being laid off, a few technologies which are either potentially defense/civil dual use, or in some sense "surplus" to the defense contractor's strategic plans, and to bring these together in an internal incubator on or near company premises. If this venture is successful, the company will win in terms of possible future license revenue, a retained subcontractor base, and community goodwill. The local economy will have gained from retained jobs and a retained technology infrastructure.

The technology transfer intermediaries play a key role in this project. For the project to succeed, entrepreneurial business resources must be added to the technical resources of the defense contractor. Both public- and private-sector seed funding must be found, to augment the defense contractor's resources. And, critically, the would-be entrepreneurs must be screened and trained in terms of survival in the commercial world. All these areas are appropriate to technology transfer intermediaries.

6. Education and Training

Technology transfer intermediaries can serve as effective agents of change, particularly with respect to the larger federal laboratories. Few organizations, no matter how excellently managed, can produce major cultural change (e.g., from a technology-driven to a market-driven outlook) driven totally from within. Technology transfer intermediaries, as a result of what they learn by working both sides of the street, can help produce the desired change. Technology transfer intermediaries can also provide training in technology marketing to federal laboratory personnel, again by virtue of their understanding of both the commercial and federal lab cultures.

7. Technology Distribution Channel

Every successful new venture, as part of its basic strategic/business plan, must ask and answer the question: "How am I going to distribute my products and/or services?" Federal laboratories must do this as well, as part of their general strategic planning. (1) In many cases the federal lab may elect to market its technology directly to all the end users—the analog of a "direct mail" distribution strategy for a private business. In this case, the role of the intermediary is limited to providing general advice on markets and marketing. Or the federal laboratory may choose to engage a technology transfer intermediary to help the lab identify specific customers and negotiate terms for its technology, while still taking responsibility itself for the actual details of the transfer. In this case, the analogy might be to a "manufacturers' representative." Finally, as alluded to earlier the federal lab might choose to license its technology in toto to an intermediary for relicensing—the analog to the "wholesale distributor" mentioned earlier. Flow in these models is illustrated schematically in Figure 5.

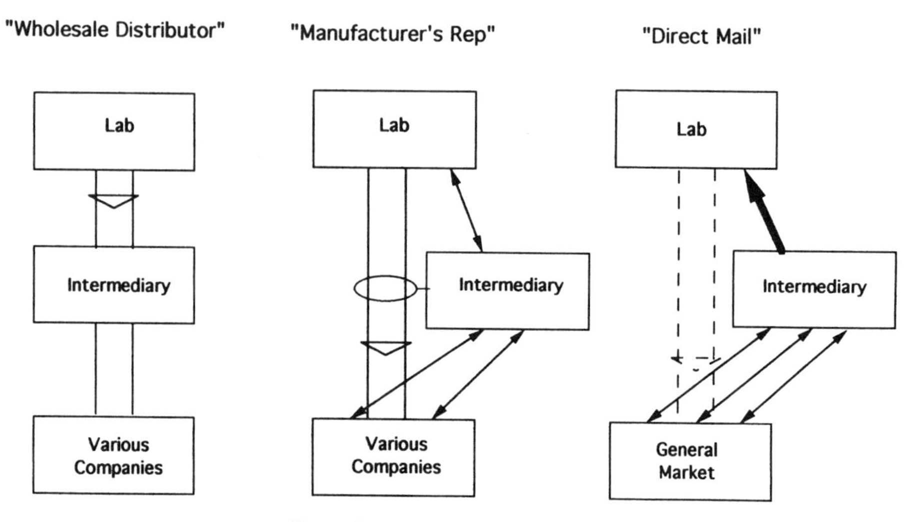

Figure 5. Distribution of technology.

The major points here are that a federal lab must make specific choices based on its overall strategy and resources, and that technology transfer intermediaries can play a major role in that strategy and can be one of the resources.

TYPES OF TECHNOLOGY TRANSFER INTERMEDIARIES

There is a variety of internal and external technology transfer intermediaries, both public and private sector. For our purposes, it is convenient to group them into five categories:

1. Federal lab internal tech transfer offices
2. University-based intermediaries
3. State-agency-based intermediaries
4. Federal-agency-based intermediaries
5. Private sector consultants

1. Federal Lab Offices

By law, all federal laboratories must have, as a minimum, an internal individual designated as the lab's technology transfer officer, often called an ORTA (Office of Research Technology Assessment). These officers are technology transfer intermediaries only in the sense that they serve as brokers between the laboratory's technologists and the outside commercial world. They are, in effect, the lab's marketing arm. While many of these offices have become fairly large and elaborate, with several tens of people, most of them were put in place when technology transfer was of decidedly secondary strategic importance to the lab. Now that technology transfer is a key mission of most labs, the offices are often struggling to keep up with demand. This is partly because of cultural left-overs, and partly because of resource limitations in even the larger laboratories; in spite of the recent growth of technology transfer offices, most must still function from a resource base that is small compared to the resources a private sector company would expend on aggressive marketing. There is thus a clear opportunity for external technology transfer intermediaries to supplement the internal resources.

2. University-Based Intermediaries

These not-for-profit external technology transfer intermediaries are sometimes associated with the technology transfer offices maintained by most research universities, and are sometimes quite independent. Two such offices the author is intimately familiar with, having founded both, are the Alliance for Photonic Technology (APT) at the University of New Mexico (UNM) in Albuquerque and the Colorado Institute for Technology Transfer and Implementation (CITTI) at the University of Colorado at Colorado Springs (UCCS). APT focuses on the transfer of technology in a limited technical area, photonics (fiber-optics, optoelectronics, lasers, etc.) from the Los Alamos and Sandia National Laboratories (DOE), from the Air Force's Phillips Lab, and from UNM's Center for High Technology Materials. APT functions as a seller's agent—it is supported by the labs and university to help market their technologies. It has no direct ties to UNM's technology-transfer or sponsored-research offices. CITTI, on the other hand, has multiple roles. As the local campus office of the University of Colorado's system-wide Intellectual Resources Program, it serves as a seller's agent in brokering the university's technology to the world. As an institute, it does sponsored research (usually supported by state and/or federal grants) in the theory and practice of technology transfer. And, acting in the role of a buyer's agent, it brokers technology deals among the university, regional federal laboratories and private-sector clients.

An example of a CITTI-brokered deal-in-the-making is one involving a small start-up we shall call the New Venture X Corp (NVX). The technology in this case is a proprietary, high-density silicon semiconductor memory-cell technology devised by two engineers who spun out of a small semiconductor company. NVX is now faced with the classical catch-22 of technology transfer; they cannot license their technology or get capital to develop it themselves until they have at least a rudimentary proof of principle that the technology works as advertised, but they cannot afford to do the necessary prove-in demonstration until they raise more capital. CITTI's goal is to break this vicious circle by making arrangements for access to university design facilities and federal lab wafer-fab facilities on a quid pro quo basis. In addition, CITTI is helping NVX identify sources of capital, based on the expectation of university and federal lab partnering. If this works as planned, the university will have a new CAD capability in place, the federal lab will have a new technology of value to its core mission, and the two entrepreneurs will have their proof of principle and start-up capital.

3. State-Agency Based Intermediaries

Wanting to catalyze technology-driven economic development within their borders, many states have formal state-wide technology transfer intermediaries funded by general state tax revenue. Two of these the author is familiar with are the Colorado Advanced Technology Institute (CATI) and New Mexico's Technology Enterprise Division. CATI's main role has been to disburse seed money to help fund, in Colorado's universities, research and research centers deemed to have a long-term impact on economic growth. More recently, CATI has begun to disburse a larger fraction of its resources to support technology transfer infrastructure (advanced data and telecommunications, business incubators, etc.). The Technology Enterprise Division in New Mexico on the other hand, has recently refocused its resources entirely from a focus on technology creation through research centers, to technology distribution through such programs as the State Technology Assistance Resource System (STARS) program, which is aimed at helping the front-line state business-support people get prompt access to the technology they need to help their private-sector client companies. In general there seems to be a country-wide shift by state agencies away from the creation of new technology (which is often viewed as paying off only in the very long term, if at all, in terms of economic growth) towards support of the infrastructure necessary to do an effective job of distributing available technology. Typical of such programs is that at the Colorado Technology Transfer Center, which has a contract with the National Renewable Energy Laboratory (NREL) to assist the latter in marketing some of its technologies.

4. Federal-Agency Based Intermediaries

Typical of these is the new National Technology Transfer Center (NTTC) and its network of six regional technology transfer centers (RTTCs). These are partially supported by NASA as part of its very broad technology transfer program, and in part by user fees from private sector clients, primarily to the RTTCs and their affiliates. The NTTC and (to a lesser extent) the RTTCs provide training in technology transfer methodology. They also provide an infrastructure giving the private sector access to federal technology. They act in this case as buyer's agent.

5. Private-Sector Consultants

Backing up the public sector intermediaries cataloged above is a rich collection of private sector entities, which can be lumped into three broad categories:

a. For-profit licensing firms, such as University Technologies Corporation, or Research Corporation Technologies which, historically have specialized in licensing technologies from universities for subsequent relicensing or, rarely, internal development. There is no intrinsic reason why these firms could not do for the federal labs what they do now for private and public universities, the barriers seem to be cultural rather than legal.

b. For-profit companies specializing in on-line electronic databases of available technology and R & D people resources. Typical of these are Knowledge Express, which has focused on licensable technologies from universities and federal labs, Teltec, which provides a consultant referral service based primarily on private-sector and university researchers, and Best North America, which maintains records of faculty research interests and licensable university technology. These organizations are all buyer's agents, in the sense that they derive most of their revenue from private-sector companies which sign up, on a fixed-fee or transaction-fee basis, to gain access to the respective agents' databases of available technology and people. While federal laboratory technology has been reasonably well represented in these databases, the capabilities and facilities associated with federal laboratory researchers have been largely absent. This has transpired in large measure because federal laboratory directors have been understandably reluctant to subject their researchers to a "flood" of calls for help, many of them perhaps, of dubious quality. A knowledgeable technology transfer intermediary can counter this concern by serving as an effective initial (but *only* initial) screen between the researcher and the customer. Also, in many cases, the transaction fees charged by the private-sector database companies discourage frivolous use of the database by their clients.

c. For-profit, private-sector consultants, often individuals. Taken altogether, individual technology-transfer consultants represent a very powerful resource. In particular, many of the public-sector technology transfer intermediaries, such as CITTI, rely on a "stable" of such consultants to work with them and their customer companies sometimes on a pro bono basis, sometimes on a reduced-fee basis, and sometimes on a full-fee basis as circumstances dictate.

CONCLUSIONS

There is a growing cadre of technology transfer intermediaries who can make an efficient market in technologies from a range of research institutions. But federal laboratories have generally been slow to take advantage of these external intermediaries. Part of this reluctance stems from a culture rooted in two historical realities: some government technology was secret and one could go to jail for transferring it to *anyone*, and the balance of government technology developed at the tax payer's expense, belonged to the people, and had to be shared with *everyone*. It has been understandably difficult for many federal laboratories to adjust to a world in which technology transfer (from a federal lab) has gone from being *forbidden*, through being *encouraged*, to, today, being *mandated* as a requirement for the lab's survival.

It is difficult for someone brought up in the government tradition of public service to adjust to the reality that successful technology transfer implies that someone in the private sector is making a profit at the taxpayers' expense. But that is exactly what technology transfer is all about. Clearly, successful transfer has occurred only when the transferred technology has resulted in a product, service or process that is so successful that it is profitable. To the extent that the underlying technology was developed in a taxpayer-supported federal laboratory, we are clearly talking about the taxpayers' expense. This whole process is justified by the expectation that the combination of taxpayer and private-sector investment results in substantial benefits to society in terms of more jobs, better health, higher living standards, etc. The prospect that some one particular firm makes a profit commercializing a federal technology is justified as long as that firm's competitors had a reasonable chance to do likewise ("fairness of opportunity"), but lost out through inaction or in a fair fight. There is a great temptation to try to deal with the built-in opportunities for unfairness of opportunity and with the possibilities of conflict of interest by putting in place a rigid web of rules. But legislating away the possibility for unfairness of opportunity and conflict of interest probably also legislates away some possibility of meaningful technology transfer. Instead, it seems best to set up a system which can police itself at the lowest level and which deals with every case on its merits.

Given the difficulty the federal laboratories experience in adjusting to the whole concept of technology transfer, it should come as no surprise to external technology-transfer intermediaries that they have difficulty gaining acceptance by the federal laboratories.

Another factor in this reluctance to embrace intermediaries is the view widely held by many technologists in both private and public sectors, in federal labs and academia, that selling (or even marketing) is somehow a parasitical activity. ("Every dollar spent marketing a product, service or technology is a dollar we don't have to make it technically better.") This view, while understandable, is not defensible. We are all familiar, of course, in the commercial world, with products or services where the resources put into selling the product have gotten out of balance with the resources dedicated to improving it. In the case of technology from the federal labs, it can be argued that the pendulum is still way over on the other side—the resources now being dedicated to getting this technology to the marketplace are dwarfed by the technical resources being deployed to develop new technology, and by the potential value of the nascent technologies which already exist in the laboratories.

By taking advantage of the many services available from public- and private-sector technology transfer intermediaries, incorporating these into their long-range strategic plans, federal laboratories can efficiently leverage their own limited resources. By putting the profit motive to work for them, carefully, instead of fighting the concept, they are likely to be much more successful in getting technology quickly to market. This will surely involve risk. However meaningful technology transfer cannot occur in a risk-free environment. That is just as true in the federal sector as it is in the private sector.

REFERENCES

1. Gurney, B., & Anderson, L. K. (1993). *Benchmarking best practices in technology transfer.* Presented at the 1993 Technology Transfer Society Annual Meeting, Ann Arbor, MI.
2. Radosevich, R. (1993) A mixed strategy model and case example of federal technology transfer. To be published in *International Journal of Technology Management.*
3. Eye, K. (1993) *1993 Colorado technology survey.* Available from University of Colorado Business Advancement Center (303–444–5723).
4. Botkin, J. W., & Matthews, J. B. (1992) *Winning Combinations.* New York: John Wiley & Sons.
5. Anderson, L. K. (1992). Technology transfer from federal laboratories — An industrial perspective. *Proceedings of the Technology Transfer Society 17th Annual Meeting and International Symposium* Atlanta, GA.
6. Lundquist, G. M. (1993 February) *Defining value: Translating from the technical-ese.* Available from Market Engineering International, Inc.

III

MECHANISMS AND PROCESSES

INTRODUCTION

The future of the U.S. position in the international economic arena depends heavily on the increase in the utilization rate of federal technology by private enterprises in the U.S. and on the success of the universities and state and local governments as intermediaries in the technology transfer and commercialization process. In this section, we provide a number of mechanisms and processes that support improved technology commercialization activities. The previous section described the relationship between the roles played by various participants in technology transfer and the mechanisms and processes used. The purpose of this section is to examine in greater detail the mechanisms that are being used and proposed as well as the processes that accompany those mechanisms. As the terms are used here, "mechanisms" means the devices or entities that are involved in the "processes" (the set of activities which result in technology transfer and commercialization).

TRANSFER AND COMMERCIALIZATION MECHANISMS

Several studies have suggested that technology transfer and commercialization is a multi-stage process (especially if new enterprises are established in the process) and that the effectiveness of a given mechanism depends upon the stage in the process.[1] This implies that it may be best to employ a combination of mechanisms, with changes being implemented as the transfer progresses. Although not all-inclusive, the array of mechanisms described below demonstrates the relative effectiveness of various mechanisms at different stages.

Early Stage Transfer Mechanisms

The earliest stage of technology transfer and commercialization is often described as the prospecting or "hand-shake" stage during which the technology source and the recipient become aware of each other. On occasion this stage begins with a general interest in learning more about the capabilities and needs of each party. Initial interaction may also be triggered by awareness by one party of a particular technology or application available from or sought by the other. The probability that the two parties will become aware of the transfer opportunity can be enhanced by a number of mechanisms as discussed below.

Technical Advisory Boards and Panels. An increasing number of public and private organizations are relying on external technical advisory boards to provide information about the existence and characteristics of technology developed by other organizations. Technical advisory boards are very useful for start-up ventures needing not only advice on technical solutions to problems but also information regarding desired functionality by users. Local governments often need scientific and technical advisory boards and panels not only to help them evaluate technical alternatives but also in planning economic development programs with technical components. Similarly, the federal government requires technical advice on matters ranging from science policy to choices of technical approaches for specific programs. All these requests for technical expertise result in opportunities for laboratory scientists to increase awareness of laboratory technology and capabilities—an essential step in the early stages of technology transfer.

Technical Visits. Since the late 1980s, when federal laboratories were ordered to increase technology transfer, the increased interest of many firms has been manifested in visits to laboratories by technical persons from industry. Although mechanisms such as personnel exchanges exist within most agencies to foster increased knowledge by laboratory personnel of industry needs, visits from laboratories to firms have not approached the volume of the reverse flow. A number of mechanisms have been tried to make those involved in directed research and developmental activities to increase their understanding of potential users' needs. For example, scientists in China have been required to spend time in common labor in the factories and fields to understand better the productive processes which could be improved by their work. Incentives for scientists from federal laboratories to visit industrial firms rarely exist.

Professional Meetings. Most university and federal laboratory scientists receive the bulk of their understanding about industry problems from their industrial counterparts while attending professional meetings. As professional societies become more actively involved in technology transfer and commercialization, this mechanism may become even more important.

Published Materials. Reports, articles, books, etc. and other published materials, either on specific technologies or general capabilities, have historically been heavily used by technology sources to stimulate interest in their results. Because newer mechanisms such as CRADAs reserve data rights to the collaborating partner, there exists some concern that opportunities for publishing in scientific journals may be reduced. Even if this occurs, it may not be injurious to U.S. international competitiveness, since many experts feel that publishing scientific results in the open literature provides equal opportunity to competitors.

Technology "Inventories". Various government agencies have tried to produce formal inventories of technologies with commercial potential. Inventories like the *NASA TECH BRIEFS* or laboratory documents like *Sandia Technology* or *Los Alamos Science* publicize specific technologies for which proprietary intellectual property exists as well as demonstrating general competence in a particular field of science or technology in order to attract potential research partners. To date, little systematic matching of these inventories to commercial lists of industry interest has taken place.

Technical Conferences on Specific Technologies. When federal laboratories develop a technology that appears to have strong commercial potential, it is not uncommon to arrange regional conferences to broadcast the technology's availability and describe its functionality. This "show-and-tell" mechanism has been used very effectively by the regional committees of the Federal Laboratory Consortium, especially in creating outreach to small businesses.

Test Demonstrations. It is not uncommon for federal agencies to support demonstration projects embodying technologies developed in their laboratories, especially when they see opportunities for widespread diffusion of the innovation rather than selective transfer to one or few recipients. This has been successful historically in fields like agriculture and alternative energy. Specific laboratories may be designated demonstration sites or, occasionally, a portable demonstration is developed so that the convenience of geographic proximity induces more potential recipients to attend.

Data Base Compilation. Research results, in the form of reports or published materials, are usually sent to a central repository by the technology source for inclusion in a data base accessible to a wide audience. Access to a few data bases (especially within the Department of Defense and the Department of Energy) may require special status, such as that of a vendor or contractor. While some data bases are commercial and include extensive options for search services, those sponsored by the government are more traditional in their operations. Some, like the National Technical Information Service, cross all agency boundaries, while others are agency-specific, like the National Energy Software Center or the DOE Office of Scientific and Technical Information.

Extramural Service. Technology source workers, especially those inclined to be gatekeepers or boundary spanners, frequently serve on boards, committees and task forces of businesses, governments or other organizations. In those instances in which laboratories are committed to local economic development, one often discovers laboratory personnel extensively involved in planning task forces, business incubator boards or advisory committees to local and state governments. Such involvement not only presents opportunities to interact with potential technology recipients, but also helps the development of local infrastructure to support related assistance services.

Technology Maturation Stage

At this stage of the transfer process, further development toward commercial application is achieved, either by the source or recipient or perhaps, most expediently, is done jointly. Although one occasionally finds almost complete products or processes in public-sector technology sources, one is usually fortunate to find a fully-functional prototype. Agricultural "products" or practices may be well tested before the government mounts an extensive adoption/diffusion campaign, and products like high-energy-density capacitors, routinely built into nuclear weapons, may find immediate applications in medical devices. More commonly, several years of development and testing is required before specific applications of public-sector technology is market ready. A variety of mechanisms are employed to aid this maturation process.

Laboratory/Vender Relationships. Contracts from laboratories to industry for products or services not currently available frequently result in the provision of technical assistance to vendor to develop the desired purchase. In those instances in which vendors do not have the technical capacity to deliver products with the desired qualifications, some laboratories have defined collaborative development programs to improve vendor capabilities. In one case, when no U.S. vendor had suitable capabilities and the Department of Energy did not want to depend upon foreign suppliers, substantial effort has been made to support the technology competitiveness of U.S. industry through development programs, technical assistance and purchase contracts.

Laboratory/Industry Personnel Exchange. Temporary exchanges of personnel between industry and laboratory is a possibility in most agencies but is under-utilized by both sides. The least common mechanism, exchange from laboratory to industry, may be the most useful to accommodate technology transfer. Since technology transfer is commonly described as a people-to-people phenomenon, personnel exchanges could be used to create greater substance in the interpersonal contact. However, it appears that better incentives are necessary to induce laboratory personnel to be interested. Although not an official transfer mechanism, the permanent transfer of laboratory personnel to industry may become a more common occurrence if federal laboratories (especially defense-related ones) are subjected to substantial reductions in force. In this case the personnel exchange may also assist the commercialization stage as well as the maturation stage.

Joint Research Projects. Since the advent of the cooperative research and development agreement (CRADA), federal laboratories have dramatically accelerated their collaborations with industry. Historically, joint research projects have been undertaken principally for the advancement of science or technology by augmenting local expertise with that existing elsewhere. Collaboration with consortia has increased dramatically since the 1984 legislation authorizing cooperative precompetitive research; however, there are few short-term applied results which will affect international competitiveness from this sort of research. Collaborative work with other laboratories and universities has usually been at the basic or applied research phase and therefore has not furthered the laboratory's capability to solve specific industrial problems or define new products and processes.

The formation of strategic alliances between complementary organizations is likely to become an increasingly important technology commercialization mechanism. James Botkin's paper in this section of the book, *How to form, manage, and evaluate effective strategic alliances*, (p. 225) describes how strategic alliances help organizations speed up the innovation process, compete in global markets and convert from the military emphasis of technology to an economic emphasis. He also prescribes the steps necessary to form and maintain an alliance that benefits all the parties concerned.

Jana Matthews's paper titled *Forming effective partnerships to commercialize public sector technology*, (p. 233) discusses changes in the federal laboratories' missions needed to manage the formation of strategic alliances that are more in line with the technology commercialization goals of the laboratories. She also presents a model of the technology transfer process that is used in commercializing manufacturing software.

David Gibson and Everett Rogers, in their paper, *The evolution of technology transfer at MCC*, (p. 257) further describe the transfer mechanism of consortia. They use the case of the Microelectronics and Computer Technology Corporation (MCC) to describe the work of an R&D consortium and evaluate its performance through its initial years in its attempt to manage and transfer technology.

Bruce Winchell discusses the ingredients for success in strategic alliances in his paper, *Partnerships are a people business*, (p. 239). Apart from finding the right partner with the right motivation and creating an arrangement that is beneficial for all parties, success needs to be defined in terms that are applicable not only in the commercial world but also to the federal laboratories.

Ray's paper, *Some observations on laboratory-industry alliances* (p. 243) reports on a survey conducted to determine how industrial firms acquire or develop technology. It points to a concern with the differing objectives that companies have vis-à-vis laboratories. Companies want to build a relationship quickly and cheaply, whereas the laboratories have a different objective in mind.

Barquinero, in his paper, *Targeted technology commercialization through value-added facilitators* (p. 247), describes several mechanisms for technology commercialization including the NASA version of joint research programs. The novelty of this mechanism suggested by Barquinero is the use of facilitators that help accelerate the process of commercialization. As he describes it the use of facilitators has increased effectiveness in two new NASA initiatives—not only the joint-sponsored research program but also NASA's technology commercialization centers. The activities described elaborate the concept of "targeted technology commercialization," which focuses the effort of the federal agency (in this instance NASA) on its commercialization efforts. Barquinero describes the activities of these two programs, as well as two examples of recently completed commercialization efforts facilitated by commercialization experts.

Contract R&D, "Work for Others. Federal laboratories have the capability to perform specific contract R&D for private firms, especially when unique facilities, equipment or technical expertise exist at the laboratory. There are at present two issues that must be addressed before this "work for others" becomes a widely-used transfer mechanism. The first is the current concern of federal agencies that public facilities should not be used to compete with the private sector. Such activity would place a private facility at a serious competitive disadvantage with a public institution. It is very difficult for a laboratory to ascertain when there is a private entity with the capability to perform the work being proposed by a work-for-others agreement. Secondly, due to high overhead rates charged by many federal laboratories, private sector organizations are less likely to contract with these laboratories for work, especially if the potential client is a small business. Most agencies and their laboratories have now instituted programs to provide short-term, free services to small businesses, but there is still a need to define special considerations in the cost structure of reimbursable activities for longer-term small business assistance.

Technology Commercialization Stage

In many industries, it is difficult to separate the design and development of devices and processes from manufacturing, marketing and other commercialization activities. In VLSI (Very Large System Integration) circuits applications, for example, device designs depend heavily on production technology. In such cases as these, it is important to include additional mechanisms in a staged transfer and commercialization process. Some useful alternative processes are discussed below.

"User Facility" Programs. Many federal laboratories and some universities have formal programs allowing private firms to have access to special facilities and equipment on a basis which does not interfere with the institution's primary mission. The Los Alamos National Laboratory, for example, has over thirty formally-declared user facilities. LANL has prepared a brochure describing the program and each facility, including contact points to arrange access. Such facilities are useful to both the laboratory and the private user. Both parties become aware of common research interests through involvement in the program. The laboratory can better justify expensive equipment and facilities by increasing the utilization rate through this program. The private sector need not invest in special-purpose, limited-use equipment if it can share public resources for one-time applications. The cost structures that define user fees vary considerably between agencies universities and laboratories. For example, some have special considerations for small businesses in their cost reimbursement schedules.

Laboratory Personnel Serving as Consultants. In many of the federal laboratories, technical staff members are allowed to serve as personal consultants to private industry after securing approval by laboratory management. Since the consulting activities take place on personal time, financial arrangements depend upon the individual consultant. In some instances, these individuals have been very flexible—even taking stock, for example, in a small business in exchange for continuing consultation. Firms located near major laboratories should investigate the policies and practices of the laboratory regarding personal consulting, as this source might have an expert who can expeditiously solve the firm's technical problem.

Institutional Technical Assistance. Consultation services or technical problem-solving by the laboratory is a primary mechanism for technology transfer. Unfortunately, the number and kind of these events are hard to capture for formal statistics. One laboratory transfer official judged the number of incidents per year at his laboratory to be in the tens of thousands. In one incident at this laboratory, a single phone call resulted in a suggestion by a laboratory scientist which saved the caller over $6 million, according to a report by the firm. Often, federal laboratories have formal programs for short-term free technical assistance, especially to small or minority-owned firms.

Intellectual Property Acquisition. Each federal agency has different policies and practices with respect to licensing intellectual property. If a laboratory is a GOCO (government-owned, contractor-operated) entity, the contractor's procedures will also influence the process of acquiring rights to intellectual property. Some laboratories have had extremely limited budgets for securing patents and, therefore, have a limited amount of intellectual property. Others have sizeable "inventories" and formal procedures for searching and acquiring licenses. Any interested person can determine availability directly from the laboratory or through intermediaries such as the Federal Laboratory Consortium or agency offices such as the DOE Office of Scientific and Technical Information. The sophistication level of intellectual property management has been rising rapidly at many federal installations. Most laboratories have procedures for granting waivers for intellectual property rights to inventors who are laboratory employees especially when the technology is not relevant to the laboratory's mission. The employee may then negotiate the transfer of rights as an individual.

Entrepreneurial Spinoffs. The incidence of start-up companies based on laboratory technology varies considerably from laboratory to laboratory. An internal study by Los Alamos National Laboratory revealed more than thirty spin-off companies. Oak Ridge National Laboratory has also had a significant number of local firms started by laboratory employees. Since both these laboratories are located in small communities, this activity has been vital to local economic development.[2] Entrepreneurial activity by federal laboratory employees in most locations has not been a widely-practiced mechanism in spite of its inherent advantages. Support and incentives for such activities have generally been lacking in most agencies and laboratories. Given the typical level of resources available to federal laboratory scientists and the usual degree of job satisfaction, an increased level of incidence will require concentrated efforts by laboratory managers should they decide to invoke the regular use of this mechanism.

Although it does not include all the mechanisms that could be usefully employed to achieve technology transfer and commercialization, the array presented above should serve to illustrate the point that different mechanisms have varying efficacies depending upon the stage of the process. It should also be noted that the use of multiple mechanisms may be necessary even within a given stage, to achieve the desired probability of success. In conclusion, it should be observed that continued innovation in transfer mechanisms is necessary for the evolution of the commercialization process. Several such innovations are discussed in the last section of this book which describes prescriptive models.

TECHNOLOGY TRANSFER AND COMMERCIALIZATION PROCESSES

The technology transfer and commercialization process has many forms depending upon such factors as the technology source, the type of recipient and its characteristics, the possible use of a variety of intermediaries and agents, and the stage of development of the technology. Wilhelm's article in this section, *The impact of federal technology transfer on the commercialization process*, (p. 253) differentiates between process and product technologies in his discussion of the potential impact of federal technology. Process technologies improve processes, whereas product technologies produce new and innovative products. The mechanisms that create product technologies are centered around the act of innovation. Wilhelm describes this process in his paper.

Most descriptions of technology commercialization processes describe them in terms of a series of stages or phases. Although there is little common agreement upon the stages, some combination of those delineated below is likely to encompass most processes.

1. Value creation through the means of discovery, invention or advancement of technology.
2. Determination of applications by synthesizing the needs of potential users with the improved functionality of the technology.
3. "Outreach" by the technology source and/or "inreach" by the recipient or its agent until interest is established and developed into an alliance.
4. Maturation or continued development of the technology through a series of prototyping and testing, including possible joint work between the technology source and one or more recipients.
5. Consummation of legal relationships among all the parties involved in the transfer and commercialization process.
6. Physical transfer of the technology, including people transfer, legal rights, drawings, blueprints, parts lists, prototypes, etc.
7. Continued technical assistance from the source or an intermediary to the recipient.

Each of these stages will be described briefly.

Value Creation

The popular concept of technology or inventions having intrinsic value is disproved by statistics such as the very small percentage (some estimate at less than five percent) of patents that have any economic return to the inventor. The scientific and technical advances that take place in the vast majority of work in universities and federal laboratories are at such an early research stage that commercial value is generally small relative to the expense of generating the technology. True, some patents have returned very substantial royalties when licensed, especially in early laser work and biotechnology, but the creation of technology is only the first step in the usual process of creating value through innovation.

From the perspective of industry, technology is frequently used as a competitive edge through the establishment of a proprietary position. Until recently, however, when the source was a university or federal laboratory, a very small sum of money was available to establish a patent. The treatment of "know-how" as trade secrets by the technology source was also difficult because of the lack of experience by laboratories in trade secrets and also by a lack of regulations and policies protecting technology generated with public funds from public scrutiny. For example, it wasn't until the CRADA was established as a formal mode of collaboration with industry that five-year exclusive data rights were defined for the industrial partner. Additionally if the laboratory did use some of its very limited resources to establish a proprietary position, policy in many instances made it difficult to award exclusive rights to industry.

Janis, in his paper *Creating commercial value*, (p. 209) defines the problem as that of $16 billion and 200,000 scientists having little effect on this utilization. He suggests that for an increased return on this investment, this country needs to manage the flow of "know how" better and to remove some of the barriers that slow down this flow. For results, he points out that many mechanisms exist, yet they all depend on interaction between the source and the recipient of the technological advances. The obstacles that make these advances underutilized must be removed and replaced by effective relationships between technology sources and potential users. Interaction takes on many varied forms and measuring the effectiveness of each of these different mechanisms is at the core of the plan this country needs to adopt to increase the benefits of creating value from research and development.

McKinley supports the conclusions presented by Janis and provides another important factor in this effort. Her paper, *Building the knowledge asset* (p. 221) explores the idea of improving the way that experts communicate their knowledge so that it is used in the commercial world. For that to happen McKinley suggests managing the human resource aspect of our federal labs in a more proactive manner. Required activities include training participants, providing incentives and top level commitments, and measuring the performance of the workers in the area of technology transfer.

Application Identification

A first critical step in translating "raw" technology to a position which represents substantial commercial value is to identify one or more applications. Frequently, this entails the recognition of user needs that can be met because the technology offers better products, services and processes. To the extent that federal laboratories perceive their missions to be generally restricted to basic and applied research, researchers have been discouraged from maturing the technology in terms of specific applications. This represents a major concern for those laboratories that are executing significant numbers of CRADAs to perform basic and applied work. It also means that joint work with consortia restricted to "pre-competitive" research will provide little incentive for the laboratory researcher to seek applications. In contrast, those laboratories with missions requiring delivery of specific hardware, software and services to operating agencies occasionally have dual-use technology that can be easily modified to meet certain needs in the private sector. An example is the high-energy-density capacitor developed for the nuclear weapons program at Sandia National Laboratories, which also has applications in the implantable medical devices markets. From the same laboratories, the semiconductor bridge technology developed for the weapons programs has potential applications in the private sector such as air bag initiators. These private-sector applications are more obvious since the functionality is very similar in each dual use.

Obviously, those researchers who interact regularly with their counterparts in industry are more likely to think about potential applications of their work. For example, presentations at technical meetings which have high private-sector attendance have occasionally brought ideas for applications from discussions following the presentation. Matching the technology advancement with private-sector needs seems to be largely a serendipitous process, although various creativity-inducing techniques have reportedly been used by industry in improving the rate of applications from captive R&D efforts.

Outreach

In 1990, the Department of Energy conducted a survey of its 34 laboratories to categorize the range of outreach mechanisms in use. The survey found the laboratories were using dozens of mechanisms, varying from newsletters to electronic bulletin boards and technology-specific conferences and demonstrations. Innovation in new mechanisms appeared to be fairly high. Outreach activities became a major concern of most agencies as policies were formulated to assure "fairness of opportunity" or "special considerations for small and disadvantaged businesses," as examples. Most mechanisms in wide use employed the broadcast method, such as advertisements in the *Commerce Business Daily*, in which specific firms were not targeted. As efforts to reach specific targets were required by new policies, mailing lists had to be developed or alliances formed with third-party organizations representing a class of targeted firms. Many laboratories did not know which form of media would reach selected audiences. For these laboratories, collaboration with trade and industry associations has generally provided insights into appropriate mechanisms and media to reach their targets. Several laboratories have even employed advertising agencies as consultants to help them deliver their messages.

In the past, outreach usually meant holding bidders' conferences or mailing "technical notes" about specific technologies. Increasingly, federal laboratories are employing brochures and newsletters aimed at demonstrating their capabilities. Some of these promote areas of technical competence, such as materials science, while others are more general, such as promotions of techniques for industry collaboration (for example, a brochure describing user facilities and procedures to take advantage of them). The rush for CRADAs has greatly increased the volume of these promotional devices.

Inreach has also increased as industry has become aware of federal programs for collaboration. In most federal laboratories, the volume of visits by industry has shown significant growth. As U.S. industry adopts processes more commonly used abroad, such as the appointment of Chief Technology Officers involved at a strategy formulation level, one can expect that inreach mechanisms and efforts will be expanded. Many of the more successful U.S. corporations have traditionally been active in surveying external technology resources, although many of these have focused on other private organizations and universities in the past. If the number of new information brokers seeking to help in the search through federal laboratories is a valid indicator of industry interest in the federal laboratories, the volume of laboratory/industry collaborations will continue to grow. Intermediaries such as the Federal Laboratory Consortium (see the article by Beverly Berger—*Technology transfer in a time of transition* p. 29) and the National Technology Transfer Center (see the article by Lee Rivers, *The role of the researcher*, p. 103) have designed a major portion of their activities to assist industry trying to inreach into the laboratories for specific technologies.

Maturation

Because the state of the technology to be transferred from a federal laboratory in most instances is premature with respect to direct incorporation in industrial products, processes or services, continued development is necessary. If a specific application has been identified in collaboration with a potential industrial recipient, the necessary development may best be accomplished by joint work between the source and the transferee. In this way, the recipient learns more about the technology in the process of designing prototypes, testing and evaluating them, and redesigning for continued improvements. Continuing development can be performed exclusively within the laboratory through a funds-in agreement if the industrial user so wishes; however, development work in most federal laboratories is expensive and the know-how transfer to the recipient is minimized.

A number of the transfer mechanisms discussed above can be used to facilitate joint development of technology. If the technology recipient wishes to involve the source in long-term development, including applied research as well as development related to multiple applications, an exchange of scientists may result in a productive relationship. If specific work is desired with short-term goals, the scope of work defined in a CRADA may well be the best mechanism.

Legal Relationships

Each agency, and each laboratory in the case of those operated by contractors, has different approaches to the establishment of legal relationships defining the technology transfer. In government-operated laboratories, inventors may receive waivers providing them with exclusive rights to their inventions (with the exception of the rights retained by the federal government). In most other instances, legal relationships are most often defined by licensing agreements, the terms and conditions of which may vary depending upon the source. Mechanisms like CRADAs define certain exclusive data rights (up to five years) and funds-in agreements can provide even greater exclusivity. As the federal laboratories gain experience in technology transfer, model legal agreements have evolved and the procedures for establishing them have been expedited. Even so, the use of an experienced agent or intermediary can, for those potential recipients approaching a federal laboratory for the first time, often relieve some of the frustration of dealing with a government bureaucracy.

Physical Transfer

The actual physical transfer of technology can take many forms, from the movement of people in whom the technology is imbued to the transfer of rights from the sale or licensing of technology, depending on the choice of mechanisms to use. Contrary to popular belief, individual incidents of transfer by federal laboratories have exhibited significant innovativeness. In the case of one spin-off company, the laboratory was able to declare prototypes as surplus property and transfer them to the company. In transferring the technology associated with an implantable medical device, one laboratory provided two man-years of support services, parts lists, drawings, process descriptions, supplier information, etc.

Since technology transfer has often been described as a "people-intensive" activity, the transfer of people has been critical to several successful efforts at technology commercialization, especially if the mechanism used is spin-off entrepreneurship. In one start-up company, five laboratory people left to join the company under different arrangements, each arrangement meeting different personal needs. One person simply resigned, but the others were allowed to work part-time at the company and part-time at the laboratory or to take leaves of absence as they desired. Temporary transfer of personnel from the laboratory to the technology recipient is a highly effective mechanism, but unfortunately is generally underutilized.

Continued Technical Assistance

Too often the process of transfer and commercialization is considered a short-term, finite relationship. In some instances, the continued relationship between companies and laboratories has resulted in new research programs in each organization. In the instance of the spin-off company from Sandia National Laboratories, SCB Technologies, Inc., continued joint technology development resulted in product improvements for the company and additional research (with incremental funding) for the laboratory. Some of the federal laboratories, especially those involved in programs like nuclear weapons that require extraordinary reliability, have learned to design systems and components with much more scientific understanding behind each design than could be afforded by industry. As a result, continued technical assistance is very valuable because the technologist at the laboratory may understand the underlying science to a level which permits rapid insights into problem solving or adaptations to the technology. This phenomenon was vividly demonstrated by a laboratory that had transferred designs for a vertical-axis, wind-turbine design to the private sector. When several commercial machines failed, laboratory scientists and engineers were able to diagnosis the problems much more quickly than their industrial counterparts.

Small companies need much more "hand holding" during the transfer process from federal laboratories or universities, and the extent of continued assistance required is often grossly underestimated.[3] Nevertheless, once the technology is successfully acquired and understood by the small company and the firm is able to accumulate sufficient internal technical capacity the innovative milieu of the small company may result in better exploitation of the base technology, thus benefiting the technology source through greater royalties.

THE SPIN-OFF: AN EXAMPLE MULTI-STAGE PROCESS

Because of the extensive initial support and continued technical assistance required of a technology source to increase the probability of success of a spin-off firm, the process should usually take a number of years and use a number of different mechanisms in each of the process stages described above.

In the value creation stage, the technologist must be aware of the possibility of creating a proprietary position in the technology by helping the laboratory with the proper disclosure of inventions. A number of early-stage transfer mechanisms can be used by the inventor to help the patent attorneys establish the prior art. Existing literature and the industry practices learned at professional meetings can help in this process.

The determination of applications is also fostered by interaction with industry. Outreach mechanisms can tell industry about the availability of the technology and its functional advantages. Consulting to industry in the application of the technology may reveal business opportunities which may be the basis of the new company. Since strategic alliances between small companies and large companies with substantial market presence are described as highly mutually beneficial in the article by Botkin, these relationships are also likely to be useful after the spin-off is launched.

Technology maturation can take place in either the laboratory or the private firm. To the extent that the spin-off occurs early in this cycle, there will be less apparent conflict of interest if the inventor benefits from the commercialization of the invention. Mechanisms like the user facilities program will allow the firm to gain access to the laboratory equipment needed to advance the technology to a mature form without capital commitments to expensive facilities. CRADAs and work-for-others agreements can also be effective maturation assistance for specific pieces of technical work to be done in the laboratories. Extensive work using these mechanisms is unlikely to be useful because the expense of large laboratories defeats the advantage of "garage" overhead in the start-up firm. Using laboratory colleagues as technical consultants has proven very effective (especially if they are willing to provide support in exchange for equity).

The actual physical transfer of people, equipment, prototypes and documentation embodying the technology can be accomplished through the use of mechanisms described above such as leaves of absence for the inventor/would-be entrepreneurs. If the technology transfer signifies the termination of laboratory interest in the advancement and application of the technology, it may be possible to transfer related equipment and prototypes without any apparent conflict of interest.

Continued technical assistance may well be critical to the success of the spin-off, especially if there has been a substantial team effort within the laboratory and if members of the team elect to remain at the laboratory rather than joining their colleagues in the spin-off.

CONCLUSIONS

Depending on the characteristics of the technology source, recipient and intermediaries (if any), the processes of technology transfer and commercialization involve a number of stages. Each stage may usefully apply a number of facilitiating mechanisms. This paper has described a representative sample of mechanisms and their potential application. It should be noted that new mechanisms are being created, although probably not at the rate necessary to promote sufficient innovation in this difficult process. It should also be noted that the staffs and resources devoted to technology transfer and commercialization at most federal laboratories have expanded dramatically in recent years and the practices and procedures being implemented involve increasing experimentation with existing and new mechanisms.

The purpose of section III of this book is to examine in greater detail the mechanisms being used and proposed, as well as the processes that accompany those mechanisms. It is hoped that articles such as those included here increase individuals' awareness of practices being employed by others, and therefore enlarge our perspectives on the possibilities for elaboration of each of our experimental efforts.

REFERENCES

1. Souder, W. E., Nashar, A. S., & Padmanabhan, V. (1990 Winter/Spring). A guide to the best technology-transfer practices. *Technology Transfer*, 5–16.
2. For a study examining one local economic development program based on start-up commercialization of technology from federal laboratories see: R. Radosevich (1988). Employing the indigenous startup strategy using spinoff technology from federal laboratories: A seven-year New Mexico experience. *1988 Proceedings, Technology Transfer Society.*
3. Goldhar, R. S. & Lund, R. T. (1983). University-to-industry advanced technology transfer: A case study. *Research Policy, 12* 121–152.

CREATING COMMERCIAL VALUE

The Untapped Federal Resource-Technical Know-How

F. Timothy Janis

ARAC Inc.
Indianapolis, IN 46204

OVERVIEW

During the 1980s, a historic transformation of policy and practice occurred concerning commercial exploitation of federally sponsored research and development. Earlier, the primary concern had been with developing and inserting technology into specific U.S. government-required applications. Technology transfer, while not repressed, was not encouraged. Today, technology transfer has become a major vehicle for increasing U.S. economic competitiveness.

The technology transfer enabling legislation of the '80s has generated programs supporting dissemination, application and technical assistance. Generally, the targets have been small businesses although emerging in the '90s is the concept of strategic partnering that has produced more interaction with large firms. While competitiveness has been the driver, two quantitative outcomes have become emphasized—licenses (and royalty income) and cooperative research and development agreements (CRADAs). A flurry of processes has resulted, producing an increasing number of these outcomes.

This paper, while not discounting the value of these outcomes proposes that current initiatives do not fully exploit the most plentiful resource available, namely the government's technical know-how of its scientists and engineers. If this know-how were used more by businesses, U.S. technological (and concomitant economic) competitiveness would be greatly improved.

Focusing only on the federal laboratory system (the major 700 labs) one finds that the government employs almost 200,000 scientists engineers and technicians. Annually, the skill and knowledge base of this work force expands, through work experience, formal training opportunities, symposia attendance, etc. In many areas, this know-how is on the cutting edge of a technology. The issue to be examined here is whether this technical know-how base could be managed more effectively, resulting in more efficient laboratories and more successful U.S. businesses.

One interesting question is whether asset value can be assigned to a technical work force. Models yielding viable quantitative data apparently do not exist, or have not been implemented. Industry however, is beginning to examine this topic. A simple approach using salaries can be defined for the federal work base. This yields an estimate of an asset of over $100 billion, a staggering value. Unfortunately, measuring the return on this asset is speculative and difficult.

From Lab to Market: Commercialization of Public Sector Technology,
Edited by S. K. Kassicieh and H. R. Radosevich, Plenum Press, New York, 1994

Private-sector analogs can be examined to elucidate estimates of market need and economic value. Teltech, Inc., Best America, NERAC and ARAC are four private firms set up to market know-how. The results of their efforts are only beginning to be realized. All indications are that know-how is marketable and useful to U.S. businesses.

The federal technical work base has made direct contributions of its know-how internally and to industry. Most researchers do publish and have their knowledge disseminated. Many respond to technical inquiries, but usually on only an ad hoc basis. No organized network for promoting and facilitating know-how movement exists, although the National Technology Transfer Center is beginning such a program.

The time is right for mobilizing technical know-how in support of U.S. business needs. It has been said that the cultural differences between the labs and business preclude effective exchange. In general however, technical people love to be challenged by difficult technological requests. Furthermore, workplace differences disappear when it comes to discussing technical problems.

Barriers do exist. Limitations on the time and resources of federal researchers, lack of incentives to participate, limited management recognition of need/commitment, and nonexistent job requirements for sharing/transferring know-how are but a few. Legislation will be needed to overcome these workplace barriers to facilitate use of this national asset. Job descriptions will need to include technology transfer. Mechanisms for internal and external sharing of know-how will need to be instituted and supported. To make the process function, a database of available resources is needed. Management of the asset will require recognition of its asset value. Doing so will make its management more essential.

LEGISLATIVE BACKGROUND

The literature contains many discussions of federal legislation affecting technology transfer. [1,2,3] From Stevenson-Wydler in 1980 to the Omnibus Trade and Competitiveness Act, technology transfer has been legitimized and mandated in the federal system. Of particular interest for this paper are the mechanisms initiated and the metrics advanced by the General Accounting Office. [7]

Alan Schresheim, the director of Argonne National Laboratory, states that "The national laboratories are a reservoir of scientific and technological talent that can help America compete in international markets." In his paper [4], he describes the mechanisms being employed. These and some others are summarized in Table 1.

With the exception of training for patent disclosure, all these mechanisms address one of the most important guiding principles of technology transfer: interaction.

CURRENT MEASUREMENT PRACTICE OF TECHNOLOGY TRANSFER

Measurement of technology transfer activity has been difficult and has produced mixed results. One significant problem is what to measure: the process or the outcome? The former is easier, the latter more difficult but more desirable.

A review of the literature indicates that most measurement has been done on the technology transfer process. The results, however, do not appear easily correlated to offer a consensus view of best practice. Gibson and Smilor's [5] analysis of the Microelectronics Computer Consortium (MCC) suggests the following four critical variables:

- Communication interactivity
- Cultural and geographic distance

Table 1. Commercialization Mechanisms for Federal Laboratories

Mechanism	Description
Patent development	Training and assistance for inventors in preparing disclosures
Licensing	Passive and some proactive approaches to seek adopters of federal technologies
Cooperative research and development agreements (CRADAs)	Federal laboratory/industry (and university) partnerships for technology development
"Work for Others"	Contracts for work performed for someone other than the sponsoring agency
Consortia	Symbiotic match of needs and capabilities
Staff exchanges	Industry personnel at federal laboratories (and vice versa)
Facility use laboratories	Access to unique facilities operated by federal
Consulting	Some laboratories permit staff to consult
Technical information	Research in progress reports stored in specialized as well as generalized databases (NTIS)

- Technology equivocality
- Personal motivation

These are all process-addressable variables. With consistent measurements, better assessments of best practice might be forthcoming.

Chapman [6], in some landmark work for NASA and ARS, has shown that the spin-off potential (secondary uses of technology), in terms of sales or savings that have been realized, is over $150 million per case. This data, however, is anecdotal and, while directly obtained from adopters of technologies, is hard to substantiate.

The current measurement in vogue is the number of CRADAs, licenses and royalty returns. Since 1989, when CRADAs became firmly established mechanisms, the numbers have grown rapidly. For example, DOE CRADAs went from 80 to almost 300 by the end of 1992. Comparative data by agency were not found.

Summarized in a recent GAO report, *Barriers Limit Royalty Sharing's Effectiveness* [7], were the number of inventions licensed by agency and royalties derived from these licenses. Between 1981 and 1986, the yearly average over all agencies was 69. For the period 1987–91 it went up to 87. From 1987 to 1991, about $8.7 million in royalties was obtained from US inventions ($76,000 per invention). John Preston of MIT has compared MIT's rate of return to that of DOE. [8] He showed that for an R&D budget of $700 million, MIT generated 410 patent applications, and obtained 80 licenses with an average royalty value of $202,000 per license. DOE, from $3,000 million R&D expenditures generated 748 applications and 125 licenses, with an average return of $25,000 per license. The comparison is not flattering. However, the federal laboratories have not been in the business as long as, nor do many have the reputation of, MIT. It is interesting to examine the ratio of the return on investment (royalties/R&D). For MIT it is 2.3% and for the DOE 0.1%. Again, MIT fares better; however, neither result seems impressive.

The conclusion to be drawn is that the measurement of technology transfer is imprecise, and that statistics, therefore, can be misleading (e.g., the 6-foot statistician who drowned in a creek whose average depth was 3 feet). The decision by the GAO and sponsors within the government to focus measurement on "hard" technology transfers is not a complete measure of technology transfer success.

THE HUMAN RESOURCE BASE

The federal government sponsors over 700 R&D laboratories. There are almost 200,000 technical workers employed at these laboratories. For the purposes of this paper, we will use the data given in Table 2 which is reproduced from the GAO report, *Barriers Limit Royalty Sharing's Effectiveness*[7].

As shown in the table, there are 177,000 agency scientists. Department of Labor salary data yield:

Average scientist's wages [27] = $50,400

Thus, the annual human capital investment (in wages only) is $8.9 billion for scientists or over 67% of the intramural budget given in Table 2. It has been estimated that federal scientists have over 2 million years of accumulated experience (approximately 11 years per scientist). [9] This translates into a human capital (wages) investment of almost $101 billion.

Table 2. Agency Financial and Human Resources

Department/agency	Federal obligations for intramural R&D [1]	Total number of agency scientists
Agriculture		
ARS	551.5*	3,100
Forest Service	130.1	10,300
Commerce		
NIST	126.8	1,600
NOAA	209.6	6,400
Defense		
Air Force	1,533.4	10,700
Army	2,200.3	38,200
Navy	3,088.6	62,700
Energy [2]	427.3	400
Health and Human Services		
ADAMHA	241.5	500
CDC	87.5	2,300
FDA	118.3	3,000
NIH	1,401.6	2,400
Interior		
Bureau of Mines	69.1	1,500
Fish and Wildlife Service	62.8	600
Geological Survey	277.2	6,100
Transportation		
Coast Guard	15.9	1,400
FAA	129.2	7,500
NHTSA	14.4	100
EPA	129.3	7,500
NASA	2,573.0	13,000
TVA	25.3	2,200
Total	**13,412.7**	**177,000**

*In millions of dollars (fiscal year 1991).

While human capital does not show up on any statement of account, for the purposes of this paper, will it be getting full value for this asset? The answer is probably no and is certainly no if asset value is measured using current technology transfer measures.

THE KNOWLEDGE BASE

What other measures might be used to ascertain the federal technical knowledge base? In R&D institutions, one measure of performance is number of publications. What is the US output? The author was not successful in finding in one location any succinct summary of output by type and agency. Some of the data gathered are summarized in Table 3.

By linearly extrapolating the NASA database records (1.8 million) to all federal R&D agencies, adjusting for size differences, one obtains a database of almost 10 million citations. Assuming a redundancy factor of one third, one arrives at almost 40 publications per government scientist. If one normalizes by average man-years [11], one gets 3.5 publications per scientist per year.

The National Technical Information Service (NTIS) should be the repository of all documents; however, the database contains only about two million citations. This is partially explainable on the basis that not every technical report results in a publication and the fact that, in the past, NTIS did not receive documents from every agency. The Technology Preeminence Act of 1992 requires all agencies to submit a copy of any material generated to NTIS.

How well is the government doing? Again no general measure was found; however, continuing the empirical approach, one arrives at the cost of a single report as approximately $14,000. According to the NTIS [12] over half its documents have been unrequested (sold). Using this, one concludes that a federal report costs $28,000 to generate (human capital expenditure only). Thus, the inescapable conclusion is that on simple measures such as CRADAs, royalties and publications, the US ROI is not very high.

KNOWLEDGE GROWTH

Another important factor not to be overlooked is the growth of the human capital asset. The discussion to this point has costed the human capital asset simply based upon number of scientists, average wages and years of service. To be a little more accurate requires including experience. For example, the Oak Ridge Institute for Science & Education is the training arm of DOE. Tina McKinley, who is responsible for much of the training, stated that "ORISE offers over 3,000 courses to DOE employees."[14] These run the gamut of subject areas, but emphasize pragmatic requirements. The Office of Personnel Management offers hundreds of

Table 3. Federal Publications

Repository	Type	Number
DTIC [10]	Technical Report Database Work Unit Info. System Database IR&D	1,590,000 records 240,000 records 128,000 records
NASA RECOM [11]	Patents Citations	5,235 citations 3,500,000 records total 1,800,000 records NASA 10,000 new titles annually
NTIS [12]	Documents	2,000,000 in archives 70,000 new documents
DOE [13]	Citations RIP Projects	56,000 added annually in '92 33,000

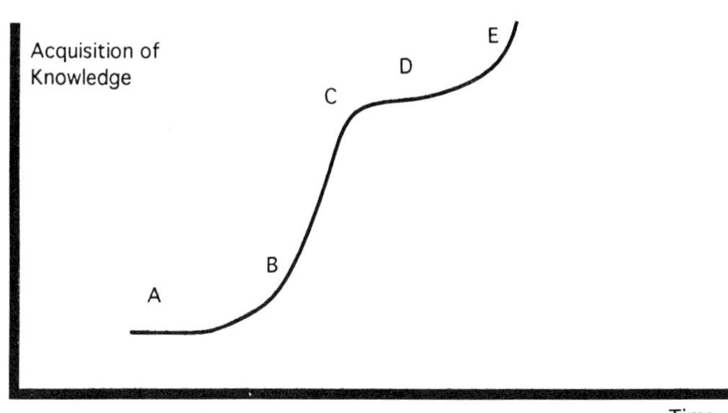

Figure 1. Stages of knowledge accumulation.

courses annually. Their subject focus is managerial as opposed to technical. During the last couple of years each agency has offered extensive programs in Total Quality Management. The conclusion is that the federal technical asset is growing through formal training.

What about on-the-job experience? Knowledge growth is discussed in a paper by Reisman and Xu [15]. They give the analogy that "the growth of knowledge in a given field may also be linked to a chemical reaction." They suggest that the rate of growth can be depicted by a rate diagram which is reproduced as Figure 1.

The slope represents the rate of knowledge growth. Note that overall knowledge accumulation is predicted. In the early stages of an activity, the rate may be slow due to inhibiting forces (e.g., lack of direction, lack of support, lack of interest). As the area develops so does the rate, because the forces (internal and external) are reinforcing (e.g., critical personnel mass, support, training). As the idea matures, equilibrium is established and the forward rate decreases until catalyzed by a new development.

The obvious question is: Does this accumulated knowledge lead to greater technological developments? Marchetti [16] developed various models to which he fit data from several hundred technological developments. His empirical results indicated that the growth of knowledge contributed to new technologies; however, there was a time lag between the development and the application. Reisman and Xu [15] conclude "Technology can be considered as a form of knowledge, which is based either on practical experience or on scientific theory. Knowledge in a certain field involves many technologies; the growth of knowledge, in turn, provides knowledge bases for the development of new technologies."

The federal technical staff can therefore be viewed as a dynamic growing asset base. Can it be quantified? Cost models are methods [17] used to approximate real situations for estimation or illustration. The following simplified model is suggested for the federal scientist/engineer work force. The output can be viewed as the estimated "value" of this technical asset.

$$\mathrm{UV} = \mathrm{S} \times \mathrm{No}$$

where UV = unadjusted value, S =salary, and No = number.

$$\mathrm{AV} = \mathrm{UV} \times (\mathrm{Dp} + \mathrm{Ap} + 1)$$

where AV = adjusted value, Dp = depreciation, and Ap = appreciation.

$$\mathrm{HCA} = 50{,}400 \times 177{,}000 = \$8.9 \times 10^9$$

Dp estimated to be −20% by assuming the life of a new idea is 5 years.
Ap estimated to be +50% by assuming knowledge growth is compounded.

$$AV = 8.9 \times 10^9 (-.20 + .50 + 1) = \$11.6 \times 10^9$$

$$HCA = 11.6 \times 10^9 \times 11 = 1.3 \times 10^{11}$$

The federal human capital asset current value is estimated to be $130 billion.

KNOWLEDGE UTILIZATION

In a summary paper [18], Adams, Spann and Souder summarized the barriers reported by sponsors, developers and adopters in technology commercialization. On a scale of 0–5, the highest ranked barrier identified by adopters was unawareness of a technology. Pinelli et al. [19] showed that personal knowledge is the primary source used by managers and nonmanagers to solve technical problems. Joe Schuster [20], in testimony before the House of Representatives, observed that “Smart, world class competitors are constantly looking for leverage, and nothing provides more economic leverage than technical knowledge—NOTHING”.

Souder et al. [18], while stating that adopters were unaware of technologies, did not address the corollary: were they interested in becoming more aware? Dennis Hogan was a co-author of an internal Alcoa document entitled *Knowledge Transfer* [21]. The primary problem area addressed was that “throughout Alcoa, we are not adequately exploiting the full potential of technology required to achieve our vision, values and milestones”. During their study the authors concluded “Technology Transfer is more aptly described as Knowledge Transfer.” They defined three desirable forms of knowledge transfer:

- Concept to market
- Proven solutions
- Individual to individual

To implement these mechanisms, Alcoa concluded that the company had to develop a knowledge transfer infrastructure, had to work on changing culture, and had to provide incentives and measures.

The importance of this information is that conflicting variables are at work. Knowledge is the driver for decisions, but the corporate infrastructure (and culture) has not yet been developed to facilitate knowledge movement, both internally and from external sources. Hence even if the federal system develops a highly effective internal and external infrastructure, the process still may not work. The reason is simple: having a car in 1910 that could go 100 miles per hour wouldn’t have been very useful because the road system required to handle such a speed didn’t exist. Thus, a well-maintained and effective federal technology transfer highway will become congested unless it is connected to an industrial superhighway.

A SOLUTION

Recently, F. Sherwood Rowland, President of the American Association for the Advancement of Science (AAAS), sounded the following warning for scientists at an AAAS meeting: “Scientists have contributed enormously to human welfare since World War II, but to regain public support for their efforts, they must do a better job of explaining what they do.” [22] He went on to state that the common lament of scientist is, “If only you understood.” This lament is now commonly issued by the federal laboratories regarding technology transfer. Specifically, labs would like to be conduits for public technology flowing to the private sector. Only, as pointed out earlier, the process is not working well. Gillespie [23] has stated that, even with all the statutory restrictions existing in the federal system, "the NBS (NIST) shoulder-to-shoulder contact at the bench level has provided an extremely important conduit for technology transfer."

While the 1986 Federal Technology Act (FTTA) mandates technology transfer programs at national laboratories, few of the labs make the transfer a workplace requirement. Frank Penaranda [24] reported that the NASA Marshall Space Flight Center has instituted technology transfer job performance requirements for all employees at the center. Penaranda went on to state that over NASA's existence there have been many attempts to develop a NASA management instruction (NMI) on technology transfer. To date none has been implemented, but indications are that an NMI on technology transfer may be forthcoming in the near future.

What is needed is a mechanism that addresses understanding and culture and yields a better ROI for federal R&D. To achieve these objectives it is proposed that there be developed a Federal Extension Assistance in Science and Technology (FEAST) program. The objective of this program would be to enable federal scientists and engineers to transfer their knowledge to U.S. businesses.

How might FEAST operate? If every federal scientist and engineer's job performance criteria included technology transfer, each would need to report his/her activities annually. Many of the activities are currently being performed; others may be new or extensions of current functions. The critical element is that FEAST would be enabling, and thus require management to implement mechanisms to facilitate activities like those given in Table 4.

Why implement FEAST? Because it would address a major issue—top management support. NASA scientists frequently claim that they support technology transfer, but their job is to produce specific science or technology. Anything that diverted attention from that objective has been a potential management problem. If the lab and its employees were to be measured on technology transfer activity, necessary management support would be forthcoming.

Will it cost more? This is a hard question to answer. On the surface it would appear that it would cost more. For example, if everyone were required to do four hours a week of tech transfer, and assuming all were new activities, a 10% decrease in weekly productivity might be expected. However, productivity will increase, especially if internal technology transfer is also facilitated. Another idea is to follow the university approach and make one day a week available for consultation (paid and unpaid).

Table 4. Some Technology Transfer Activities

Activity	Audience(s)
Publications, reports	Internal,external
Conferences, workshops	Internal,external
Consultation	Internal, external
Technology commercialization	External
Public technical service	External
Training	External
Politics	External
Public marketing	External
Public education	External

Table 5. Some FEAST Advantages

- Enhanced communication
- Cultural changes (internal and external)
- Conduit for know-how movement
- Increased public (business) awareness
- Enhanced contributions to business competitiveness
- Greater ROI in public investment in R&D
- Greater job satisfaction
- Stimulus for technological advances

Why do it? In Table 5 are cited some of the advantages that would emanate from FEAST if it is implemented.

Roessner and Bean (25) summarized the results of an IRI study entitled *Industry Interacting with Federal Laboratories*. Reproduced from that report is a figure (Fig. 2) describing the respondents' experience with federal labs. The overwhelming deficiency is a lack of interactive experience.

- Of particular note is that very few firms have had employee exchanges. What interactions do occur are at professional meetings. FEAST could greatly improve that statistic. Would it be important? In the same article, Roessner reports that successful interaction was influenced by:
- Person-to-person contact
- Flexibility in approach
- Support of middle and top management

FEAST can produce these interactions.

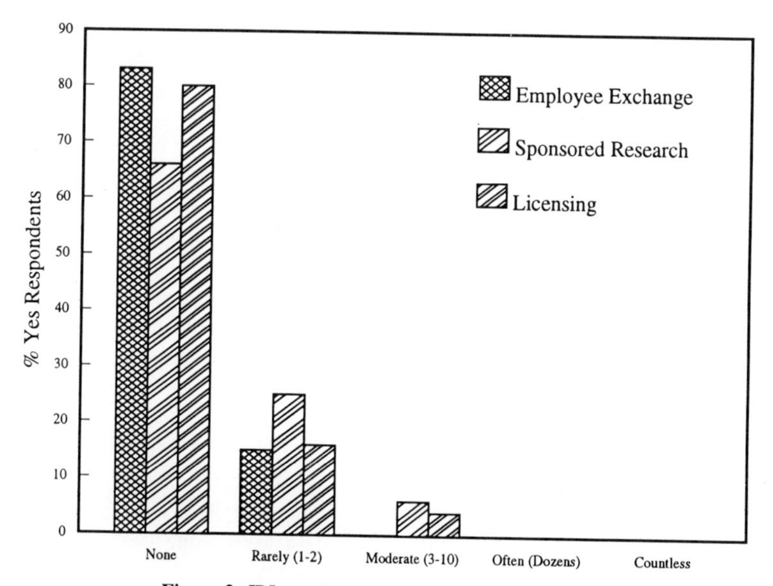

Figure 2. IRI members' experience with federal labs.

What if it is not done? At a recent NCAR meeting the program focused on the federal technology transfer system. Irwin Feller [26] stated the "creating of new institutions specially designed to commercialize new technology itself represents technological innovation—and risk—but times warrant risk taking." He went on to say: "Attempts to redeploy the federal labs may prove fatal." To date, policy decisions have been on transformations and alliances. They could easily shift to reconstitution.

FEAST will provide greater opportunities for federal scientists to share their knowledge with business. This can be mutually educational (i.e., address cultural differences) and mutually rewarding (i.e., transfer knowledge). The interaction does not have to be formalized (i.e., work for hire, consultant), but can be simply responses to inquiries requiring accumulated knowledge.

Will business be receptive? For small companies, Teltech, NERAC, BEST and ARAC facilitate access to knowledge. While all are relatively small (< $20 millon), all are growing and their primary support is from industry. As Joe Schuster stated in testimony to the House [26]: "Many believe that Japan's economic miracle happened because they excel in commercializing research. This is partially true, but I believe that their greatest genius is in funding relevant technologies in the first place...they clearly understand that access to research is as good as doing the research...and one heck of a lot cheaper."

If FEAST is implemented, the ROI on the knowledge asset can be quantified. It is not possible with the data currently available; however, consider the following scenario:

GIVEN 177,000 federal scientist whose knowledge asset is worth $\$101 \times 10^9$ or \$570,000 per scientist. There are approximately 360,000 small manufacturers in the US.

ASSUME Each scientist provides 4 hours/week or ~ 200 hrs/year or ~10% of his knowledge capital assets.

THEN Each small manufacturer could receive almost \$28,000 of world-class technical knowledge annually.

In reality, only a fraction (say one third) would take advantage of this opportunity, thus select businesses could receive over \$80,000 of know-how annually. This ought to help those firms be more competitive.

CONCLUSION

To prepare for the year 2000 and beyond, U.S. businesses will need to become technologically more up-to-date and competitive. Various studies have shown this does not necessarily require being at the technological cutting edge, but it does require incremental advances along the technology curve. Most U.S. businesses have found it difficult to move as rapidly as their international competition. To help these businesses, many programs have been implemented or proposed.

The focus of this paper has been that the federal government has built up the world's most formidable technical knowledge asset, but it is not adequately developed for U.S. business use. The various technology transfer acts have enabled transfer, but have focused on the movement of U.S.-owned "hard" technologies. Little emphasis has been placed upon mobilizing the technical knowledge asset.

A Federal Extension Assistance in Science and Technology (FEAST) has been suggested, to address the problem of increasing the return on the U.S. knowledge asset. It is proposed that job performance measures should include technology transfer. This requirement will enable increased direct interaction between federal employees as well as with U.S. businesses. It is suggested that doing this will help change the culture, increase the flow of knowledge, and, in the process, improve both federal and business technical competitiveness.

Acknowledgement

Dr. George Launey, Economics Professor at Franklin College and Mark D. Janis, ESQ. of Barnes & Thornburg, Attorneys at Law provided important economic and technical input.

REFERENCES

1. Departments of Agriculture, Commerce, Energy, Transportation and NASA (1992, October). *National technology initiative summary proceedings* 6.
2. U.S. Department of Energy (1993). *Technology transfer 92/93.* DOE/ST-005p DE93003623.
3. Grissom, F. G., & Chapman, R. L. (1992). *Mining the nation's brain trust* (Chapter 3). Reading MA: Addison-Wesley.
4. Schresheim, Alan (1990–91, Winter). Towards a golden age for technology transfer. *Issues in Science & Technology*, 52–58.
5. Smilor, R. W., & Gibson, D. V. (1989). Key variables in technology transfer. *Journal of Engineering & Technology Management, 8* (3–4) 287–312.
6. Chapman, R. L. Private conversation, data taken from projects conducted for NASA and USDA.
7. *Technology transfer barriers limit royalty sharing's effectiveness.* GAO/RCED-93–6.
8. Preston, J. (1993, February). Presentation to Office of Personal Management training course on Technology Transfer, Denver Executive Seminar Center.
9. *Using federal laboratories to improve the competitive position of U.S. industry* (1993, February). OPM training course, Denver.
10. Lesser, Barbara, DTIC (1993, January). Personal communication.
11. Heiland, Walter, RMS Associates (1992, December). Personal communication.
12. Quality campaign reinvents structure and mission of the National Technical Information Service. (1993, February). *Technology Access Report* 16–18.
13. Jordon, Sharion, DOE (1993, February). Personal communication.
14. McKinley, Tina, ORISE/DOE (1992, December). Personal communication.
15. Reisman, A., & Xu, X. (1992). On stages of knowledge growth in management sciences. *IEEE Transactions on Engineering Management, 29* (2), 119–128.
16. Marchitti, C. (1990). A personal memoir: From terawatts to witches. *Technological Forecasting & Social Change, 37*, 409–414.
17. Keith, E. (1992). Smashing the cost barrier: The only technical roadblock to space. *Space Commerce, 1*, 215–228.
18. Adams, M., Spann, M. & Souder, W. (1993). *Summary report on measures of and barriers to technology commercialization among sponsors developers and adopters.* To be published.
19. Pinelli, T., Barclay, R., Glassman, M. & Oliu, W. (1990). The value of scientific and technical information. *Proceedings of European Forum "External Information: A Decision Tool.* Strasbourg (France).
20. Shuster, Joseph (1990, September). *National technical extension service.* Testimony before Congressional hearing on House Bill 4659.
21. *Knowledge transfer* (1991). Alcoa Company internal document on technology transfer.
22. Heylin, M. (1993, February 22). AAAS president sounds warning for scientists. *Chemical & Engineering News* 5.
23. Gillespie, G. (1988, Fall). Federal laboratories: Economic development and intellectual property constraints. *Journal of Technology Transfer* 20–26.
24. Penaranda, F. (1992, December). Personal communication.

25. Roessner, J. D., & Bean, A. (1990, Fall). Industry interactions with federal laboratories. *Journal of Technology Transfer*, 5–14.

26. Odza, M. (1992, October). Research leaders debate: Reshaping existing institutions to commercialize research or create new ones. *Technology Access Report* 10.

27. Department of Labor (1992). *Labor statistics, occupational outlook handbook.* Bulletin 2400.

BUILDING THE KNOWLEDGE ASSET

Tina McKinley

Oak Ridge Institute for Science and Education
Oak Ridge Associated Universities
Oak Ridge, TN 37831

How can we translate the extraordinary knowledge base defined by Tim Janis [*Creating commercial value*, p. 221] into performance in technology transfer? How can we build on this asset in a way that extends its usefulness in technology commercialization? And having done so, how do we measure, judge, and modify this performance in the future so that we can continue to optimize the technical knowledge asset? We know what we have to work with, and it is indeed impressive. The experts in our federal R&D system need to share what they know more effectively.

I would like to reflect on some lessons learned in the past and describe how training, as one approach to building the knowledge asset, can help us gain the greatest benefit from the technical know-how that is so critical to the successful commercialization of federal expertise and technologies.

TECHNOLOGY TRANSFER AS A CROSSCUTTING PERFORMANCE FUNCTION

We want technical personnel, administrative support personnel, and managers to perform differently from the way they have in the past. We want to move technology transfer higher up on their priority list and make it evident in the criteria they use (and their supervisors use) when evaluating their work.

Technology transfer is one of the newer items on the list of organizational crosscutting functions that are being recognized and managed in new ways across the federal community. However, it is by no means the only one. We can learn some lessons from some of the other items on this list, and I feel certain we will add our own words of wisdom as we continue to integrate technology transfer into routine research, development, and application operations.

In the Department of Energy system, we have spent much of the past four years recognizing the critical importance of environment, safety and health as crosscutting functions that must be incorporated into all phases of our operations.

At the same time, the federal system has embraced total quality management (or some variation of this term) as a new way of doing business. In the course of adopting and implementing these crosscutting functions, we have learned some things:

From Lab to Market: Commercialization of Public Sector Technology,
Edited by S. K. Kassicieh and H. R. Radosevich, Plenum Press, New York, 1994

- Routine activities that traditionally have been performed in informal ways must be formalized. Policies, procedures, and standards for conducting operations must be established which emphasize the importance of staff functions, such as ES&H (Environment, Safety & Health) and technology transfer, as well as line operations.
- The signal that ES&H and quality management (and technology transfer) are important must be sent by the highest levels of management. Top-down commitment defines the critical path leading to the greatest behavior change.
- At the same time that signals are sent from top-level management, ES&H and total quality management stress the importance of individual responsibility. Each worker must be held accountable for the outcome of the activities performed. This concept is the pivotal point in training; the content must provide what the worker needs to know, and the worker must come away with a greater commitment to technology transfer.
- The customer is the most important player. In technology transfer, it is sometimes difficult to define who the customer is (the taxpayer has a way of getting mixed up in all this), and in many organizations, different operations perceive different customers. But the bottom line is that the customer must be acknowledged and listened to at all stages of planning and implementation.

Let us not have to relearn from ES&H, quality management, and the other "new ways of doing business" these lessons that we have all experienced. If we can be smart enough to design technology transfer efforts in a way that builds upon what we already know, we stand a far better chance of succeeding more quickly.

TRAINING AS AN INVESTMENT

Training is one way to build and direct the knowledge asset. We define training as the acquisition and transfer of site- and job-specific skills and knowledge required by a worker to perform a job (as distinguished from the educational process the same worker experienced before obtaining the job). Training may be provided in a classroom or on the job; regardless of the setting, it requires the preparation of printed and audiovisual resources such as curricula, checklists, instructor guides, and similar documents. In short, "good training" requires the same systematic assessment and approach (followed by objective evaluation) that is expected from other operations that prove their value to the organization time and time again.

The Clinton/Gore technology policy describes training as an investment in human infrastructure; labor economists define education and training as one way to help optimize human capital. The intended result of training is to increase what workers know and are able to do, and to measure that increase by improvements in their performance.

Who can benefit from training associated with technology transfer? There are two major groups: (1) technical staff and their management and (2) technology transfer personnel and the administrative support personnel with whom they work. For the purposes of this discussion, I would like to confine my remarks to the first group.

How do we know what should be included in the training content? We find there are three major drivers for the training we design and deliver in the federal system: requirements, technology, and the worker. The first, requirements based on legislative and other regulatory guidance, is the easiest to define and measure. Compliance-based training in ES&H, for example, can often be traced to a specific requirement directed toward a particular target (for example hazardous waste operations training required by OSHA). Not only is this training fairly easy to develop because there are specifications but also its outcome can be measured effectively with respect to changes in work performed, decreases in accidents, increased use of protective measures, and similar other indices. Unfortunately (for trainers, anyway), in the technology transfer world we have few established requirements for training. There are some general statements about the need for training, but little direction as to who should be trained, in what areas, and how frequently. In fact, the little direction that does exist is usually found in agency implementation of legislative mandates.

So much for requirements. Another driver for training design is technology. What are the particular characteristics of the technology and the technical research area that a potential user or a new application should consider? How is this technology or technical area different from others, broadly speaking? What value can the technology make available to the buyer? How can the uniqueness of this technology be communicated without getting so technical that it is impossible to understand? The technical staff provides guidance in answering these questions, and the instructional designers take the subject-matter expertise and frame it in sound approaches that communicate as effectively as possible (as measured by testing, post-training evaluation, and similar instruments). The transferability of training content designed to address technology, by definition, is limited because of technical and site-specific content.

Usually, we consider the worker to be the most important driver for training design. The instructional designer must use the knowledge base the worker brings to training as the framework for defining the learning objectives and the instructional approach to be used. Learning objectives are designed based on front-end analysis that defines what the worker must know and do to perform the desired tasks in a measurable, acceptable manner. The bottom line is worker performance, and worker performance is a direct function of skill and knowledge base.

WORKER PERFORMANCE IS THE KEY

If training is to change performance, it must be based in the real world. We characterize reality for the workers who will receive the training by performing job/task or some other form of front-end analysis. Usually, we do this by identifying expert performers at the top of the curve and involving them in an interactive, analytical process. We define what the expert performers are doing and what they believe they need to know to perform at an expert level. This information becomes the definition of desired performance. If we are fortunate enough to have access to documented success stories, we analyze them to understand the vital knowledge and skills that helped to create their success.

Next, the assessment process looks at what the targeted training audience must do and know to perform these tasks. This information becomes the baseline. When the baseline and the ideal are established we can design training to get from here to there.

Recently, we spent a day completing what we call a tabletop job/task analysis with the Federal Laboratory Consortium executive committee. Collectively, the 20 participants had amassed almost 300 years of experience in federal technology transfer. By the end of the day, we had defined a series of technology transfer tasks performed by technology transfer policy makers, laboratory technology transfer managers, and technical program managers. We also described the skills and knowledge the committee members believed necessary to perform these tasks in a competent manner.

Now what? We are compiling that data and will undertake a modified Delphi process with committee members and others who are willing to spend some time thinking about the details of what they know and do. The data will remain fluid because the expected performance will evolve as experience is gained in federal technology transfer. And even as they change, the data will provide a foundation for the Federal Laboratory Consortium and other interested parties to design training that will lead to an increased ability to perform these tasks.

How will we know the training is effective? How will we recognize desired performance when we see it? Metrics in training, as in technology transfer, can be difficult to define. They often concentrate on process rather than outcome, because process is easier to measure and less time needs to pass before measurement can be taken. Furthermore, evaluating training can become complicated if both the content and the desired behavior changes are complex; this is certainly true in technology transfer, where we are dealing with highly intellectual target audiences.

There are a number of metrics which we routinely use to measure the value of training; the most important ones (and the hardest to get) are assessments by the trainees' supervisors as to how job performance has changed after training. There are very extensive models for evaluating training, such as the one proposed by the Nuclear Regulatory Commission, but most of these models were developed for training that is targeted towards skill enrichment (knowledge enrichment can be much more difficult to measure.)

The actions we expect technical personnel to take to address their new technology transfer mission may be very different two years from now (and are certainly different from two years ago). The analogy of pollution technology, where we have expanded the limits for detection of pollutants to such a point that we are now pushing expectations for control way beyond anything conceived just a few years ago, may be very appropriate in our technology transfer world.

We can define categories of metrics, and increasingly more agencies and organizations are doing so. We can use the standard outcome measures that have been used in the past to assess the results of applied research (papers, patents, presentations, and the like) and we can count the steps taken by technical personnel to work actively with outside partners (phone calls, visits, personnel exchanges, lab open houses, consulting arrangements, SBIR participation, etc.). Most important, as we change expectations for worker performance we must translate those expectations into operational indices. At a minimum we must consciously include expectations for performance in the job descriptions and appraisals of all laboratory technical personnel, administrative support personnel, and technical managers. Certainly we will continue to use the outcome measures we know now. But we must also recognize the specific types of activities which work best at different sites. For a defense lab, this may mean luncheon discussion sessions on potential commercial applications of a defense technology; at an energy research lab, this may mean a concerted effort to bring in potential partners from other than the usual industrial base; at a more application-oriented laboratory, this may mean far more interaction with both the local school system and local small businesses than has been evident in the past.

We have the opportunity now to define the measures against which we will be judged and to train to them; let us take advantage of the chance to do so.

HOW DO WE GET THERE FROM HERE?

Training can help change both individual and organizational expectations and performance in technology transfer. Training directs the technical knowledge base in new directions while at the same time increasing its value. It is our responsibility to take the same systematic approach to designing and evaluating training that we need to use in all other aspects of implementing and evaluating technology transfer.

We must define our customer, encourage commitment from the highest levels and the grass roots, and be honest in defining the metrics against which we will be judged. Finally, we must continue to concentrate on showing initial successes to motivate our experts to do more, but we must also acknowledge the importance of time (and this is particularly difficult when we have little of it to spend) in making the kind of cultural change we are talking about in the federal R&D community.

THE COMMERCIALIZATION OF PUBLIC SECTOR TECHNOLOGY

How to Form, Manage, and Evaluate Effective Strategic Alliances

James Botkin

Technology Resources Group
Sante Fe, NM 87501

STRATEGIC ALLIANCES

What are the reasons for forming strategic alliances or, more generally partnerships?* Why should we be considering alliances, when we speak about national and federal laboratories and their future?

President Clinton recently noted that all the new jobs in the last decade have come from small businesses, many of them start-up ventures. Unsaid but clear in his statement is the implication that large companies need to become more entrepreneurial in order to survive, create jobs, and strengthen the U.S. economy. Seldom said is that, while America is a leader in entrepreneurship, it could do even better—the failure rate of new start-ups is about 80% in the first five years, a rate which we would find intolerable in other areas, like quality of cars or dropouts from schools.

Entrepreneurship is being challenged. Small companies, the so-called garage shops, are often not up to the technological complexity or global markets of the day. Large companies have global reach and technical sophistication, but they are slow to bring new ideas to market.

Public sector R&D laboratories face an even greater challenge. Many don't yet know how to commercialize their technologies, and most lack customer focus—an orientation that is central to the success of private business. Commercialization and customer focus have not heretofore been part of the mission of laboratories; they haven't had to learn this behavior, and they haven't hired people who have these skills and talents. Commercialization and customer focus were not congruent with the past mission of labs.

---

* I will use *partnerships* and *strategic alliances* interchangeably. *Laboratories* will be used to refer to federal and national labs such as Los Alamos and Sandia. *Innovation* should be understood to mean the successful commercialization of a discovery or invention, and *entrepreneurship* is the process of founding a startup business to promote an innovation.

From Lab to Market: Commercialization of Public Sector Technology,
Edited by S. K. Kassicieh and H. R. Radosevich, Plenum Press, New York, 1994

Reasons for strategic alliances:*

1. *Accelerate innovation, and commercialization of a new technology, product, or service.* Such commercialization implies new business and is known as entrepreneurship, especially when referring to start-up companies, and sometimes as intrapreneurship when referring to existing companies.
2. *Competition in global markets.* Large U.S. businesses have learned to form alliances to share development risks, reduce production costs and extend marketing reach and distribution systems. Airlines form international links to meet the needs of worldwide travelers; Continental-SAS is an example. Automotive companies like GM link to Saab, Volvo to Renault. Computer companies are said to do business in a web of more than 100 separate strategic alliances.
3. *The shift in emphasis from military to economic national security with the end of the Cold War.* The laboratories are rich in capacity for science, basic and applied research, and an ability to develop new technologies. Examples in manufacturing include biotechnology techniques for genetic engineering, or transportation, satellite and fiber optic knowledge and resources, not to mention advances in electronics and computing. These capabilities could be applied to such civilian activities as environmental services, weather monitoring energy and conservation, and even educational technologies.**

In the private sector, large companies are at a watershed. They need to become more innovative, behave more like entrepreneurs, and compete more effectively in the global marketplace. To increase innovation they have followed a number of internal strategies such as investing more in R&D, intrapreneuring, internal ventures, and skunk works. They have also practiced several external strategies, such as joining university centers, partnering with other large companies, mergers and acquisitions, and linking up with small companies.

Small companies, too, are at a watershed; they always are. They have plenty of innovations, but they need access to more money, marketing, management assistance, and the fruits of basic research, which they can rarely afford.

In both the private and public sectors, a new principle will be necessary to realize the changes that partnering can bring—collaborating in order to compete.

VALUE CHAIN ASSESSMENTS: WHO SHOULD PARTNER AROUND WHAT?

In *Winning Combinations*, the authors developed the idea of using value chain assessments, first as a descriptive way to focus on where to partner, and secondly as an action technique to determine which type of partner to seek. The value chain starts with basic and applied research, prototype development, and design. For convenience, we term this "phase one" of the value chain. The next link in the chain is production and fabrication, which we term "phase two." "Phase three," then, is the marketing, sales, distribution, and service of an end product. Figure 1 shows a value chain and its three phases.

* Many of the ideas in this paper are based on J. Botkin and J. Matthews, *Winning combinations: The coming wave of entrepreneurial partnerships between large and small companies.* New York: John Wiley & Sons, 1992.

** See *Schoolhouse Manifesto: The Ten Ways Business Will Revolutionize Education*, by Stan Davis and Jim Botkin, Simon and Schuster, NY forthcoming 1994.

The process of assessment begins with listing an organization's strengths and weaknesses along the value chain. In the commercial world we often add another component to the chain—the ability to hand off from one phase of the value chain to the next, because many large companies with strengths in all three phases typically find their problem is difficulty in moving from design to production, or sometimes from manufacturing to sales.

After listing strengths and weaknesses, the next step is to list problems and potential partners along the dimensions of the value chain. One goal is to enable the organization to move as quickly as possible from one end of the chain to the other. Sometimes this involves "moving forward; " an organization is strong in basic research and needs to learn to move quickly toward production and ultimate sales. This is probably the case for most laboratories.

In other cases, one may want to "move backwards." Organizations with a strong marketing capacity may search for a product or technology that can take advantage of their sales capacity. A familiar example is Dell Computer, which is strong in an innovative approach to selling computers and servicing by telephone.

The point of the assessment is that partnerships enable companies to compensate for weaknesses. Partnering companies may gain special expertise, or get access to new markets. We may summarize this by saying that *complementary dissimilarities lead to successful collaborations*. Too much similarity can be dysfunctional in a partnership. The quid pro quo is often technology innovation, speed and agility from one partner in exchange for capital and market access from the other.

For a laboratory, the question arises as to whether it is a large or small potential partner, or if it is something else. That labs are supported by public dollars and responsible to a public mission also differentiates them from commercial companies.

Sometimes labs are like small companies. They can be like garage shops that have invented a new technology, a new product, or a new use for an old technology. Certainly, compared to the size of a giant corporation like GM or Du Pont, most labs appear small. IBM, for example, was until recently spending about \$6 billion annually for its own research, while the Los Alamos total budget was closer to \$1 billion. To IBM, LANL appears small.

At times, labs are like large companies. To a small company, say one with total revenues of \$10 million, a billion dollar laboratory appears big. Its size is magnified by the fact that labs are both public and in part classified, which to a small company translates into bureaucratic

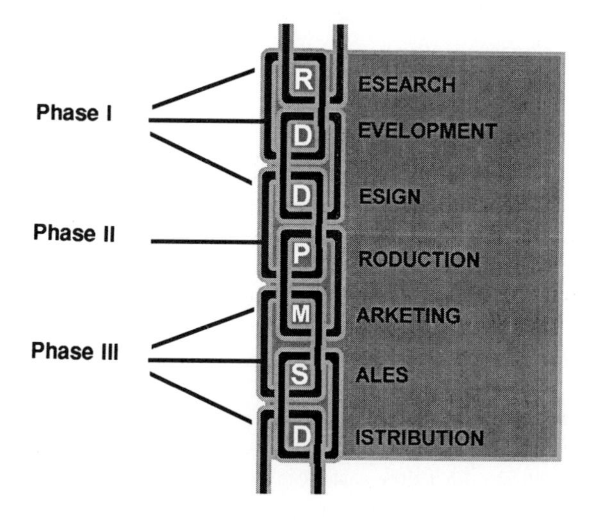

Figure 1. Value chain.

slow, and—worse yet—open to public debate and media attention. These are characteristics that small companies see in large companies. But unlike large commercial organizations, labs seldom have any strength in marketing or distribution, nor are they a potential source of venture capital as some large companies are.

In truth, to both small companies and large alike, laboratories would not be high on the list of potential partners were it not for their cache of technology know-how.

A laboratory has at least five possibilities for partnering:

1. It can be a source of technology which it markets to a large company to incorporate into its products.
2. It can be a source of technology which it licenses to a small company to transform from military use into commercial products and market.
3. It can be a source of technology which a small company transforms and then turns over to another company to market.
4. It can be a broker, by linking a small technology company to a large marketing company. Often this function is performed by a small networking company.
5. It can create its own internal venture by setting up a subsidiary to develop technology, where the subsidiary develops its own marketing capacity, perhaps even setting up its own internal incubator for more subsidiaries.

Until recently, laboratories seemed to gravitate to the first partnering arrangement, where lab offices of industrial liaison link up with well-known large corporations, often with older industrial products. What has been less characteristic is to link up with newer high-tech companies, which now sometimes are larger than their older industrial counterparts. (MicroSoft, for example, is now larger than General Motors in terms of book value of the company.) If we were to make a matrix of possibilities, I suspect that most lab partnerships would fall in the upper left-hand quadrant, but I think perhaps the best opportunity for growth is in the lower right.

	Older Industrial	**Newer Hi Tech**
Large	X	
Small		?

PRINCIPLES OF SUCCESSFUL STRATEGIC ALLIANCES

Twelve principles characterizing successful commercial partnerships can be grouped into three categories:

- Finding a partner
- Creating a contract
- Managing the partnership

Finding a Partner

An organization should follow a stage-by-stage strategy. In practice this means such things as making sure the goal of strategic alliances is part of the business plan, part of the mission statement, contained in vision statements, and the like. One thing we've learned about following stage-by-stage strategies is that they always take longer to execute than anyone thinks.

Who should be assigned to find a partner? We have learned to put a motivated networker in charge of the search. For small companies, this is seldom the founder, though it could be. For large firms, this is seldom a director of R&D, though there are exceptions.

Essential to finding a partner is developing a profile of the partner(s) one seeks, using the value chain assessment. Don't link up with the first company that expresses interest, find the one that really fits your mission and goals.

To find a suitable partner, the best strategy is to contact multiple candidates. One experienced executive we know expressed it like the fairy tale: "You've got to kiss a lot of frogs before one turns into a prince." In many commercial cases, it's only one to five candidates out of 100 who prove to be suitable.

Creating a Contract

There are four principles for success:

1. focus on mutual benefits
2. start simple
3. set benchmarks by which you can assess whether the partnership is working on target and on time, and
4. something particularly hard for a public agency, involve the lawyers—LATER. The best commercial partnerships are built on trust. In fact, mutual trust and putting both parties at an equal psychological level of respect is probably the most critical success factor. Lawyers are clearly necessary to protect intellectual and other property, and to design legal safeguards and security. But the legal propensity for adversarial relationships can be quite contrary to the spirit of partnership.

Managing the Partnership

Again there are four principles:

1. *Emphasize the partnership mentality at every opportunity.* You're trying to create something that is quite beyond and different from the usual customer-vendor relationship.
2. *Develop a team of champions.* In many organizations, partnerships don't happen naturally. It takes a champion to nurture and protect them. The problem with champions is that they are often rising stars, who have a tendency to rise right up and out of an organization. That is why having a team of champions is a good idea, so that when the protector moves on, the partnership stays in place.
3. *Frequent communication.* Both parties and all players need to know what's going on and how it's performing.
4. *Think long-term but deliver short-term successes.* This is a little more difficult for most managers, some of whom think only short-term—especially when linked to scientists who think mostly long-term. For alliance-seeking labs, this aspect may prove to be a challenging balancing act.

A summary of the twelve principles is below.

Twelve Principles of Successful Partnerships

- *Finding a partner*
  - Follow a stage-by-stage strategy
  - Put a motivated networker in charge of the search
  - Develop a profile of the partner(s) you seek
  - Contact multiple candidates
- *Creating a contract*
  - Focus on mutual benefits
  - Start simple
  - Set benchmarks
  - Involve the lawyers—later
- *Managing the partnership*
  - Emphasize the partnership mentality
  - Develop a team of champions
  - Communicate frequently
  - Think long-term but deliver short-term success

How do we know if a partnership is being successful? What types of measures can be used? Important for organizations used to public policy measures is to get away from the tendency to measure inputs—such as numbers of partnerships, numbers of meetings, amounts of resources etc.—and to concentrate instead on outputs. Did the partnership produce income and make a profit? Do the participants come back for repeat business? Did the partnership lead to new uses of technology that consumers relate to the originator?

POTENTIAL PITFALLS

What can go wrong with partnerships between large and small companies or laboratories? We've grouped the issues under the following headings:

- Trust and liability
- Control and failure
- Differing perceptions of time
- Value and compensation

As noted above, the trust issue is one of the most important underlying a partnership. A contract works on compliance, a partnership works on commitment. Lack of trust is the most frequent stumbling block to successful alliances.

When working with small companies, labs must realize that entrepreneurship is messy and unpredictable. Most executives, in contrast, prefer predictability and order. Entrepreneurial business is inherently risky, while executives and directors try to control and minimize risk.

New businesses, like scientific developments, take a long time to mature. Large companies and political agencies, however, often want short-term results. Furthermore, managers measure themselves by the size of their organizations and budgets, while entrepreneurs measure themselves by results and profits. For entrepreneurs, large staffs represent overhead, not power.

It is also important to build in exit strategies in the beginning: how to plan for dissolution of the partnership and how to determine who owns what.

In summary, partnerships are founded on a new way of thinking: cooperating in order to compete.

CONCLUSION

Finally, there are questions for the future, which many laboratories may need to grapple with as part of their plans to form strategic alliances to commercialize public sector technology:

- *In technology transfer, is the concept of transfer incomplete?* In partnering, one party understands the technology, the other understands the market. Each brings something to the table. Together they begin thinking in new ways, creating a creative team, making a new concept that is better than either party could do alone. Transfer suggests a one-way transaction, it doesn't always allow for two-way interaction. "Technology transfer" should be multidimensional; it should anticipate a transformation. If you simply transfer to someone, you have a customer-supplier relationship, not a partnership.
- *Can laboratories become learning organizations?* Strategic alliances are opportunities for one partner to learn from another. But if the learning is limited to certain individuals, the opportunity for organizational learning and change is lost. The change is one from the individual as an autonomous learner to team learning, so that organizations can deal with their purpose and mission, and become agile organizations sharing a common vision of the future. Labs and universities often find it difficult to grasp the idea of a learning organization. The usual method has been adversarial scientific inquiry; a new way is to develop the synergy of organizational action and shared mental models. This means getting the best from the team as a group rather than from only the scientifically brilliant individual.
- What will it take to change traditional lab cultures from a focus on R&D to a focus on applications and markets? Will alliances provide ways to encourage these changes?
- Who are the labs' customers for customer-focused programs?
- Where are the venture capitalists in lab alliances?

FORMING EFFECTIVE PARTNERSHIPS TO COMMERCIALIZE PUBLIC SECTOR TECHNOLOGY

Jana B. Matthews

Center for Entrepreneurial Leadership
Ewing Marion Kauffman Foundation
Kansas City, MO 64112

INTRODUCTION

Moving technology developed in the federal labs to the marketplace (i.e., commercializing public sector technology) has clearly become an imperative for both management and researchers in federal laboratories. Two major problems negatively affect the labs' abilities to do this efficiently and effectively:

1. Labs have focused on the needs of one market, the military, to the exclusion of industrial, commercial markets.
2. Labs do not know how to commercialize technology.

The labs' traditional role has been focused at the front of the Value Chain (see Botkin, Figure 1), research, development and design used by military contractors. Since technology transfer was not an integral part of the labs' mission, they have not placed a high priority on this function and only a few have hired people with this kind of expertise. Therefore, the labs are faced with a dilemma regarding how to respond to pressures to transfer technology to the marketplace:

- Should they reconstitute their mission, reorganize and expand along the Value Chain—move into prototype development, production marketing, sales and distribution of products based on the technologies developed in the lab?

or

- Should they maintain their mission and depend on other organizations and companies to develop and commercialize technology developed by the labs?

It is fairly obvious that the second alternative will result in more and faster technology transfer. Hence, I suggest a reformulation of the problem facing labs: how can labs facilitate the transfer and commercialization of technology by other companies and organizations?

From Lab to Market: Commercialization of Public Sector Technology,
Edited by S. K. Kassicieh and H. R. Radosevich, Plenum Press, New York, 1994

REASONS FOR FORMING ALLIANCES

Let us review the reasons for forming strategic alliances:

1. Accelerate innovation (faster commercialization of a new technology, product, or service).

2. Facilitate competition in the global marketplace by sharing the risks of development, reducing the costs of production facilities, extending market share, and using existing distribution systems.

Strategic alliances between labs and the private sector make more sense than for labs to try to commercialize technology themselves. However, in order for alliances to work, two things will be required:

1. Technology identification: someone needs to determine what technologies the labs have developed that can be commercialized.

2. Partnership identification: labs need to determine what kinds of partnerships will be most effective in getting the technology out of the lab and into the marketplace.

ALTERNATIVE PARTNERING ARRANGEMENTS

When the technology is identified, partnering becomes a proactive strategy to move it to the marketplace. Thus, the focus of this paper is on identifying and developing effective partnerships. Let me suggest four types of partnership that the labs might consider:

Option 1 With large, established companies. Those labs that have tried to transfer their technology through partnerships appear to be seeking partners from among the Fortune 500 companies. Many of these companies, however, seem to be commercializing "old technology" rather than new, high-tech, or emerging technology.

Option 2 With a small company that takes technology from the federal lab and completes all aspects of the Value Chain, turns the technology into a product, process, or service, then markets, sells and distributes it. There are few examples of this kind of partnership between a lab and a small company.

Option 3 With a small company that takes technology from the federal lab, completes the next step of the Value Chain, then partners with another company to commercialize the technology and take it to the marketplace. There are even fewer examples of this kind of partnership between a lab and multiple partners.

Option 4 With a small company, to identify technology at one or more labs that is readily marketable, and let the small company become the exclusive licenser of the technology. This involves building performance requirements into the contract, and expecting the company to make the deals that will result in technology transfer.

While all four options are viable, I think the labs will benefit most and transfer technology more quickly by working with a group of companies and organizations that have differentiated roles to play in the technology transfer process—Option 3.

CHANGES IN PERSPECTIVE

There are a few changes that federal labs need to make in order to transfer technology effectively through a partnership arrangement:

Think long-term but act short-term. I predict that labs have three years at most to begin showing the results of technology transfer. One of the reasons for partnering is to accelerate the transfer and commercialization process. In the short term, politics, not global competition, will drive labs to transfer their technology as quickly as possible—if they want a continuing funding stream from the federal government.

Think small company, not large company. Note that in Options 2, 3, and 4 a small company is the recommended partner to which the technology is directly transferred. In essence, my message to federal labs is: "Partner with a small company if you really want to transfer technology quickly." Thinking of small companies as feasible partners will require a change in the mind-set of many people in labs, but the benefits will be enormous. Small companies move more quickly than large companies; they create most of the new jobs in this country; and there is less of the "Not Invented Here" syndrome to work through than you will find in large companies.

Think national economy, not national defense. In the past, the labs' customers have been those concerned with national defense. Now the labs need to understand that their customers are those whose products and services will improve the national economy.

The Clinton Administration is recognizing the importance and value of resources at the federal labs and is promising to increase their budgets by millions of dollars. These increases should not be interpreted as affirmation of the status quo, but as inducements to change behavior and transfer technology. The best way to maintain the flow of dollars into the labs is to begin moving technology out of the lab—as soon as possible.

Therefore, a high priority in the labs should be to sift through the research done at the labs and identify commercial applications for the R&D that was undertaken for military (or defense) purposes. If labs do not have the staff to do this themselves, they should hire or partner with other companies or organizations.

Finally, over the long run, the labs need to seek out and undertake more research in response to commercial customers' needs.

Think multiple partners, not single partner. Develop alliances with more than one company or organization focused on commercializing innovation and getting products to the marketplace quickly. For example, seek out and license technology to a small company that can move technology through the idea to the prototype stage, and then help it link with larger companies to manufacture, market and distribute the resulting products. An alternative, which can be implemented at the same time as the partnering strategy, is to contract with a technology broker—a company or organization whose sole purpose is to market technology to others.

The bottom line for labs is: find ways to get your technology to the market place as fast as possible, using multiple partners. The window of opportunity is becoming increasingly narrow for labs to show results from transferring technology. Working with multiple partners can enable labs to act more quickly, to move on several fronts, and to cover more ground.

What kind of technology transfer model could be developed that embodies these perspectives and facilitates technology transfer?

MODEL OF TECHNOLOGY TRANSFER

Let me describe a model of technology transfer that we developed at the Kauffman Foundation's Center for Entrepreneurial Leadership (CEL). The Center was established in 1992 to be a catalyst for developing, supporting, and accelerating entrepreneurship in America.

The target market is entrepreneurs in growth companies. Although service companies outnumber technology-based companies, the latter tend to grow faster and create higher-paying jobs with higher multiplier effects. For this reason, technology-based companies are prime candidates for CEL's programs. Companies commercializing federal lab technology are of interest to us. Consequently, the center helped develop a model for transferring technology from a federal lab to the marketplace.

For the past two months, representatives of six organizations have worked together to develop a proposal describing how we would commercialize a technology developed by a federal lab—in this case manufacturing software. The "partners" in this project are two federal labs, a university, a business incubator, a not-for-profit organization whose mission is technology transfer, and CEL.

The model we have designed for transferring technology from a federal lab to the marketplace, uses the Value Chain:

- Phase I. R, D, & D

Research. A federal lab, Wright Patterson Air Force Base (WPAFB) undertook the research that produced the innovation—manufacturing software.

Development. Faculty members at the University of Missouri/Rolla identified this software and worked with it enough to recognize its potential. It could save small manufacturing companies, of which there are 4500 in the Kansas/Missouri area and 100,000 in the U.S., millions of dollars and make them considerably more efficient.

Prototype Design. UM-Rolla, two federal labs (Sandia and WPAFB), and several small companies have agreed to test the software and prove the feasibility of the concept.

- Phase II. Production

A company will be created: Manufacturing Software Company, Inc. (MSCI) to commercialize the technology. MSCI will be housed within the Center for Business Innovation in Kansas City, an outstanding incubator that just received an award from the National Business Incubation Association as the Outstanding Incubator of the Year. The software will be produced by MSCI.

- Phase III. Marketing, Sales and Distribution

MidAmerica Manufacturing Technology Center (MAMTC), one of the National Institute of Standards and Technology (NIST) centers, will play a key role in the marketing, sale and distribution of the software. MAMTC works with hundreds of small manufacturing companies in a multi-state region and will play a key role in the distribution of the software, as well as in education and training concerning the use of the software.

In this process, our foundation will play a catalytic role. For example CEL is committed to doing the following, should the proposal be accepted by the National Technology Transfer Center:

1. Since we helped design and describe this technology transfer model, we will allocate funds to the project to make certain that the process of technology transfer is sufficiently documented so that it is replicable.

2. We will provide general professional and financial support to the Center for Business Innovation. We will provide additional support to the entrepreneurial team developing MSCI to increase the odds that they will be successful in starting up a spin-out company.
3. Two of us will serve on an advisory board comprised of representatives of all the partners, plus one or two additional people representing potential customers of the product. This board will be instrumental in keeping the project on track and investigating other ways that the "partners" in this project can collaborate on other projects.
4. We will host a day-long workshop midway through the project to review progress and suggest mid-course corrections. We will bring the best brains and experts together to focus on the technology transfer model being developed through this project.
5. We will provide assistance with the dissemination of the technology transfer model after it is developed.

We expect this project to produce several outcomes. First and foremost we expect to demonstrate the successful transfer of technology (in this case, manufacturing software) from a federal lab through the creation of a spin-out company to market and distribute the technology. In effect the lab becomes a partner with a small company to commercialize the technology it developed. Of course, there are many other partners in the picture, as well.

Second, we believe this is a viable model for technology transfer and commercialization that will be sufficiently documented, analyzed, and evaluated to provide valuable new information and insights concerning how to transfer technology from a federal lab to the marketplace via a spin-out company.

Third, we believe this project will provide information and insights about the startup and operation of a consortium of partners—diverse types of organizations that can shape a bold new partnership for technology transfer and commercialization.

Since the technology has already been identified and tested, we believe we can move quickly and develop a product that will be ready to take to the marketplace in 14 months.

The transfer and commercialization of federal laboratory technology is essential for American competitiveness. Partnering is a proven and effective strategy to accelerate this process. The model described here provides an innovative, timely and mutually beneficial way for federal labs to structure alliances with multiple partners to move technology out of the laboratory and into the marketplace quickly and successfully.

PARTNERSHIPS ARE A PEOPLE BUSINESS

Bruce Winchell

Martin Marietta Corporation
Bethesda, MD 20817

FINDING THE RIGHT PARTNER

Forming effective alliances for the commercialization of public-sector technology is the goal and the mission of both today's federal laboratories and the businesses seeking a profit from the sale of products and/or services resulting from the technology. Success will require an understanding of how the federal laboratories operate as well as an understanding of how the private sector must satisfy market demands for products and/or services in order to make a profit and thus survive.

Certainly the twelve principles of successful partnerships outlined by Dr. Botkin [*Commercialization of public sector technology*, p. 225] are very important in terms of finding the correct partner, creating an agreement and managing that partnership to a success. For a private-sector business to find the right partner in a federal laboratory setting will require investigating the available laboratories and their various technologies. There are technologies worth commercial exploitation in many of the different federal laboratories.

In shifting a public-sector technology into a commercially successful operation, it is essential that both or all parties to the alliance feel it is a win-win arrangement for all of them. A federal laboratory can be a good partner because the federal laboratory usually receives royalties only when the private sector partner is successful in selling products and/or services. This means that neither party wins anything unless the private sector partner is successful. The federal laboratory is a committed party without the competition that sometimes results between two private sector parties.

For effective alliances, it is important that the strengths and weaknesses of the various parties to the alliance are complementary to one another in a way that results in a strong alliance with all of the necessary skills to take the public-sector technology all the way to the market place to meet a specific market demand. The parties need experience and expertise in: R&D portion of the public-sector technology, producing a product and/or a service, knowing potential private-sector markets, how to formulate an effective strategy to provide the product and/or service to those markets and lining up the financial backing to accomplish the desired end result. For instance, it is very important to understand whether a large capital investment will be necessary to take the public-sector technology to the market place or whether the swiftness and boldness of a small firm without substantial available capital could accomplish a job in a much more expeditious fashion. Additionally, the federal laboratory can use consortia

From Lab to Market: Commercialization of Public Sector Technology,
Edited by S. K. Kassicieh and H. R. Radosevich, Plenum Press, New York, 1994

to bring together whole industries to compete more effectively in international market places through leveraging the R&D dollars necessary to commercialize public-sector technology in the private sector market place.

CREATING THE ARRANGEMENT

For an alliance to achieve success in commercializing public-sector technology, it must be constructed in such a way that all parties clearly understand their expectations and responsibilities. The private-sector entity should find out at a very early stage what kinds of agreements the laboratory can enter into, in order to understand possible approaches to effect the arrangement.

This might involve, for instance, a straight licensing transaction wherein the federal laboratory licenses its intellectual property to the private-sector business entity for commercialization in some identified market. It could also include, in addition to the license agreement, additional technical assistance, perhaps in the form of work-for-others agreements, or cooperative research and development agreements or CRADAs covered by the Stevenson-Wydler Act.

There may be small business innovation research (SBIR) agreements directly from the federal agency. Today, many of those agreements, particularly for Phase Two funding, require that a small business show how it would produce from the Phase Two development project a marketable product and/or service. This may require that for Phase Two SBIR funding, the small-business entity would, under federal acquisition regulations, engage in some kind of teaming agreement with a large private-sector business entity or government contractor in order to assure the agency that the technology will actually be produced in the form of a product and/or service for an identified market place.

While there are a large number of options for forming effective alliances to commercialize public-sector technology, the private-sector business entity must become familiar with the operation of a federal laboratory, so that the arrangement can proceed more quickly and conserve its assets for the real task of commercializing the public-sector technology.

Small businesses, in particular, find it very difficult to spend the time necessary to work with a large business entity or a federal laboratory, which may proceed at a more deliberate pace in terms of finalizing the arrangements. The involvement at an early stage of lawyers who are knowledgeable about these various mechanisms, as well as intellectual property, can help the parties 1) find the right mechanisms for the given arrangement, and 2) clarify their respective rights and obligations in a fashion which will enable them to manage the alliance effectively upon completion of the contractual arrangements. Early involvement of lawyers can also be beneficial, 1) for the large corporation, in avoiding arrangements that ultimately would not be approved or executed by the corporation, and 2) for the small business, in effectively and quickly completing an arrangement which will maximize potential profits while minimizing potential risks. (The small business should be aware that most large corporations do involve counsel early, as do the federal laboratories).

While arrangements should always focus on the mutual benefits of the parties they must also deal with potential problems and liabilities, to maintain a realistic attitude toward accomplishing successful commercialization of the public-sector technology.

DEFINING SUCCESS

Most private-sector business entities define success in the commercialization of public-sector technology in terms of profits and sales of products and/or services. For the federal laboratory, success can have a significantly broader meaning in accomplishing the mission of a laboratory by leveraging its R&D efforts with outside business entities. Laboratories also gain significantly from interaction with business entities with different goals and a different way of approaching many of the problems attendant to taking an idea to the market place. There are substantial differences in the cultures and the environments of federal laboratories and private-sector business entities. Mixing the two can be an exciting process which is rewarding for both the laboratories and the business entities.

In terms of technology transfer and the commercialization of public sector technology, the laboratories might well consider the prospect of interagency transfers of technology, to help maximize the yield for the federal government from any given technology that may exist within the federal laboratory system. There are many technologies existing today in federal laboratories that have more than one end use, often in both the public and private sectors.

SOME OBSERVATIONS ON INDUSTRY–LABORATORY ALLIANCES

Joseph W. Ray

Great Lakes Industrial Technology Center

The Great Lakes Industrial Technology Center (GLITeC) is one of six regional technology transfer centers established by NASA in January 1992, with a mission to help industry locate, acquire, adapt, and utilize technology and related capabilities from the federal laboratories. GLITeC is operated by Battelle, and its lead staff members have an average of 15 years experience in developing and applying technology for industry. In addition, the center has worked with over 700 companies in the Great Lakes region in the past year. The following observations about alliances between industry and the laboratories are based on that experience.

WHY ALLIANCES?

Although there are many reasons why specific industry-laboratory alliances are pursued, there are some basic drivers underlying most of them. For industry, the basic driver is competitive advantage—technology as a business opportunity. Effective use of technology continues to be important to global competitiveness, and more and more companies realize that it is shortsighted to focus solely on internally generated technology. Nevertheless, acquisition of external technology is a business undertaking. As one of GLITeC's clients put it recently, "These relationships require an investment—dollars, engineering and management time, opportunity cost—and they must have a measurable return."

For the laboratory, a key driver is fulfilling its mission. Legislation and executive directives over the past several years have supported and enabled the transfer and deployment of technology by federal laboratories, and have for all practical purposes made it part of their mission. Alliances with industry have become a measure of laboratory performance, and in some cases part of a strategy for survival—a way to replace lost budgets as primary missions decline.

OPPORTUNITIES

With both potential partners interested, there should be many opportunities for alliances, and this is the case. GLITeC is conducting a survey of manufacturing firms in the Great Lakes states to identify key technology areas and assess knowledge and attitudes toward federal

From Lab to Market: Commercialization of Public Sector Technology,
Edited by S. K. Kassicieh and H. R. Radosevich, Plenum Press, New York, 1994

technology. Although survey responses are still being received, preliminary results are available based on the first 383 responses.

Figure 1 summarizes responses to a question related to how companies meet their technology needs. The two main groupings are "develop" and "acquire". The three approaches to acquisition are equally popular, and about half of the respondents have used each approach—licensing, joint venturing, and purchasing a company. In contrast, essentially everyone develops technology in-house, but there is significant variation in the utilization of various external development partners. About half the respondents have done some development with a university and nearly three out of four with an independent R&D organization, but fewer than one in five have worked with a federal laboratory.

This result is not surprising, but the picture improves when we examine responses related to how companies plan to meet technology needs in the future. As shown in Figure 1, everyone seems to be saying, "More technology", and nearly everyone seems to say, "More from outside the company." Of particular interest here is a tripling in the number of companies which say they plan to develop technology in alliance with a federal laboratory. This indicates that opportunities for industry-laboratory alliances will be plentiful over the next few years. It also suggests that the next few years will be critical. If these new customers come away from the experience both satisfied and supportive, then this 50 percent may increase to 80 or 90 percent. However, if these new customers are dissatisfied, then we are likely to return to the 10 or 20 percent we have seen in the past. We need to create useful alliances and make them work—and this includes understanding and working on the obstacles.

OBSTACLES

The fact that there are obstacles to these alliances should not be surprising. Many companies have as much trouble "partnering" with their own internal R&D laboratories as they do with an outside organization. It sometimes seems as though there are as many ways an alliance can fall apart as there are potential alliances. Three major obstacles are:

- *Alliance mechanisms.* There are many types of relationships which industry can use to gain access to technology and technical expertise from the federal labs, ranging from

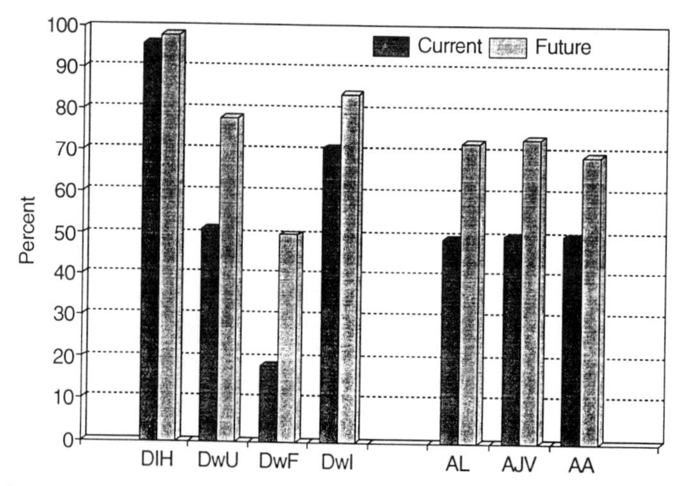

Figure 1. Meeting technology needs—current practice versu future plans. DIH: develop in-house; DwU: develop with a university; DwF: develop with a federal laboratory; DwI: develop with an R&D organization; AL: obtain via licensing; AJV: obtain via joint venture; AA: obtain via purchasing a firm.

customer-supplier relationships to near-partnership relationships. These include quick-response consulting, contract R&D (personnel, facilities or both), staff exchanges, straight licensing, cooperative projects (CRADAs), and various combinations of these. Although this is not an inherent obstacle, there is a danger of focusing on one mechanism and trying to make it all fit situations, or possibly of ignoring opportunities which don't fit the current mechanism-of-choice.

- *Different objectives.* The underlying objectives of the alliance are likely to be different for the company and the laboratory. To industry, it is an investment which should be completed quickly and cheaply—and also quietly (confidentially). However, the laboratory may not have similar motivation. Indeed, there may be reasons to keep it going, make it as lucrative as possible—and publicize it.
- *Ownership of intellectual property.* A company looking for a business opportunity and a competitive advantage wants control of the technology—ownership or exclusive license. Anything that appears to threaten or compromise this control will represent a potential obstacle. This is rarely an insurmountable obstacle, but it is frequently an issue that needs attention.

CONCLUSION

We can anticipate a growing industrial interest in federal technology, with an accompanying increase in opportunities for industry-laboratory alliances. The next two or three years will be a critical period in the commercial application of federal technology. Recognizing differences in basic objectives and potential obstacles to agreements will help us design alliances which meet the specific needs of individual opportunities.

TARGETED TECHNOLOGY COMMERCIALIZATION THROUGH VALUE-ADDED FACILITATORS

Kevin Barquinero

Office of Advanced Concepts and Technology
NASA Headquarters
Washington, DC 20504

INTRODUCTION

NASA has been conducting an experiment with value-added facilitators for targeted technology transfer to accelerate commercialization of NASA-developed technology. The hypothesis is that, by bringing commercialization expertise directly to NASA technologists, the probability of successful technology transfer will be increased. This paper will discuss two new NASA activities—the Joint Sponsored Research Program and the Technology Commercialization Centers at Johnson Space Center and Ames Research Center—both of which, taken together, will test the value-added facilitator hypothesis.

The phrase "targeted technology transfer" was coined by Dr. Jerry Creedon of Langley Research Center and his special initiatives team on technology transfer. This team was chartered by NASA administrator Daniel Goldin in May 1992 to review and make recommendations to improve NASA's processes to transfer and commercialize its aeronautics and space technology. The report to the administrator in December, 1992 identified two categories of technology transfer activities: non-targeted and targeted technology transfer[1].

The Creedon committee defined non-targeted technology transfer as "technology which is transferred via 'acquisition and dissemination' of information and is used 'as is' or is extended by the user without further NASA assistance." Most of NASA's 30-year technology transfer program falls into this category. While noting that these mechanisms to acquire and disseminate technical information are important, the committee believed that relying solely on such processes would not enable the agency to fulfill its potential contribution to the nation's economic competitiveness. Instead, the team recommended that NASA become active in its efforts to transfer technology to the private sector, which they called targeted technology transfer.

Targeted technology transfer involves NASA's conscious involvement in collaboration with industry to commercialize its technology. The team broke down this category into two subcategories: primary and secondary targeted technology transfer. Primary targeted technology transfer occurs when "the technology is part of NASA's primary mission and is developed from the outset with the purpose in mind of transferring it to an identified aerospace user."

From Lab to Market: Commercialization of Public Sector Technology,
Edited by S. K. Kassicieh and H. R. Radosevich, Plenum Press, New York, 1994

NASA's entire aeronautics program represents this category. Newer programs like the Centers for the Commercial Development of Space, are examples from the agency's space program.

Secondary targeted technology transfer refers to "technology originally developed for a NASA mission...extended by NASA to meet the identified needs of a specific user for a non-aerospace application." The committee noted that NASA dedicates very little effort or resources to this category, although it is this area, the broader U.S. economy, which offers greater opportunity for transfer of NASA technology. This is the only area where the Creedon committee recommended that NASA increase its budget.

The Creedon committee report is important because it affirms the need for NASA to be more active in its efforts to transfer its technology. The challenge facing NASA is how to accomplish this mission when its vast technical talents lie in developing technology for its aeronautics and space missions—not in collaborating with industry to commercialize this technology. This is a knowledge gap which thwarts our best intentions to transfer technology. The premise behind using facilitators is that they fill the knowledge gap between NASA's technology and the know-how needed to target the technology's transfer to industry.

VALUE-ADDED FACILITATORS FOR TARGETED TECHNOLOGY TRANSFER

For successful commercialization of any technology, four elements must somehow come together: a useful technology, a dedicated champion with business acumen, market demand for the technology, and available working capital to bring the technology to market[2]. In the case of technologies developed in NASA laboratories, only one of these elements always exists—the technology. The other elements do not commonly reside in the field center. More importantly, few NASA technologists embody all these qualifications, and it is at the technologist's level that the process of technology transfer begins.

The purpose of employing value-added facilitators for targeted technology transfer is to introduce the missing skills of the third-party nonprofit expert discussed in this paper, who works directly with the NASA technologist, legal staff, and central management. The added values these experts bring to NASA are knowledge of how to bring a technology to market, business expertise, deal-structuring skills, and facilitation know-how. Of the four, facilitation know-how is the most important and the toughest to quantify.

Facilitation know-how is essentially a bridging function, linking the sometimes conflicting cultures of government science and technology, which is driven by mission and schedule, with business, which is market-driven. Industry technologists and government technologists are different, almost by definition, because of the self-selection process that sends some people into government service while others enter the world of commerce. In short, there is a cultural gap between these groups. Bridging this gap is a requirement for successful targeted technology transfer; being this bridge is the value added by the facilitators.

JOINT SPONSORED RESEARCH PROGRAM FOR PRIMARY TARGETED TECHNOLOGY TRANSFER, AMERICAN TECHNOLOGY INITIATIVE, INC., FACILITATOR

The Joint Sponsored Research Program is a NASA, industry and university partnership to promote dual-use technology development, accelerate technology transfer and commercialization, and leverage R&D resources. The truly unique aspect of this program is the formation of formal collaborative R&D agreements between NASA, a company and a university (or other nonprofit research institution) under the authority of the National Aeronautics and Space Act of 1958, as amended, otherwise known as the Space Act. The Space Act authorizes the NASA administrator, at his or her discretion, to enter into "other agreements" to fulfill the agency's mission. In March 1992 the NASA administrator delegated this authority for the specific propose of entering into joint sponsored research agreements (JSRAs).

While authorization exists to enter in these new agreements, only a tiny fraction of NASA civil servants have the facilitation know-how to craft a JSRA. To fill this gap, NASA has entered into a cooperative agreement with American Technology Initiative, Inc. (AmTech), a California nonprofit corporation, to conduct research on how to form these agreements and, when appropriate, to facilitate the formulation of a JSRA.

AmTech's efforts have resulted in a detailed understanding of all the elements that need to be covered and the process to follow in crafting a JSRA. Its focus is to identify common areas of interest between NASA research and industry. In return, NASA receives technology rights for government use and the company receive exclusive rights for commercial use of the technology. The university owns the intellectual property[4].

Figure 1 depicts the process AmTech uses to structure a JSRA. AmTech first works with NASA to identify candidate R&D. It is important to note that this is R&D that NASA is already pursuing to meet some agency-approved mission. If a JSRA cannot be crafted, NASA will pursue this work unilaterally. With candidate R&D in hand, AmTech seeks potential industry partners. The initial criterion is a company which is pursuing similar R&D for its own purposes. A match is made when NASA and a company agree to cosponsor a specific research project at a university. At this point, Amtech facilitates the negotiation of the agreement among all three parties. The end product is a JSRA that details the research to be conducted at the university, the money, people and equipment NASA and the company will contribute to the effort, and the intellectual property rights. AmTech has no implementation role in the agreement other then being available to facilitate future communication between the parties.

Both NASA and the company benefit in similar ways from participating in joint sponsored research:

- Leveraged resources through partnership, sharing the R&D risks while simultaneously producing more research
- Promotion of dual use technology development through the identification of commercially relevant R&D
- Acceleration of technology transfer and commercialization
- Intellectual property rights negotiated upfront, thereby providing NASA rights for government use and the company with clear commercial benefit
- The intangible benefit of knowledge transfer resulting from NASA's and industry technologists' close collaboration

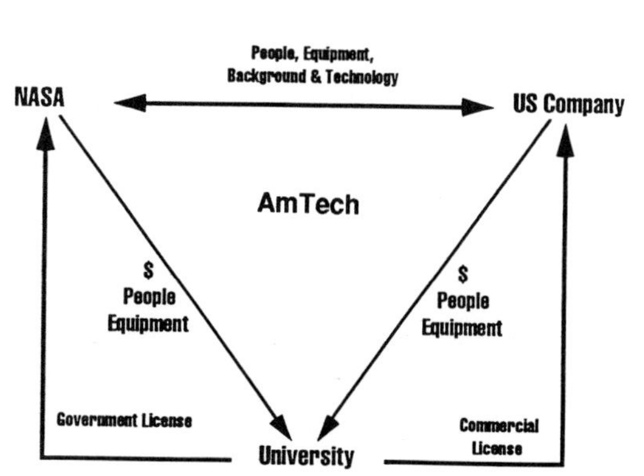

Figure 1. Process AmTech uses to structure a JSRA.

While AmTech is not required to participate in the formation of JSRAs, it would be difficult for others to serve this facilitation function without the equivalent knowledge residing at AmTech.

TECHNOLOGY COMMERCIALIZATION CENTERS AT JOHNSON SPACE CENTER AND AMES RESEARCH CENTER FOR SECONDARY TARGETED TECHNOLOGY TRANSFER, IC^2 INSTITUTE UNIVERSITY OF TEXAS AT AUSTIN, FACILITATOR

The Technology Commercialization Centers (TCCs) at Johnson Space Center and Ames Research Center are a fundamentally new way to transfer technology from NASA field centers. The innovation is in establishing the value-added facilitator as an institutional resource at the field center that focuses on identifying secondary commercial applications of that center's technology. The concept of the TCCs is an amalgamation of two successful technology commercialization models—the entrepreneurial deal-making and technology-licensing model that exists at universities like MIT and Stanford, and new technology business incubation as practiced by the Austin Technology Incubator.

The IC^2 Institute of the University of Texas at Austin, the operator of the Austin Technology Incubator, is the facilitator. IC^2's job is working with NASA technologists to identify technologies that can be commercialized quickly, in from one to two years. For those technologies which affect products serving an existing market, the goal will be to aggressively license the technology to an established company already serving the market. IC^2's role here is to identify these opportunities and hand them over quickly to the field center's legal staff.

If a technology innovation represents a brand new business opportunity, the technology will be commercialized by incubating a start-up company to bring the technology to market. The value added goes beyond the management of an incubator to a more intangible asset —linking NASA technology with regional entrepreneurs, business expertise, capital and market knowledge. NASA lawyers are responsible for intellectual property protection and subsequent licensing of the NASA technology to these start-up companies residing in the incubator.

Figure 2 lists the success factors critical to the technology incubation function of the commercialization centers[3]:

• On-site business expertise in management, marketing, business planning and accounting

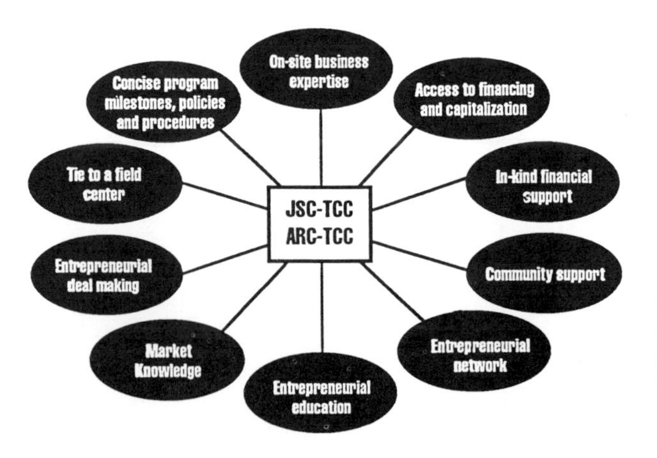

Figure 2. Critical success factors to the technology incubation function for the commercialization centers. IC^2's role.

- Access to financing and capitalization
- In-kind services, such as secretarial and administrative support facilities and equipment
- Community support of efforts to diversify, create jobs, build indigenous companies, and public relations
- Entrepreneurial networks that establish links between individuals and institutions, promote and sustain new ventures, and create new opportunities
- Entrepreneurial education to develop skills and prepare entrepreneurs to learn about technology commercialization
- Perception of success, an intangible element that helps position incubator tenants in their community
- Entrepreneurial deal-making that creates win-win linkages between NASA and industry
- Ties to the NASA field center
- Concise program milestones with clear policies, procedures and metrics for success

These factors are the innovations that make the TCCs a technology-transfer experiment. Once successful, the benefits to NASA will be significant. First the existence of full-time technology transfer experts at the field center will infuse commercialization know-how and enthusiasm into the NASA culture. This in turn will promote dual-use applications of NASA technology as the civil servant technologist becomes more knowledgeable about the importance of technology transfer and the value-added services available through the facilitator. With the technologists' involvement, the transfer and commercialization of technology should be accelerated. The subsequent collaboration between the government and industry technologists will result in the intangible, yet extremely valuable, benefit of mutual learning. Finally, as NASA technologists work with companies, an institutional benefit is afforded the field center in particular and the agency in general—linkage with the regional economy.

SUMMARY

The purpose of these experiments is to demonstrate that it is possible to improve technology commercialization from a NASA field center through employment of value-added facilitators. The facilitator's unique expertise should accelerate the process of technology transfer and commercialization, promote dual-use technology development, and contribute to national and regional economic competitiveness. The measures for success are: leveraged economic development, technology transfer to existing companies, technology transfer to new firms, and knowledge transfer. If these experiments are successful, NASA will transform itself from its past role as a civilian fixture of the Cold War to a national technological engine for economic growth, through the accomplishments of its aeronautics and space missions.

REFERENCES

1. Creedon, J., (Chair), (1992, December 21). *Special initiatives team on technology transfer report.* National Aeronautics and Space Administration. A copy of this report can be obtained from the author.
2. Jennings, J. (1992, July). *Technology transfer models, methods and summaries of existing innovative programs.* Internal NASA paper, National Aeronautics and Space Administration. A copy of this paper can be obtained from the author.
3. Kozmetsky, G., Gibson, D., & Kilcrease, L, (1992). *NASA (Field Center based) technology commercialization centers: Value added technology transfer for U.S. competitive advantage.* Austin, TX: IC^2 Institute, University of Texas at Austin, 13.

4. Shariq, S. (1992). American technology initiative: Competitiveness through R&D joint ventures. In D. Gibson, & R. Smilor, (Eds.) *Technology transfer in consortia and strategic alliances*. Lanham, MD: Rowman & Littlefields Publishers, Chapter 20.

COMMENTS ABOUT *The Impact of Federal Technology Transfer on the Commercialization Process*

James P. Wilhelm

Mid-Continent Technology Transfer Center
College Station, TX 77843

Much has been said about technology transfer and little about technology commercialization. My comments will focus on the commercialization of public sector technology by industry.

Technology transfer occurs when a technology is transferred from a federal lab to a company. This process is facilitated by such intermediaries as the Federal Laboratory Consortium (FLC), the National Technology Transfer Center (NTTC) and the six Regional Technology Transfer Centers (RTTC).

There are two types of technology to transfer. Process technology (e.g., software, tools) produces manufacturing and business process improvements that can result in a higher quality and/or reduced cost product. Process technology is often transferred by technical assistance from the federal lab technologist to the company engineer. On the other hand, the transfer of product technology (e.g., a new battery) often involves the transfer of intellectual property (e.g., patent license) from the lab to the company.

Technology commercialization occurs when the company uses the technology to develop a new or improved product which is, ideally, sold to the market for a profit. The RTTCs offer technical, market and business services which help companies, especially small businesses, successfully commercialize a federal technology.

A product technology must go through the innovation process to be commercialized. The Mid-Continent Technology Transfer Center (MCTTC) uses the innovation process model (Table 1) as a guide. This model is similar to the value-chain and ladder-cycle process models presented earlier. The innovation process model is more comprehensive, using an eight-stage process from concept development to commercial maturity, with technical, market and business steps to be completed at each stage. Inventors tend to race through the technical steps from stage to stage, giving relatively little attention to the market and business steps until they are blocked from going to the next stage. At that point, the inventor must know his market and must develop market and business plans to obtain the resources necessary to implement his plans and move on to the next stage of the innovation process.

Most company start-ups, including those that spin out of a federal lab, university, or defense/aerospace contractor, need help through the innovation process. For further information about the innovation process, the reader is referred to the source document for the model *From Invention to Innovation: Commercialization of New Technology by Independent and*

From Lab to Market: Commercialization of Public Sector Technology,
Edited by S. K. Kassicieh and H. R. Radosevich, Plenum Press, New York, 1994

Table 1. Innovation Process Model

Technical Steps	Market Steps	Business Steps
Stage 1: Concept devlopment		
Technology search	*Market needs assessment*	*Development requirements assessment*
•Database search •Literature review •Contact experts	•Explore market •Explore product uniqueness •Explore competition	•Capacities assessment •Needs assessment •Innovation process status evaluation
Stage 2: Concept Analysis		
Technology application assessment	*Market survey*	*Pre-development venture feasibility*
• Evaluate critical barriers • Evaluate applicability • Select technology	• Commercial product requirement • Other products on the market and their suppliers • Description, specifications, price and shortcoming of each product	• Review intellectual property protection procedures • Preliminary capital requirements analysis
Stage 3: Predevelopment		
Working model	*Formal market analysis*	*Economic feasibility*
• Production feasibility • Performance data • Product or process designs • Safety and environmental factors • Plans for next stage	• Detailed market research • Identify specifics: • Market size and segments • Customers • Prices • Distribution channels • Competition	• Economic feasibility study • Identify seed capital • Form advisor team
Stage 4: Early Development		
Engineering prototype	*Strategic market plan*	*Strategic business plan*
• Construct prototype • Identify critical materials • Fabricate pilot process • Design iterations • Conduct tests • Develop manufacturing methods and reliability	• Identify marketing team participants • Finalize marketing strategy • Field test product	• Obtain patent • Determine organization structure • Determine financing method • Write business plan
Stage 5: Final Development		
Production prototype	*Market management planning*	*Business formation*
• Complete preproduction prototype • Determine production process • Select manufacturing procedures and equipment	• Complete formal marketing plan • Establish formal marketing relationships • Beta test product	• Comply with legal requirements • Establish business functions • Obtain seed capital
Stage 6: Commercial Start-up		
Pilot production	*Formal marketing*	*Business start-up*
• Limited production • Qualification testing • Running changes	• Contact customers • Commence distribution • Seek product endorsements • Follow-up sales • Advertise • Publish in technical journals	• Hire and train personnel • Acquire facilities and equipment • Execute contracts • Obtain second stage capital
Stage 7: Commercial Growth		
Full-scale production	*Sales and distribution*	*Business management*
• Commercial level design Quality control • Minor evolutionary modifications • Production maturity	• All of the previous plus • Expand distribution • Analyze competitor response	• Refine business strategies and practices • Obtain third-stage capital

Table 1. *Continued*

Stage 8: Commercial Maturity		
Production support	*Market development*	*New innovation product market search*
• Maintaining maximum value of production or process through engineering improvements	• Develop and implement market retention and expansion strategies and tactics	• Identify new market opportunities OR • Exit

Table 2. Innovation Process

Stages	Technical steps	Market steps	Business steps
Concept phase			
Stage 1: Concept development	Technology search	Market needs assessment	Development requirements assessment
Stage 2: Concept analysis	Technology applications assessment	Market survey	Predevelopment venture feasibility
Development phase			
Stage 3: Predevelopment	Working model	Formal market analysis	Economic feasibility
Stage 4: Early development	Engineering prototype	Strategic market plan	Strategic and business plans
Stage 5 : Final development	Production prototype	Market management planning	Business formation
Commercial phase			
Stage 6: Commercial start up	Pilot production	Formal marketing plan	Business start up
Stage 7: Commercial growth	Full scale production	Sales and distribution	Business management
Stage 8: Commercial maturity	Production support	Market development	New innovation, product, market search or exit

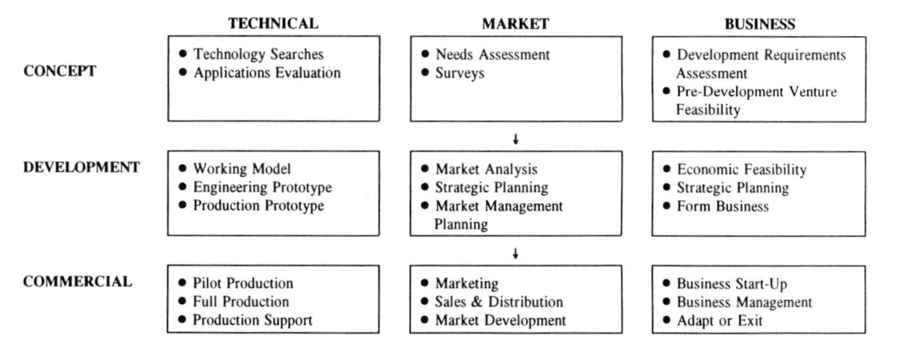

Figure 1. Innovation process steps.

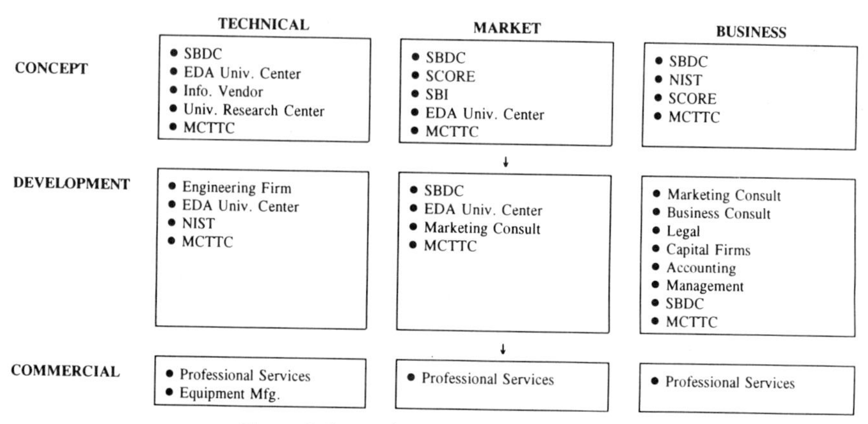

Figure 2. Innovation process service provider.

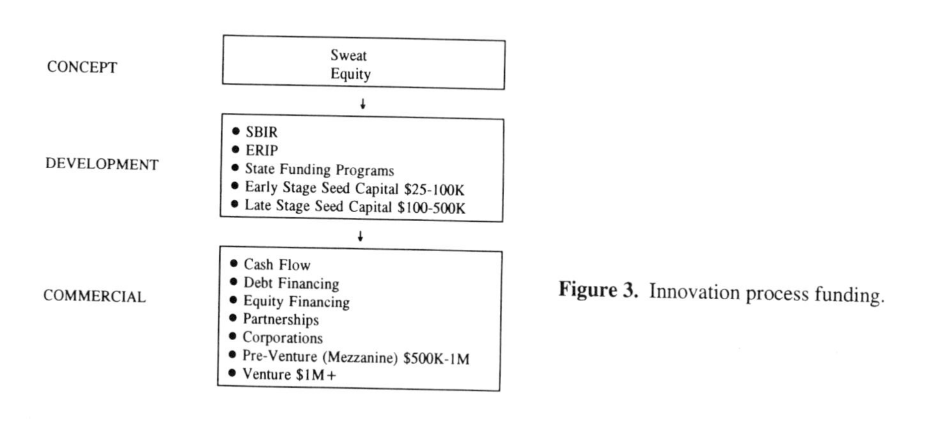

Figure 3. Innovation process funding.

Small Business Inventors, (DOE/NBB-0087), prepared for the U.S. Department of Energy by Mohawk Research Corporation, May 15 1989.

The innovation process can be divided into three phases: concept phase, development phase and commercial phase (see Table 2 and Figure 1). Realizing that companies need outside assistance through some of the technical, market and business steps in these phases, the Mid-Continent Technology Transfer Center links each company with the appropriate service provider(s) listed in Figure 2. The center also helps companies understand that the type of funding they are likely to need will change as they progress through the innovation process (see Figure 3).

The innovation process is complicated and difficult. There are many technical, market and business hurdles that an inventor-turned-entrepreneur must overcome to commercialize a federal technology successfully. As federal technology-transfer personnel better understand the innovation process, we can become more effective at facilitating the transfer and commercialization of public-sector technology from the lab to market.

THE EVOLUTION OF TECHNOLOGY TRANSFER AT MCC*

David V. Gibson[1] and Everett M. Rogers[2]

[1]IC^2 Institute
University of Texas/Austin
Austin, TX 78705

[2]University of New Mexico
Albuquerque, NM 87131

The Microelectronics and Computer Technology Corporation (MCC) is an important experiment in R&D management and technology transfer. This consortium is the first major U.S. effort to get private corporations to collaborate in research while they compete in the marketplace. In 1983, MCC challenged U.S. antitrust law and a long-standing belief in the value of free-enterprise, unfettered competition. In 1993, MCC is still on trial as a symbol of an emerging paradox in U.S. economic development—the belief that for U.S. corporations to compete effectively in the international marketplace, they must collaborate with their domestic competitors.

Despite dramatic changes in MCC's structure, organizational culture, and research activities, the consortium's stated mission has remained fairly consistent: to strengthen and sustain the competitiveness of member companies who share common elements of a vision in information technology.

MCC's strategic intent has been to gain competitive advantage—to collaborate to win. Although a private, for-profit consortium, MCC has traditionally emphasized technology development and transfer, not profit. Shareholder or associate member companies use MCC technologies to produce products and services of their own design and to compete in markets of their choice.

MCC's most dramatic operational and strategic changes have occurred during three distinct phases of leadership: (1) a honeymoon era, from 1983 to 1986 under the leadership of retired Admiral Bobby Inman, (2) a reassessment phase, from 1987 to 1990, under the leadership of Grant Dove, a 28-year veteran of Texas Instruments, and (3) a major restructuring phase from 1990 to present, directed by former DARPA (Defense Advanced Research

* For a more extensive and detailed account of the evolution of technology transfer at MCC, see Gibson & Rogers, (1994) *R&D collaboration on trial.* Boston: Harvard Business School Press.

From Lab to Market: Commercialization of Public Sector Technology,
Edited by S. K. Kassicieh and H. R. Radosevich, Plenum Press, New York, 1994

Programs Agency) executive, Dr. Craig Fields. During each era, MCC has faced different challenges and opportunities. Important lessons have been learned about public/private alliances and about learning how to collaborate in research and related activities in order to compete effectively in the world's marketplace.

The challenges faced by MCC and its member companies are similar to those now confronting participants in inter-organizational alliances (among government, universities, and private participants as well as other new organizational forms) which are concerned with technology commercialization. When it was launched in 1983, MCC was described by some as the prototypical organization of the future—i.e., a boundaryless organization composed of strategic alliances with public/private organizations.

Technology transfer, from the laboratory to product development, has been a continuing challenge at MCC. A major criticism of R&D consortia in general is that, in terms of ROI (return on investment), research results leading to successful product commercialization have been sparse, especially given the amount of funds and research talent invested. While the U.S. has excelled in basic research and in technology development, the nation's firms have not been competitive in getting new technologies to the market place in a cost-effective, timely manner. Technology transfer is especially difficult when it involves crossing organizational boundaries, as in the case of an R&D consortium and its member companies.

MCC has employed a range of creative initiatives to facilitate technology transfer to its member companies. The lessons learned are relevant for (1) firms trying to transfer technology from a research activity to the marketplace in a timely manner, (2) R&D consortia, universities, and federal laboratories, and (3) firms trying to learn how to collaborate in strategic alliances which encourage technology transfer across organizational boundaries.

THE DIFFICULT NATURE OF TECHNOLOGY TRANSFER

Two kinds of technology transfer directly affect U.S. industrial competitiveness, as shown in Figure 1: (1) spinning out technologies into start-up companies (the dashed line), and (2) transferring new technologies to established firms (the solid line). Spin-out technologies may originate in the private sector, federal labs, universities, or consortia. These spin-out companies may or may not be nurtured by an incubator. The U.S. is a successful role model for much of the industrialized world regarding spin-out technology transfer. The U.S. is not as competitive regarding technology transfer across the organizational boundaries of established firms.

MCC's shareholder companies realized that getting technologies transferred from the consortium was going to be a challenge. A 1986 survey of MCC member companies indicated that 46 percent of the firms responded that technology transfer was the most difficult task facing the consortium [9]. However neither the managers nor the researchers at MCC or at the shareholder companies understood just how difficult such technology transfer would be. "One of the fascinating elements of the MCC experience," said Admiral Inman [6], "was to see how different the companies were in dealing with ongoing research...the ones that prowl, looking for things that were new, and the majority that waited for you to deliver things to them."

Depending on the perspective of the evaluator, MCC's success at technology transfer could be based on a range of criteria. The R&D consortium has published over 2,000 technical reports, produced over 400 technical videotapes, "transferred" more than 180 technologies, and has been issued 87 patents and 182 technology licenses.

There are also less tangible and less accepted criteria for judging MCC's success at technology transfer: learning from R&D failures while having the costs dispersed over a number of firms, small incremental wins, support services provided, increased intra- and inter-firm communication among MCC's member companies, and learning how to communicate and collaborate across organizational boundaries. It would be almost impossible to specify

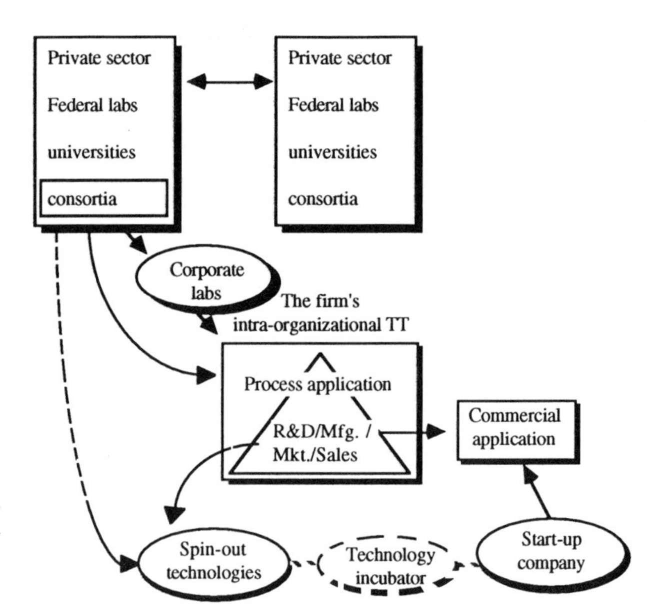

Figure 1. Two basic forms of technology transfer to commercial applications.

the ROI (return on investment) of these forms of technology transfer, but they may well be MCC's most significant contribution to U.S. industrial competitiveness.

Learning from failures is sometimes offered as a benefit of R&D consortia. If ten companies invest in a risky technology and it fails, then each company learns what not to do at one-tenth the cost. Such a strategy would certainly seem preferable to the "bet your company" type of investment in cutting-edge R&D. However, MCC researchers and managers soon realized that "learning from failures" was not the way to promote the benefits of belonging to MCC; member companies were not keen on funding research, on any scale, which produced "failures" as a measure of success.

MCC has engaged in many collaborative research projects leading to incremental improvements in member company technologies and internal processes. The consortium has achieved numerous and often imperceptible transfers of such technologies to its member companies. At times, MCC researchers did not even realize they were transferring important technologies/ideas to member company engineers, researchers, and managers. Small and continuous transfers were often embedded in larger product or process technologies. However, shareholder representatives found it difficult to justify to their superiors at their home companies that such transfers justified continued support for MCC.

Member companies have looked to MCC for support services or particular product or process applications which the member company could not, or would not, find elsewhere. MCC has hosted teleconferences on cutting-edge research for its member companies, and it has organized seminars on important emerging technology areas. MCC's international liaison office (ILO), which translates foreign technical publications and professional reports, provides valuable competitive information to MCC's member companies. U.S. firms have traditionally done a poor job of monitoring the technological advances of their foreign competition, especially when compared to the technology scanning and utilization capabilities of their Japanese, Korean, and Taiwanese counterparts. Yet it would be difficult to place a precise value on how much such information has helped MCC's member firms compete in the global marketplace.

Perhaps the most significant "technology" transferred from MCC to its member companies concerns increased communication and collaboration. MCC provided a means for its member companies to increase communication among the functions, divisions, and hierarchies within each member company as well as across company boundaries. From the board of directors to the program technical advisory panels, MCC provided a forum for cross-organizational communication on a variety of issues. As a result of such increased intra- and inter-firm communication, MCC's member companies made significant progress in learning how to collaborate while competing in the marketplace.

In the final analysis, MCC (like many other U.S. consortia) is ultimately being judged in terms of its impact on short-term, visible product commercialization. The main criticism of MCC is that the consortium has absorbed over $500 million of member company funds with only "thin" results to show for such a large investment: "Where are the significant technological breakthroughs?" "Where are the commercial wins?" "What if the member companies had invested their MCC-dedicated funds in their own R&D?"

TECHNOLOGY TRANSFER DEFINED

Many U.S. managers think of technology transfer as looking something like a barbell. One ball represents R&D, charged with creating technological advancements. The other ball represents the users of the new technology, who operate in the world of manufacturing and marketing. The technology-transfer process consists of moving new technologies from the R&D unit to the technology-using unit.

This one-way barbell model is what MCC had in mind when it began operation in the early 1980s. A "standard technology package" was offered simultaneously to all the funding companies when MCC researchers released the technology. The consortium soon learned that this oversimplified model did not work. "If you wait to transfer the technology until the researchers are finished with it, you'll have a long wait," commented Inman. Not much came out at the user end, or at least not much technology was implemented by MCC's member companies. How MCC came to recognize just how complex and difficult a process technology transfer is, and how MCC scientists and managers have evolved in meeting these challenges, is the focus of this chapter.

The concepts of technology and of transfer are defined by both theoreticians and practitioners in different ways. There is usually agreement, however that (1) technology is not just a "thing," and (2) that transfer is a profoundly human endeavor. Essentially, *technology* is information that is put to use in order to accomplish some task, the knowledge of how to do something. *Transfer* is the movement of technology via some channel from one individual or organization to another. Thus, the transfer of technology involves the application of knowledge, putting a tool to use. Technology transfer is a particularly difficult type of communication, in that it often requires collaborative activity between two or more individuals or functional units separated by structural, cultural, and organizational boundaries. Appreciation of the human component in technology transfer directs us away from thinking of simply moving technology from point "A" to point "B", as in the barbell analogy described earlier. Instead, we must think of technology transfer as an interactive process with a great deal of back-and-forth exchange among individuals over an extended period of time.

THE SERENDIPITY OF TECHNOLOGY TRANSFER

A communication-based model of technology transfer centers on information exchange between a technology source and a receptor as a two-way process. Such information exchange is typically not orderly or one-directional. It is often chaotic and disorderly. Participants are "transceivers", who exchange ideas simultaneously and continuously, thereby blurring the distinction between senders and receivers. The technology being transferred is often not a fully formed idea that can be neatly packaged and forwarded. It may have no inherent meaning or value; meaning is in the minds of the participants. Accordingly, transmitters and users are likely to have different perceptions of the same technology. Feedback helps transceivers reach convergence about important dimensions of the technology. Technology transfer to product commercialization involves an ongoing, multilevel exchange of information.

There is a serendipitous aspect to technology transfer in that researchers often make unexpected discoveries, and technology users find unintended applications. Researchers and users may combine in a synergistic way, one that could not be predicted or managed, to produce unexpected results. Such technology transfer is an example of the "garbage can" model of decision-making.[8] A transferred technology results from an unplanned mixture of participants, solutions, choice opportunities, and problems. Both problems-looking-for-solutions (technology pull) as well as technology-solutions-looking-for-problems (technology push) are encountered.

"On the technology side," commented Admiral Inman, "it isn't just a strategic planning push because you don't know in advance what's likely to evolve out of the investment in research. You're dealing with an unknown. You may think you know what you're after, but the issue for many companies is are you prepared to deal with the other events that occur, the accidental discoveries." Inman offered Xerox PARC (Palo Alto Research Center) as an example of a company unprepared to utilize unexpected findings [10]. "They had gathered an incredible array of talent in California...did much of the pioneering work in modern computing...not only didn't they use their own products, but they didn't put into place the mechanisms to sell that technology to other companies. The technology was ultimately transferred by disaffected employees who took the technology with them."

LEVELS OF TECHNOLOGY TRANSFER

MCC's experiences with technology transfer suggest four levels of collaborative activity, and four correspondingly different definitions of technology transfer success (Figure 2). At Level I, technology R&D researchers conduct state-of-the-art, precompetitive research, and transfer their results by such varied means as research publications, videotapes, teleconferences, and software computer tapes. Technology transfer at this level is a largely passive process which requires little collaborative behavior among the transceivers, although the researchers may work in teams or across organizational or even national boundaries. Traditionally, in the United States, technology users have not been involved at this level of the

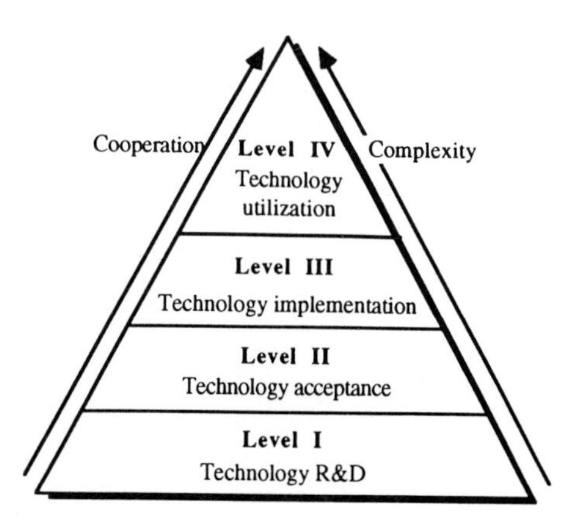

Figure 2. Technology transfer at four levels of involvement.

transfer process. Level I success is measured by the quantity and quality (usually through peer review) of research reports and journal articles. Technology transfer plans and processes are not considered very important. *Research strength* is most important. The belief is that good ideas sell themselves; pressures of the marketplace are all that is needed to drive technology use and commercialization.

During MCC's early years of operation, the consortium's scientists were measured in terms of Level I activity. Conceptions of technology transfer reflected (1) the norms and values that MCC scientists brought with them from university, federal, and corporate laboratories, and (2) the level of technology transfer mandated in the precompetitive research limitations of the 1984 National Cooperative Research Act.

Level II technology transfer, technology acceptance, calls for the beginnings of shared responsibility between technology developers and users. Success occurs when a technology is transferred across personal, functional, or organizational boundaries, and is accepted and understood by designated users. Moving from Level I to Level II technology transfer is an extremely difficult task for research organizations like MCC. Knowing the appropriate person to contact is an immense challenge as one looks at a large corporation like Lockheed or General Electric, or even at much smaller Advanced Micro Devices. A Level II perspective encourages the belief that successful technology transfer is simply a matter of getting the right information to the right people at the right time.

In Level III transfer, success is marked by the timely and efficient implementation of the technology. For Level III success to occur, technology users must have the knowledge and the resources needed to implement, or Beta-test, the technology. Technology implementation can occur inside the user organization in terms of manufacturing or other processes, or it can occur in terms of product development, such as building a prototype for commercial application. *Commercial strength* is required.

Level IV transfer, technology utilization, centers on product commercialization. Level IV builds cumulatively on the successes achieved in attaining the objectives of the three previous stages, but *market strength* is required. Feedback from technology users drives the transfer process. Success is measured in terms of return on investment (ROI) or market share. Here, we take a longer-term view. It is at Level IV technology transfer that MCC has been most criticized.

Overall technology transfer success, Levels I to IV, is difficult to measure in terms of traditional cost/benefit analysis since (1) it is often difficult to quantify financial or other impacts of a technology over time, and (2) different people involved in the process are likely to evaluate costs and benefits differently depending on their unique perspectives. In the case of MCC, the different member companies have different expectations as to which level of technology transfer they expect. Some member companies are happy with research reports, while others want market-strength products. Even within a member company, different managers and functional areas (e.g., R&D, production, marketing, and sales) evaluate MCC's technology transfer activities by different criteria. For example, a CEO might be motivated by a bold new concept, whereas a line manager in the same company wants technology of demonstrated industrial strength. For a scientist, success might be peer recognition or a journal publication, rather than the commercial application of his/her idea.

According to MCC's charter, the consortium's objective is to create and transfer generic technology that can be used by its shareholder companies to create new products. In keeping with this orientation, during the consortium's formative years, MCC scientists generally maintained a Level I perspective of technology transfer. To ensure trust and fairness, it was considered MCC's responsibility to present a new technology (the standard technology package) to all the funding shareholders at the same time. Success was defined as (1) conducting high-quality, long-range research, and (2) making the results available in a timely fashion to the shareholder companies. The shareholder companies were to use precompetitive MCC-developed technologies to create products of their own design and to compete in markets of their own choice. Market forces and competitive pressures were expected to drive the process.

Over time, it has become apparent to MCC's managers and scientists that Level I measures of success, however impressive, would not sustain the consortium. This realization intensified as responsibility for MCC funding worked down through the management levels of the member companies. Divisional and group managers increasingly demanded measurable technology benefits (Levels III and IV technology transfer). MCC faced a dilemma as technology transfer activities moved from a Level I to a Level IV perspective—the consortium exercised less control over events leading to successful technology commercialization as more collaboration was required across functional and organizational boundaries from a range of participants (Figure 3). MCC could develop a "silver bullet" technology, but if the member companies chose not to take it to industrial and market strength, MCC would be judged a failure.

We now offer an in-depth look at one of MCC's earliest and most publicized technology transfer success stories. Proteus was the first major and public product application (Level IV technology transfer) of an MCC-developed technology by a shareholder company. This case highlights the importance of collaboration, planning and serendipity, and the high level of effort required over an extended period of time across all four levels of technology transfer to achieve successful technology commercialization.

TECHNOLOGY TRANSFER AT MCC

Over the years, MCC has tried a variety of technology transfer strategies (Table 1). No one method has been completely effective. Based on interview, archival, and survey data collected from MCC's first decade of operation, we now discuss four key issues in achieving successful technology commercialization across organizational boundaries: (1) communication interactivity, (2) physical, cultural, and strategic proximity, (3) technology characteristics and (4) interpersonal motivation.

Communication Interactivity

Communication interactivity refers to the richness of exchange between technology transceivers, i.e., developers and users. Interactivity ranges on a continuum from passive, one-way media-based linkage such as research reports, videotapes, and computer tapes, to more interactive face-to-face linkages such as on-site research demonstrations and collaborative research projects. MCC has initiated a range of passive to interactive communication linkages between technology developers and users (Table 2).

Initially, MCC emphasized the use of passive modes of communication with its member companies through technology reports and video tapes. Passive communication demands less

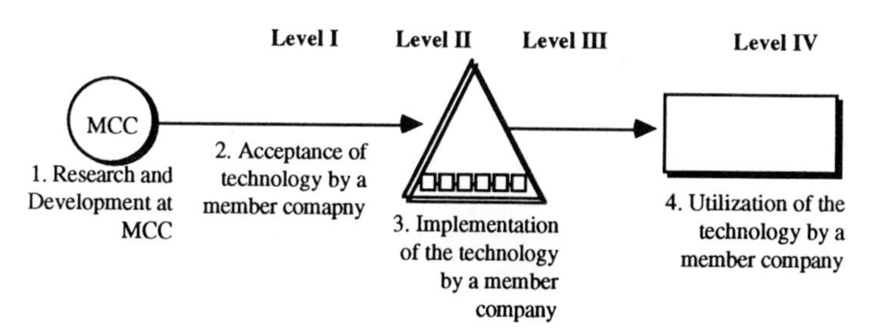

Figure 3. Four levels of technology transfer from MCC to a member company.

Table 1. Progression of Fifteen Types of Technology Transfer at MCC 1983–1993

Type #1: Shareholder Representatives: Initially, shareholder companies were to assign high-quality researchers to MCC for a period of two-to-three years. These shareholder representatives were to transfer MCC-produced technology back to their home companies. MCC accepted only about 20 percent of these company representatives and, for a variety of structural and behavioral reasons, this method of technology transfer proved to be less effective than anticipated.

Type #2: The Standard Technology Package: In support of a philosophy of trust and fairness, MCC management sought to transfer technology as a "standard technology package" to all shareholders at the same time. But MCC found that effective technology transfer is an ongoing process which must be tailored to the needs of individual researchers and technology users.

Type #3: Quarterly Meetings: Quarterly meetings were scheduled for shareholder managers and technology specialists to visit MCC and look at work-in-progress. But the visits often did not include the right people, at the right place, at the right time.

Type #4: Written Documentation: Hundreds of professionally-produced research reports were sent to shareholder companies, but it was not clear to whom the reports should be sent. The reports produced minimal feedback and technology buy-in, and they often did not get the attention of the individuals in the shareholder companies with the power and resources to turn MCC technologies into products.

Type #5: Shareholder Visits: MCC arranged shareholder visits to the consortium in order to transfer technologies as they were being developed. However, the visits did not always occur at the most appropriate time and often the most appropriate people (e.g., senior product planners) were not sent by the shareholder companies, nor were those who were sent supported over the long time-frame required.

Type #6: Multi-Media: MCC made short videotapes in which MCC researchers talked about their research results. But it was difficult to determine the best audience for these videotapes or how to obtain useful feedback from those who did see the tapes.

Type #7: Company Days: MCC instituted "company days," when individuals from a single company visited the MCC to get acquainted with work under way in the consortium's research laboratories. These visits often did not include the appropriate shareholder employees and did not occur at the right time.

Type #8: Videoconferencing: MCC experimented with transferring technology by video teleconferences, but audio feedback from the teleconferencing sites was not particularly useful. Shareholder sites were concerned with guarding proprietary knowledge. A rich exchange of information usually did not occur.

Type #9: Research Collaboration: MCC emphasized collaborative research activities between MCC and shareholder researchers on-site at the consortium. Researchers were encouraged to begin these research project collaborations as early as possible.

Type #10: Opening up and Unbundling: MCC opened up and unbundled its research programs to decrease the cost of company buy-into and to emphasize technology deliverables on the way to long-term research objectives.

Type #11: Third-Party Licensing and Vendors: MCC actively pursued third-party licensing and vendor company participation as a means of technology commercialization.

Type #12: Technology Transfer Plans: MCC emphasized technology transfer plans with start and stop dates, business impact assessments, and project management.

Type #13: Spin-Out Companies: MCC began to allow, and then encouraged spin-out technologies and start-up companies.

Type #14: Vertical and Horizontal Integration: MCC saw value in having great member (shareholder and associate) diversity to allow for vertical (e.g., vendors) and horizontal (e.g., customers) integration to facilitate technology commercialization.

Type #15: Distributed Research: MCC deemphasized the agglomeration of research talent at MCC's Austin-based headquarters in favor of being a project manager of distributed research activities.

time and expense than more interactive modes; researchers can stay "at home" while transferring technologies to potential users. A large number of individual and organizational receptors can be reached at little cost. With such computer-mediated linkages as E-mail and video-conferencing, researchers can talk directly to users about their technology. Such communication linkages diminish personal and professional risk for the technology developers, as it can be a challenging experience to go to the field to "sell" one's research. On the other hand, passive communication seldom elicits useful feedback from technology users.

MCC's program technical advisory boards (PTABs) and program technical panels (PTPs) featured formal, structured meetings to facilitate technology transfer between MCC and its member companies. At these meetings, MCC researchers gave formal presentations, tutorials, and demonstrations about specific technologies. However, the *right* information was often not given to the *right* member company recipient at the *right* time for successful technology transfer to occur. Some member companies did not even send a company representative to these technology briefings.

As Craig Fields moved MCC toward more distributed R&D in the early 1990s, he was concerned about linking people.

> One of the things we [MCC] are exploring at the moment is a teleconferencing system that links the technology development staffs at the companies with those at MCC. Remember this is a cooperation among companies; this is not a star network to MCC. I'm trying to get a little different spirit going. It's slowly happening, I don't want to claim any great revolution overnight, but at least the derivation is in the right direction. [3]

TECHNOLOGY TRANSFER THROUGH PEOPLE

At the December 4, 1982, MCC board of directors meeting in Denver, when Inman asked how MCC was going to transfer its technologies to the member companies he was told: "You shouldn't worry about that at all. We will send our research people to the consortium, and when a piece of research is done, our people will come back to their home companies and bring the technology with them." This simplification worried Inman a great deal: "From my previous

Table 2. Passive and Interactive Technology Transfer Mechanisms at MCC

Passive technology transfer	
1. Proprietary technical reports	4. Videotaped overviews
2. Refereed journal articles	5. Videotaped demonstrations
3. Newsletters	
Computer-based technology transfer	
1. Video conferencing	2. E-Mail consulting
Face-to-face technology transfer	
1. MCC/shareholder meetings	3. Shareholder representatives
• Software Technology Advisory Council (STAC)	4. Shareholders assignees
• Technical Advisory Board (TAB)	5. Visitors programs
• Program Technical Advisory Board (PTAB)	6. Shareholder site demonstrations
2. Shareholder Committees	7. Receptor organizations within shareholders
• Program Advisory Committees (PAC)	8. Shadow research projects within shareholders
• Technical Requirements Panels	9. Shareholder/MCC collaboration
• Manufacturability Panels	
• Quality Assurance Panels	

experience with the National Security Agency, I had learned that you had to have an effort to both *push* the technology out from the [R&D] laboratory, and to *pull* the technology from the outside...creating the technology is easier than getting it used." [7]

From its inception in 1982, MCC supported the view that technology transfer is a "contact sport," that people are the best means of transferring technology. An MCC visitor program allowed shareholder scientists to spend time at MCC working on a collaborative research project (as in the case of Robin Steele and the Proteus transfer to NCR). The shareholder person was expected to carry MCC know-how back to his/her company. Over the years, the most highly interactive and institutionalized method of technology transfer between MCC and its shareholder companies has been the shareholder representative.

During MCC's formative years, "shareholder liaisons/representatives" were assigned for a two-to-three- year period to a particular research program at the consortium. These representatives were expected to participate in MCC research activities and to gain in-depth knowledge of the available technologies. They were also expected to make periodic trips back to, and ultimately/permanently return to, their home companies in order to transfer this knowledge. MCC's conception of shareholder representatives as the premier means of technology transfer to the member companies has not worked out as planned. Why not?

First, MCC did not receive the expected number of quality shareholder representatives. Being assigned to MCC for two-to-three years was generally not viewed as an attractive career option for a fast-track company technician or researcher. Furthermore, shareholder companies were reluctant to give up their best researchers for a tour at MCC. As a result, shareholder representatives in residence at MCC were often not the most visible and respected company researchers, nor were they the most appropriate receptors for MCC technologies. While MCC could exercise control over the researchers it hired, the consortium had little control over a company's selection of shareholder representatives assigned to a research program. As Inman commented in 1992, these shareholder representatives often (1) knew little about their home company's technology priorities and product development, (2) were not able to contribute to MCC's research, and (3) had little "clout" in their home companies.[5] Several shareholder representatives were hired by member companies for the specific purpose of assigning them to MCC. Often these representatives were most concerned with transferring technology/knowledge from MCC to the company and were very ineffective in making the transfer a two-way process between MCC and the shareholder company. As one MCC researcher stated:

> Shareholder companies often do not send the best possible technology receptor [one who had the needed technical knowledge and shareholder power and contacts] to the consortium. Consequently, MCC scientists often wound up spending a lot of time and effort talking with the wrong shareholder representative about the wrong technologies.

Assigning a quality shareholder representative was also perceived as a threat, by some shareholders, to proprietary knowledge and competitive advantage for the parent company. For example, let us say that a shareholder representative was able to transfer an MCC technology to his/her parent company. If that company's research experiences were communicated back to MCC staff by the representative, other companies might gain valuable information or otherwise benefit from these technological advances. In effect, companies not contributing to the research effort could become "free riders" on the research activities of other companies. Such concerns with protecting proprietary knowledge began to abate in 1986. In our decade-long study of MCC, we found little meaningful scanning of a competitor's technologies for competitive advantage. In fact, the shareholder companies which participated most actively in MCC's research projects benefited the most from the technological advances.

Many of the shareholder representatives sent to MCC were volunteers who, for a variety of reasons, wanted a change of pace from their current career options. At MCC they found facilities, funding, autonomy, and research opportunities which were generally far superior to those at their home company. As one long-time MCC researcher in the advanced computing technology program said, "The result was the creation of a research environment unparalleled...offering wonderful facilities and support, as well as access to the technical resources and facilities of some of the country's leading computer corporations."

Over the years some very capable shareholder representatives have been assigned to MCC, and they have transferred technologies back to their home companies. However, even the most capable representatives found it a challenge to be knowledgeable about MCC's wide range of technologies and at the same time to keep abreast of their own companies vast research activities which might benefit from such MCC-produced technologies. As one MCC scientist stated in 1988:

> No one is competent enough to understand all of the technologies here anyway. If you sent one guy [to the MCC] who was supposed to be your 'understander ' he'd be a bottleneck. The understanding of the technology has to be accomplished by many people coming here.

Finally, the shareholder-representative approach to technology transfer emphasized the importance of the MCC technology-based contacts and deemphasized the importance of having the representative maintain strong ongoing personal contacts with key employees in the shareholder company. MCC has found that the longer a shareholder representative stayed in Austin, the more out of touch the representative became with personal networks back in his/her home company. Such contacts are critical in speeding the transfer of technology, especially to Level IV product commercialization.

PHYSICAL, CULTURAL, AND STRATEGIC PROXIMITY

A relatively small physical distance (such as different floors of a building) between individuals can decrease the frequency of interpersonal communication, despite computer-mediated networking. The worldwide spread of MCC's member-company operations certainly exacerbates the problem of targeting appropriate recipients for MCC-developed technologies. Many of the U.S. headquarters of MCC's shareholders and associate members are located at a considerable distance from the Austin consortium, in California, New York/New Jersey, St. Paul/Minneapolis, etc. (Figure 4). A travel/time analysis commissioned by Craig Fields indicated that the ideal location for MCC (based on this one criterion) would be either New York or Washington. "That's the time center, not the geographic center," says Fields. "Then there's a third and fourth-best place, and so on. Austin isn't even in the top ten."

While physical distance exacerbates barriers to prompt and efficient MCC-member-company technology transfer, MCC's technology transfer successes have not been correlated with a shareholder's physical proximity to the consortium's research activities in Austin. *Cultural proximity*—the degree of similarity between two individuals, groups, or organizations in their norms and values—is considered more important than physical distance in influencing successful technology transfer. MCC is staffed by scientists from a range of academic, government, and business institutions. Each of these researchers brought his/her particular research perspectives and styles of work to the Austin-based consortium. Cultural diversity within MCC has remained considerable despite conscious efforts early in the consortium's existence to build a common MCC culture.

MCC initially thought that shareholder researchers of comparable training and technical skill would be the most appropriate receptors for an MCC-developed technology. Effective communication would be facilitated by knowledgeable experts talking with one another, scientists talking to scientists. However such homophilious communication was inhibited as

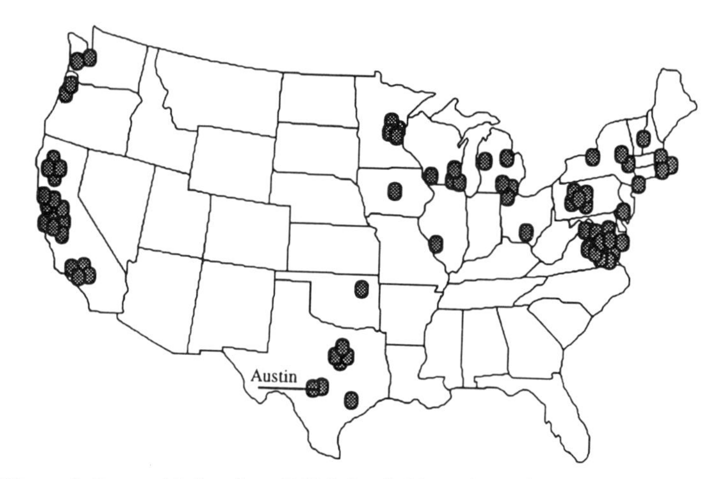

Figure 4. Geographic location of MCC shareholder and associate member companies.

MCC and shareholder scientists often competed for research funds and professional prestige. A defensive "not invented here" (NIH) attitude was also a barrier to research-to-researcher technology transfer. Another problem with MCC researchers talking to researchers in the shareholder companies was the time it took to get the information to production and marketing people in a firm. As Grant Dove observed, "...If you have to transfer the technology to a central lab and then to the product units, it's too late. It's all over with."

Over the years it has become increasingly apparent to MCC that shareholders' marketing, production, and sales personnel may at times be more appropriate receptors for MCC-produced technology than scientists and technologists. MCC's successful cases of technology transfer have often involved shareholder production and marketing people who pulled the technology out of MCC.

MCC also has to contend with strategic differences among its 22 shareholders and 59 research associates, as well as potential government and university receptors of MCC-produced technology. Diversity among MCC shareholders has created significant challenges to technology transfer.

TECHNOLOGY CHARACTERISTICS

MCC's research programs cover a spectrum from basic research to more tangible products and tools. What are the implications for technology transfer of such technological diversity? Is basic research difficult to transfer? Is software technology more difficult to transfer than hardware and other physical products? *Equivocality* refers to the level of concreteness of a technology. Technology low in equivocality is less ambiguous. MCC's first chief scientist, John Pinkston, contended that "The more a technology can be encapsulated—the more the user only has to deal with the externals (as in the case of a tool), the easier technology transfer tends to be."

MCC's high value electronics (HVE) division (formerly the packaging/interconnect (P/I) Program) has transferred more technologies to its shareholders than have all other MCC research programs combined. Most of MCC's patents have been generated in the HVE Division. However, hardware advances are easier to patent than software advances. P/I technologies are generally observable, measurable, and relatively easy to demonstrate. In contrast, the information system division's technologies (formerly Software Technology, Computer-aided Design, and Advanced Computer Architecture Programs) tend to be more idea-based, with a variety of possible applications. The superiority of a tool can be demonstrated, but the superiority of an idea is more difficult to evaluate.

During MCC's first eight years, each of the consortium's research programs approached the challenge of technology transfer somewhat differently. MCC's management did not try to force the issue and say, "This is the way to do it." The research programs most successful with Level IV technology transfer were the ones that had a steady stream of evolutionary output. The least successful research programs delivered concepts and ideas. Grant Dove [2] commented:

> Most companies are in a better position to evaluate hardware, take it, and put it to work. Software is harder because of the large and expensive gap to get to industrial and market strength. In the software world in particular, we were challenged to see if we could take it [the technology] to Beta test level (which in technical terminology means that it is pretty robust, and it does something real, and it will work reasonably well). We learned to fly in the CAD program by going to open architectures and to robust, industrial-strength code.

STP was embedded in an essentially microelectronics research environment observed program director Laszlo Belady.[1] "On the surface it looked very progressive and promised interdisciplinary work, but in reality it provided a hardware-flavored management with reduced understanding of, and sympathy for software."

According to Dove, "The shareholder mix was important in P/I program's success because they had materials, semiconductor, computer, and aerospace companies...a good mix to determine what the needs were from their [the customers'] views." Barry Whalen, P/I Program Director, had industry experience. He understood the needs and motivations of shareholder group and division managers (his customers) much better than do some of the more academic or research-oriented managers in MCC's other programs. Indeed Whalen's researchers and technicians were referred to as the "blue collar" people of MCC—with negative connotations by MCC's researchers in the other program areas.

Risk and cost are important barriers to technology transfer. The consortium's original mission was high-risk, long-term (a 7-to-10 year time-frame) precompetitive research. Such a mandate encouraged cutting-edge research using state-of-the-art equipment. This long-term goal drove staffing and research decisions. As Kay Hammer (a former MCC researcher) stated [4]:

> Much of the first three years of work was conducted on LISP workstations because of the enhanced productivity that the environment provides. There was considerable debate at the time about the choice of this platform since LISP was not a programming language in commercial use and the cost of LISP workstations ($55,000–60,000) made them an unlikely choice for a commercial workstation. However, since the research programs did not expect their work to be complete [or asked for] for years out, the pro-LISP forces argued that inexpensive LISP environments would by that time be available thereby allowing a relatively easy transfer of technology.

MCC researchers were equipped with Sun workstations years before their counterparts in the shareholder companies. While facilitating recruiting quality researchers to MCC, these advanced computer technologies also acted as a barrier to transferring research results. The hardware platforms in most MCC shareholders were DEC, IBM, or Apple computers. Even when MCC-developed technologies fit the hardware and software platforms of a particular shareholder, a lack of agreement about existing platform technologies still existed across the shareholders belonging to the same MCC research program.

The timing of technology transfer is important because of ever-shorter product life-cycles and increasing worldwide competition. Transfer must occur rapidly for a technology to be of maximum use to a receptor. Central to rapid technology transfer is the product planning strategy of the shareholder companies. For MCC it has been a significant challenge to get product planners from the member companies to be committed to MCC-produced technologies. If an MCC technology is transferred after a shareholder company has budgeted for and committed to a different technology, or while the shareholder is preoccupied with other strategic contingencies (such as a financial downturn), little serious consideration is likely to be given to the MCC-produced technology.

Figure 5 shows four stages of idea-to-product formulation. (The arrows represent new ideas or technologies and the solid curved lines represent project conceptualization.) During Stage 1, the research strength of a new idea has the most likelihood of effecting the process of technology development. MCC-developed technologies receive less resistance at the user organization at this stage. During Stage 2, ideas become more focused and walls begin to form around a preferred technology. This sorting-out process is driven by product champions and organization power and politics, as well as by technological and bottom-line considerations. Such forces inhibit technology acceptance, as depicted in Figure 2. During Stage 3, considerable time and effort is invested in building an industrial-strength technology and possibly a prototype (this corresponds to Level III, technology implementation). In Stage 4, the technology hopefully achieves market strength (this corresponds to Level IV). Psychological, professional, financial and strategic switching costs increase for the receptor site of an MCC-developed technology as one moves from Stage 1 to Stage 4.

INTERPERSONAL MOTIVATION

Researchers and managers from MCC and the shareholder companies ranked the ways to improve technology transfer (Figure 6). All respondents generally agreed that it was most important (1) to increase interpersonal communication while focusing on involving the shareholder researchers more with MCC, (2) to increase motivation by rewarding those involved in the technology transfer processes at both the shareholder companies and MCC, and (3) to change the culture or context of the reward structure by increasing awareness of the importance of technology transfer. Executing such cultural and structural changes in MCC and the shareholder companies has proven to be a significant challenge.

Despite the arguments for the importance of motivating technology developers and users to collaborate in the transfer process, such motivation (from Levels I through IV) is not sufficient to achieve technology utilization. While highly-motivated technology transfer participants can overcome passive or infrequent communication, physical, cultural, and strategic distance, and technology barriers, they do not guarantee successful technology transfer. During MCC's formative years, the consortium's scientists were initially elated when they located receptive shareholder-based technology champions. However, they often spent

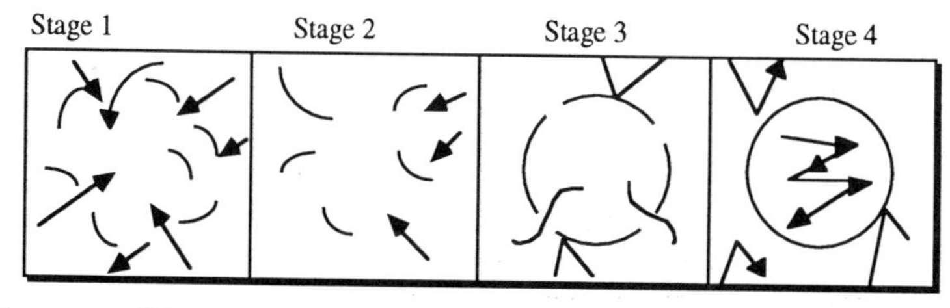

Figure 5. Four stages of idea to product formulation. Arrows represent ideas. Lines represent project institutionalization.

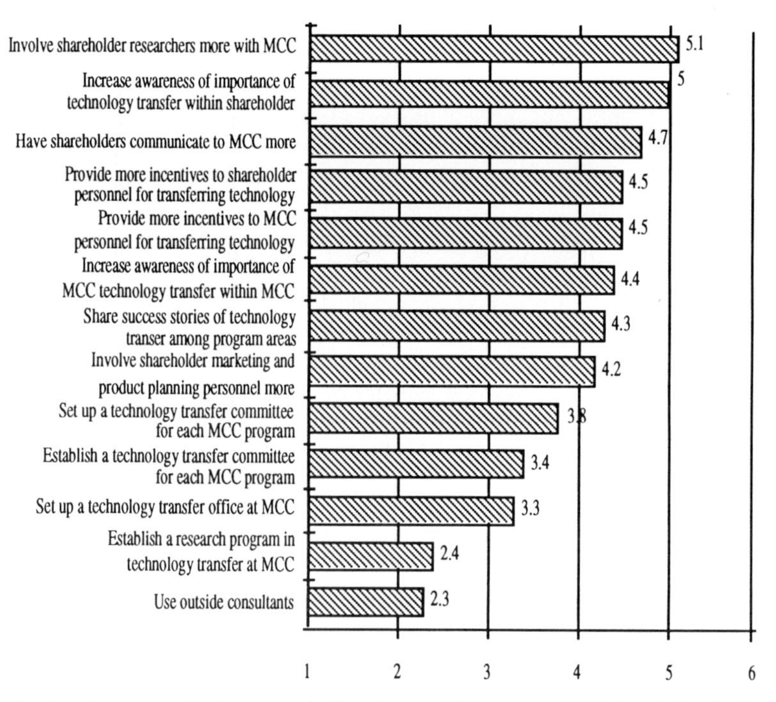

Figure 6. Ways to improve technology transfer based on a 1989 survey of MCC and shareholder company employees.

considerable time and resources transferring technologies to these receptors who, while being extremely motivated, lacked the political and resource clout to implement the MCC technology effectively in their shareholder company.

Successful technology commercialization by MCC and its member companies most frequently occurred in "win-win" situations. Frustrating such collaboration is the reality that the criteria for technology transfer success are often markedly different for research, product development, production, marketing and sales personnel. One well-placed "innovation assassin" can frustrate the successful implementation of such collaborative activity.

MCC and shareholder scientists were hired and rewarded for their research capabilities, not for their skill in technology transfer. Indeed, is it a good use of resources and talent to have highly-qualified researchers spend large amounts of time (in some cases, up to 30 to 50 percent) in technology transfer activities? Laszo Belady [1] remembered how his software researchers were instructed to market their technologies more aggressively:

> This was a new task for the entire staff. We did not anticipate the huge amounts of time, travel, and effort this marketing venture would require. It became a continuous crescendo, culminating in a sizable professional marketing staff and a huge load on all the senior people: almost daily travel and presentations. Yet it was not successful. Most prospects believed they should be able to join [STP] for just a couple of thousand dollars; they compared us to universities with regard to funding. When we created individually supported projects [unbundling and projectization] out of the program, the complexity of marketing increased, and we ended up with meager results. . . We lost many good people who were not willing to spend their creative energies for presentations and salesmanship.

CONCLUSIONS

Technology transfer from the research laboratory to product commercialization is a time-consuming, complex, and difficult process which requires a high degree of collaboration activity. MCC was formed to conduct long-term precompetitive research. Researchers were hired to conduct state-of-the-art research on state-of-the-art research equipment. It was the job of the shareholder companies to convert MCC-produced technologies into products for competitive advantage. This Level 1 technology transfer perspective (see Figure 2) eventually became unsustainable at the consortium.

As shareholder support for MCC was handled by mid-level managers of the funding companies' operating division levels, cost/benefit analyses by shareholder managers were increasingly made in terms of useful technology transferred in a timely manner—Level IV technology transfer. Such bottom-line judgments were made by division managers who funded MCC's research, and who were themselves evaluated in terms of short-term performance.

MCC scientists and managers were caught in a paradox. The more they pursued short-term research and development, the more they moved away from MCC's original motivation. To the degree that this strategy persists, MCC will be in competition for research funding and talent with the same company divisions on which MCC must rely for support. Such competition will be a significant barrier to successful technology transfer.

While precompetitive basic research can have many potential applications, MCC is now charged with transferring product-oriented, short-term technologies to a range of MCC shareholders who have different research and application needs. These diverse needs are expressed by aerospace, computer conglomerate, and semiconductor companies differing in size and in-house R&D capabilities. On the other hand, MCC's diversity of member companies has more recently been viewed as a benefit in terms of the vertical and horizontal integration needed for Level IV technology transfer (product/process commercialization).

MCC-based technology champions are often far removed from the market place. Their technology belongs to MCC's member companies and is handed off to shareholder employees for product development. Consequently, many of the rewards associated with entrepreneurship are not present for MCC scientists or for the shareholder receptors. As MCC moves to a more entrepreneurial culture, these rewards, as well as risks, are being realized by the developers of MCC's technologies. In U.S. entrepreneurial start-ups, active selling of the technology by a champion is crucial.

What motivates a researcher to devote time and energy to transfer a technology so that it is used in a profitable way by a shareholder? What is the reward for shareholder receptor personnel who must devote time and energy to receiving the technology produced by researchers in another organization? MCC's technology transfer success is measured differently by different evaluators. Some shareholder companies have supported the long-term precompetitive mission of MCC, while others want more short-term, usable research results. Some company officials have been content to monitor MCC's research by reading documents and attending shareholder meetings. Others want industrial or market-strength products. Concerning technology transfer, some shareholder personnel see themselves as full partners with the MCC, while others see themselves in competition with the consortium.

REFERENCES

1. Belady, L. (1992). Personal interview.
2. Dove, G. (1991). Personal interview.
3. Fields, C. (1991). Personal interview.
4. Hammer, K. (1991). Personal interview.

5. Inman, B. R. (1992). Personal interview.
6. Inman, B. R. (1987). Personal interview.
7. Inman, B. R. (1982). Personal interview.
8. March, J. G. & Olsen, J. P. (1976) *Ambiguity and choice in organizations.* Bergen, Norway: Universitetsforlaget.
9. Murphy, J. W. (1987). *Cooperative action to achieve competitive strategic objectives: A study of the microelectronics and computer technology corporation.* PhD thesis. Boston: Harvard University Graduate School of Business Administration.
10. Smith, D. K. & Alexander, R. C. (1988). *Fumbling the future: How Xerox invented, then ignored, the first personal computer.* New York: William Morrow.

IV

PRESCRIPTIVE PARADIGMS IN PUBLIC-SECTOR TECHNOLOGY COMMERCIALIZATION

INTRODUCTION

Federal government agencies, as well as state and local governments, not-for-profit organizations and universities, have created programs in every part of the country to encourage individuals and firms in their quest for technical and business assistance. For technology transfer to work, the popular view is that multiple channels and models must be used to put the technology and its user in contact with each other through different media. Technology transfer will succeed when and if innovative methods of furthering technology transfer and commercialization are explored and implemented. Many federal laboratories are working hard to create new incentives for technical innovation by their employees and to establish technology maturation programs to move proprietary technologies closer to the users of that technology. Cooperation with entities such as business and engineering schools, incubators and other intermediaries, is being actively pursued by many government agencies and laboratories. These models render the technology more accessible and add value to intellectual property positions.

In the quest for more technology utilization, we can use the different mechanisms, processes, models and roles that were described in the earlier three sections of this book. Some of these have been used in different contexts with varying degrees of success. At the same time, we need to look for other ways we can move technology into the commercial arena. This occurs in two different ways: first, the improvement of current practices, processes and models, and second, concentration on new paradigms that revolutionize the way we think about the problems and solutions.

The new paradigms should consider the problems and opportunities listed in section I. The problems can be alleviated and the abilities of technology transfer participants improved to take advantage of the creative models. The roles covered in Section II can produce better results with the new paradigms. So can the mechanisms and processes listed in section III. All these issues are amenable to a shift in the way we do technology transfer and commercialization.

In this rapidly evolving environment of technology transfer, many new and exciting opportunities exist for the addition of new services to the traditional technology-push mechanisms. To facilitate this new style of technology transfer and to use new models of the transfer process, we need to apply new innovative management techniques. The identification of novel applications of technology and the use of scientifically feasible, established technical advances require the creative fusion of scientific expertise and technical know-how with industrial partners wanting to solve their problems or increase their opportunities. In the world of computer communications, the cost in time and money of meetings between technical experts and the companies interested in technical advances is too great given alternatives like computer-assisted communication and problem-solving. Interaction between the user of the technology and the technical expert can be improved, in terms both of cost and actual benefits, by the use of many of the models suggested by the authors of the papers presented in this section.

Kozmetsky's *The new role of the federal laboratories* presents the case for the recent move from technology transfer to technology commercialization. It looks at the previous decades from World War II to today in terms of how America dealt with and used technology, and projects what we need to do for the 21st century in terms of the dual missions of the federal laboratories.

Kassicieh and Radosevich present an information systems model in their paper titled *A model of technology transfer: Group decision support systems and electronic meetings.* They suggest the use of electronic meetings and group decision support systems to enhance the interaction of technology sources and technology users without the cost associated with face-to-face meetings requiring extensive time and money commitments. They also list other benefits that could accrue if their systematic approach is used in technology transfer.

Carr's paper, *A proposal for a framework for measuring and evaluating technology transfer from the federal laboratories to industry*, identifies the measurement issues used to evaluate the success of technology transfer. It notes that universities use revenue from intellectual property as the measure of what has been accomplished by their technology transfer efforts. In the federal laboratories, no single measure is sufficient to capture all the issues. Carr suggests using a number of measurements to satisfy the large number of possible types of interactions.

W. O Anderson's paper *Developing effective communications with the national laboratories to commercialize applications*, suggests that increased and better communications between the laboratories and the users of the technologies produced in the laboratories will improve the technology commercialization process. He outlines specific activities to help the laboratories get the information out about their technologies.

Lundquist first paper, *Defining value: Translating from the technical-ese* provides a definition of the problem of technology transfer as being one where the customer does not understand the value of the technology, and suggests that, if customers know the value of the innovation, they are more likely to use it. His second paper, *Tech transfer and the entrepreneurial team: The need for balance*, presents the need for technical entrepreneurs and business entrepreneurs to be on the same team, working to effect a better way to implement technology transfer.

In his paper, *A free market, independent agent model for technology transfer* Bradford presents a new method of thinking about the intermediaries and agents helping the technology source and the technology user reach some form of agreement. He uses the Hollywood talent agent as a model for this intermediary in the technology transfer process.

FURTHER RESEARCH AND AGENDA FOR THE FUTURE

More research in technology commercialization is still needed. It is safe to say that all the areas covered in this book require further investigation. The research can take different forms, but longitudinal studies of the effects of applying laws, regulations, economic incentives and business thinking to technology commercialization are necessary to establish research results leading to better policies and procedures. Policy issues are central to the theme, since government is involved in the transfer of technology in many of its forms. It is important for us to understand which factors lead to success and which lead to failure, so that new policy initiatives incorporate the right incentives and mechanisms. Another area is the study of mechanisms and processes over a considerable time frame to understand the obstacles, market forces, promotional activities, and manufacturing design, and the economics of technology commercialization activities. The measurement of value added by the different intermediaries is important because of its impact on cost and incentives in technology commercialization activities. Finally, creative thinking to shift the paradigm is important for the survival of technology commercialization, given the pressures that it faces from many different constituents.

THE NEW ROLE OF THE FEDERAL LABORATORIES

George Kozmetsky

IC^2 Institute
University of Texas/Austin
Austin, TX 78705

Federal laboratories are of vital importance to America's scientific achievements, national security, and economic well-being. Science and technology, when successfully commercialized, provide opportunities and jobs across the board in research and development, design, manufacturing, customer maintenance and service, financing, and other service sectors, including personal services of all kinds.

Few question that technology has played a significant role in maintaining U.S. security and the position of our nation in a strategic peace-keeping role. In addition, U.S. scientific and research capabilities are important strategic factors in forming new jobs and promoting economic growth. America's strength has always been its ability to be scientifically creative, technologically adept, managerially innovative, and entrepreneurially daring. In these ways, we have met and will continue to meet challenges to promote the common good and support the general welfare.

The opportunities for the coming decade or more are to meet the challenges of a hypercompetitive global economic environment. To prepare for the 21st century, countries, communities, and federal labs, as well as corporations and academic institutions, must implement new strategies that account for changes in globally-competitive market systems. The next round of competition is between different forms of market-driven economies. Those who win this round will be the global leaders of the 21st century.

Transferring R&D is a recent role for federal labs. Technology-transfer policy came of age during the 1980s, with eight acts of Congress and two Presidential memoranda. Their cumulative effect is that federal labs need to alter their ways of conducting business. The message sent by Congress was that it expected federal labs, industry, and universities to collaborate in U.S. technology competitiveness.

The Clinton/Gore policy paper, *Technology for America's Economic Growth: A New Direction to Build Economic Strength*[2] and ARPA's (Advanced Research Programs Agency) *Programs Information Package for Defense Technology Conversion, Reinvestment, and Transition Assistance* [1] have changed the nature of the federal laboratories' role from technology transfer to technology commercialization. In short, I believe that the role of federal labs will be changed from being one-mission oriented to dual-mission directed. The first is their traditional mission—for example, national security. The IC^2 Institute has used the phrase "comprehensive security" to encompass both missions. Located at the University of Texas at

From Lab to Market: Commercialization of Public Sector Technology,
Edited by S. K. Kassicieh and H. R. Radosevich, Plenum Press, New York, 1994

Austin, the institute has done extensive research in the areas of technology transfer models, and has identified at least sixteen transfer models. We have learned that innovative technology-transfer mechanisms alone are insufficient for successful commercialization. Successful technology transfer requires leadership and organizations which link government, business, and academic sectors at the regional or community levels. The community provides talent (entrepreneurs and champions), technology (ideas), capital, and expertise (legal, managerial, sales) linked to a market need. The development of a smart infrastructure requires a synergy among talent, technology, capital and expertise. Existing regionally-based expertise is the key to developing and using "smart" public/private infrastructures. If one of these essential factors is missing, economic development will be slowed or stopped.

The following characteristics emphasize successful technology commercialization:

1. Customers buy benefits, not technology.
2. Market pull focuses the use of technology.
3. Successful transfer is most likely when government- and commercially-developed technologies are joined to meet market needs.
4. Technology transfer must be championed by the producing and receiving institutions as well as by key individuals.
5. Technology transfer must be embedded in the value-added chain process (e.g., Phase 1—R&D and design, Phase 2—manufacturing and Phase 3—production and marketing, sales, and distribution).
6. Technology transfer benefits from using networks that improve access to human, technology, and financial resources.
7. The probability of technology-transfer success is dependent on the quality and accessibility of (a) markets, (b) technology, (c) the management team, and (d) sources of capital.

The most critical gap in technology transfer from federal laboratories to the commercial sector is what IC^2 labeled "networking know-how", linking talent (entrepreneur/champions), technology and capital with market pull for improved U.S. industrial competitiveness. Networking know-how is the ability to leverage business or scientific knowledge in new and expanding enterprises. It is the ability to find and apply expertise, in a variety of areas central to technology commercialization, that can make the difference between success and failure. This expertise involves experience in business, manufacturing, market research, finance, distribution, sales, technologies and management. Networking know-how is a human dimension which involves person-to-person contact with conceptual understanding and shared information based upon knowledge and experience.

Federal laboratories can adjust to fill the networking gap, but it will require a willingness to experiment with new processes and organizational structures. Equitable partnerships with industry, financial institutions and universities will be required. Most important of all will be successful models that work in the short run at the community level, particularly where the laboratories are located.

RETROSPECTIVE AND PROSPECTIVE DEVELOPMENTS

Dual missions for federal labs will be a major transformation from the Cold War to competitive global markets. This is more than a cultural change; it is a fundamental change in the goals, missions and operations of federal labs. In every decade since World War II, our nation has had major transformations which, in turn, have had an impact on the role of federal laboratories.

The decade of the 1940s experienced a euphoric state of power and unsurpassed world leadership in the late 1940s. Average Americans were reinforced in the belief that, given opportunities and means, they could succeed by performing to the best of their individual abilities. Federal laboratories attracted many of the young scientists and engineering veterans of World War II. There were few civilian positions that required their wartime expertise.

In the 1950s, basic industry put people to work and achieved employment levels wherein a 4% unemployment rate was considered high. Increased emphasis on individual worker security was directly related to ever-expanding Social Security benefit laws, workmen's compensation packages, and private corporate health and pension plans.

If the times were good, what were America's fears? Politically, the major fear was one of communism at home and abroad. Economically, the only fear was whether we could provide enough human resources to meet expanding basic and technological markets. The importance of federal labs increased to meet America's fears abroad. "Guns and butter" was the national policy.

The 1960s were a decade of divisiveness. Trust in all our institutions declined, as traditional ideals that had held us together as a nation were challenged. But even in the midst of these turbulent times, three key American perceptions were reinforced:

1. All people were equal under the law.
2. American abundance was limitless for foreign and domestic needs.
3. The nation was a world leader in terms of marketing and communications.

The decade of the 1960s brought about a significant shift in federal lab R&D expenditures in constant dollar terms. The federal lab budgets were in effect "frozen" in real terms, and lab directors had to bargain hard for cost-of-living increases. High-technology commercial industry, however, was poised for growth and no longer dependent on continued federal spin-off support. Traditional financial institutions and the beginnings of the venture capital industry filled any financial gaps.

In the 1970s, the international position of the United States changed. The early '70s saw the impact on American society of the revitalization of Europe and Asia. The first indications were that the dollar as a standard could not endure. The state of American society was heavily transformed through international trade and political alliances. For the first time since 1890 the U.S. encountered an unfavorable balance of trade. This resulted in a run on the dollar at home and the build-up of dollar reserves overseas to such an extent that the financing of American industry was changed for decades to come.

For the first time in our history, we had stagflation—increased prices with increased unemployment. American business entities found themselves in competition with the government sector for funds. In this decade, there was declining support for federal laboratories, as the U.S. began to lose its consumer technology markets. Increased technology activities in computers software, energy, and biotechnology developments came from venture capital investments. This was the beginning of domestic dual technologies financed by an expanding venture capital industry.

The decade of the 1980s began with the great inflation. It saw the federal deficits increase to dimensions never before seen in U.S. history. The federal deficits for the decade of the 1950s amounted to $15 billion; in the 1960s, they increased to $63 billion; during the 1970s, they increased to $420 billion; and by the end of the 1980s, these were $4 trillion!

The 1980s were also a period of unprecedented entrepreneurial growth, as well as of small businesses. Owning their own businesses was a way for women to burst through the glass ceiling. Technology entrepreneurship, science park development, and technology incubators grew around the fifty or so U.S. research universities and their surrounding communities. However, this growth was tempered by a drop in employment opportunities for the Fortune 500 and other large companies. The drop in employment and restructuring of managerial changes in the Fortune 25 are historical precedents. It surprises many Americans that, during the 1980s and early 1990s, women entrepreneurs provided more employment than all of the Fortune 500 companies added together. This decade saw the growth of federal laboratory heavy investment in "smart" weapon system technology that ended with the U.S. cold-war victory. However, the federal labs did not participate in the growth of technology entrepreneurship, science park development, and technology incubators; these were innovations that concentrated on commercial developments. On the other hand, federal lab weapons' life cycles increased in time and decreased in costs and investments. The venture capital expansion during the 80s was one of lower returns than government bond yields.

Retrospectively, the roles of federal labs through the past five decades were viewed from their primary mission roles. When there were prolonged periods of international stalemate, real budgets were fixed. When the Cold War heated, budget allocations were extensively increased. In the early decades, the federal labs attracted the younger scientists and engineers; in the later decades, employment levels were frozen. As a result, there are pressures for cultural change to meet new expectations based on global economic competitiveness and commercialization transfer requirements of the 1990s.

PROSPECTIVE DEVELOPMENTS

The 1990s have already identified at least six major global trends that will affect future economic growth:

1. *Expansion of market-driven economies.* There is a worldwide shift from planned national economies to providing market opportunities where none were perceived a short time ago.
2. *Worldwide movement toward democracy and individual freedom.* The trend is toward entrepreneurial economics.
3. *Creation of new infrastructures.* Adequate infrastructures are necessary where a nation's economy is dependent on economical easy and rapid access to geographic areas on the one hand, and instantaneous communication on the other. Transportation and telecommunications need to be able to interact with smart facilities, value-added networks, and knowledge-creating institutions.
4. *Emergence of revolutionary technologies.* Revolutionary technologies are creating new and emerging industries that will be critical to the economic viability of the U.S. and other nations.
5. *Accountability developments.* New constraints on environment health, and energy are fostering a new sensitivity to and respect for the accountable use of technology.
6. *Emerging strategic regions.* Strategic economic regions, at present and in the future, will require multiple global relationships for trade, strategic security, and economic alliances. Each region is the city-state of the future, the economic growth of which will depend on technology commercialization. These regions are also important markets for advanced commercialization technologies.

Perspectives for the 1990s will have an impact on federal labs, as the U.S. transforms itself from a cold-war to a globally competitive economy. The overarching goal for federal labs in the 1990s and the 21st century must be one of shared prosperity domestically and abroad. Shared prosperity, in turn, requires that federal labs consider their dual missions—one for military security and the other for economic security. Such a major shift will require a major reexamination of the federal labs' impact on the domestic economy in a global context.

FEDERAL LABS DUAL MISSION ROLE

As stated earlier, a dual mission involving national and economic security will be the new role for federal laboratories. Conversion that integrates the ability to provide advanced, affordable, military systems and world-class competitive commercial products/services is difficult and complex, more so than the transfer of federal lab technology to the market place in the 1980s. Bureaucratic infighting and decision-making will be sharper in the near and mid-term, i.e., three to five years. Disputes over patents and intellectual property rights will become increasingly complex. The federal labs' task in justifying their R&D efforts, besides the primary missions mandated by Congress and the Administration, is not simply either picking the right project or choosing the right entities to collaborate with.

Federal laboratories need to pick the right projects to serve dual missions. This can be done, as set forth in ARPA's program information package, which outlines eleven broad areas for national and economic security:

1. Information infrastructure. Command, control, intelligence manufacturing, health care, education and environmental monitoring, etc.
2. *Electronic design and manufacturing.* Affordable custom electronic components, multi-chip integration packages, and novel optoelectronic module technologies and manufacturing.
3. *Mechanical design and manufacturing.* Advanced information support for design and manufacturing, flexible robotic systems integrating R&D design with manufacturing process.
4. *Materials/structures manufacturing.* Broaden military use of advanced materials extended to civilian use.
5. *Health care technology.* Health care information systems and trauma cases.
6. *Training/instruction technologies.* Digital libraries and authoring tools.
7. *Environmental technology.* Environmentally conscious electronic systems manufacturing, environmental monitors.
8. *Aeronautical technologies.* Propulsion/engine technology, fly-by-light, structures and aircraft design.
9. *Vehicle technology.* Alternative power sources, sensor & electronics for vehicle systems, and vehicle integration.
10. *Shipbuilding industrial infrastructures.* Innovative ship design and construction, propulsion and auxiliary systems.
11. *Advanced battery technology*

The federal labs' role will be more and more to initiate such programs as part of their mission planning. They will also need to coordinate their primary mission plans more closely with their commercial alliance partners. Which mission market, national or economic, is met first will affect the investment cost and operations of each. Economic security, market-based on advanced technology, will in many cases mean short-developed, short-lived cycles, while national security markets can be longer-term. Consequently national security will depend on longer-term, financially stable, commercial firms. Otherwise, federal labs might need to provide for their own production, maintenance, and service. In any event, they will need to develop better methods and tools for successful dual missions. More than ever before, federal labs need creative and innovative management.

There are possible short-term changes for federal labs. In the author's opinion, the labs must take the necessary time to set priorities to adapt to dual changes as they go through the transition from technology transfer to commercialization. Pressing current needs include employment across all skill levels, meeting foreign competition effectively, and increased productivity to help reduce federal deficits. All these factors point to the fact that the federal labs' role must be defined in short-term successes. The labs need to develop metrics for all stakeholders. In short, federal labs cannot judge proposals solely in terms of technological competence and the best price. They must also consider how to commercialize in the short term while justifying the government's investment in terms of value-added national activities as well as intellectual property asset utilization.

The role of federal laboratories will become increasingly focused on management for change. The labs will increasingly become catalyst organizations focused on linking talent, networks, and scientific and technological knowledge to identify critical issues, leverage resources, perceive markets, and continually improve organizational capabilities. Federal labs will play a role in bringing together and leveraging other institutions to help accelerate the identification of market opportunities together with plans and solutions concurrently coordinated in dual missions. This networking will help participating institutions take action faster, become smarter, and be more global in their thinking, structuring, and operating for technology commercialization.

REFERENCES

1. ARPA (1993, March 10) *Program information package for defense technology conversion, reinvestment, and transition assistance.*
2. Clinton, W. J. & Gore, A., Jr. (1993, February 22) *Technology for America's economic growth: A new direction to build economic strengths* p. 1.

MANAGING THE BUSINESS KNOWLEDGE PROCESS

Dennis M. Hogan

Enterprise Systems Consultant
Wexford, PA 15090

INTRODUCTION

Businesses today are moving through some of the most exciting times in history. The business environment today has been characterized as similar to operating in white water, with world-wide movement [1] toward free markets and reduced trade barriers. Today all facets of business can be global—product research & development, manufacturing, finances, marketing, sales distribution and competition.

There has been a great deal written about the new business paradigm, but what exactly has changed? Businesses have always had competition and many have been global for hundreds of years. They had already learned to move manufacturing to lower-cost labor markets, develop products off-shore, finance business in foreign lands and sell wherever the markets were located.

What is the difference in today's paradigm? Everyone generally agrees it is knowledge explosion and time implosion. [2] However, both these attributes have also been changing in this manner for the last thousand years. Knowledge growth over time is represented by an exponential curve. The curve is not smooth, but an integration of incremental advances or knowledge gains, some increments having more impact on society than others.

Throughout history each increment has had initially the same effect on people—anxiety. The introduction of an advance, be it a shorter land route telegraph or computers, always passes through a clearly defined cycle [3] from resistance to acceptance and finally integration into society and business.

If the mathematical model for knowledge growth is an exponential curve, the increments that are being assimilated into society must be becoming progressively shorter. In earlier times, new knowledge took many generations to be assimilated. Therefore, an advance occurring in your lifetime might not be placed in common use until you were deceased. Today, however, advances are assimilated in a few decades or even a matter of years. This is part of the time implosion.

In addition, the rate of parallel or simultaneous advances has increased. Hundreds of years ago, advances were assimilated as individual occurrences. Today, however, they are occurring simultaneously and spawning new advances through their interaction, building on each other. This is the knowledge explosion.

From Lab to Market: Commercialization of Public Sector Technology,
Edited by S. K. Kassicieh and H. R. Radosevich, Plenum Press, New York, 1994

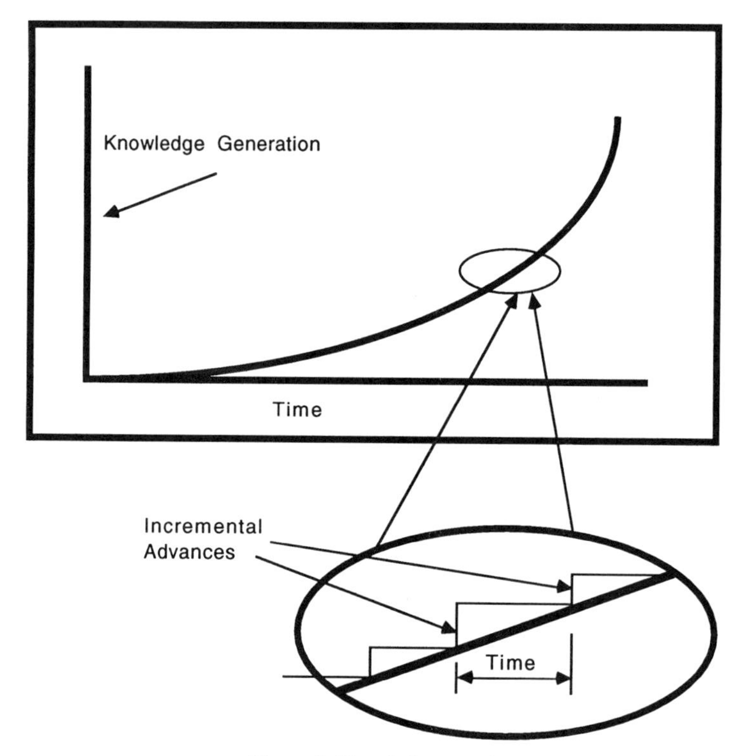

Figure 1. Knowledge growth.

The constant along this line of change—what is generating the change and why it is being generated—is people. People, over time, have progressively adapted to higher and higher rates of change. The cycle of resistance to acceptance and final integration [4] has become shorter.

Unfortunately, the rate at which people can learn to adapt to change, without being placed under severe or life-threatening pressures, isn't moving as fast as knowledge growth. Therefore, the rate of transfer and assimilation of knowledge today depends mainly on the "people variable." Things such as distance, language, and communication have become insignificant compared to people-things: culture, willingness to change, common vision or goal, the ability to establish trusting and value-adding relationships, etc. Knowledge transfer has always been a human thing. It is a social event, a sharing between people.

What does this mean to business? It means that if generating, accessing, disseminating and applying knowledge for the good of the enterprise is critical, it needs to be real-time; it is a business process that must be managed. This business process heavily involves people. Three specific types of processes [5] to be managed are:

- *Concept to market.* Creation and transfer of new concepts from research to the market place.
- *Proven solutions.* Accessing and transfer of proven solutions horizontally across the enterprise.
- *Individual to individual.* Identification and transfer of knowledge assets between people: expert to apprentice, teacher to student and retiree to new employee.

A business needs to manage all three of these processes to optimize performance. There are a number of corporations doing a good job of knowledge management. However, I don't believe any one of them would grade itself 100% in each area.

Also, success looks different at each company. This is because each company customizes its approach. It lives within a unique cultural context and operates with various types of management styles within different organizational structures. It is, therefore, hard to recognize common themes in different corporations.

After looking at a number of companies, and an equally large number of management styles, it becomes obvious there is no silver bullet. However companies which have robustly used knowledge did have some things in common. Those companies openly and formally dealt with knowledge management. They then addressed it by creating processes, building organizational structures, dedicating resources and articulating positions which created a perception in all their employees of how the leaders valued and expected knowledge to be managed. These actions broadly fell into the following areas:

- *Culture.* Shared norms as well as shared ideologies, values, attitudes beliefs and assumptions.
- *Vision.* Long term general purpose for our existence as an organization.
- *Strategy.* Long term goals and decisions concerning the means to achieve those goals.
- *Organizational structure.* Formal reporting relationships, role definition and accountability, the allocation of tasks, groupings of individuals and structural systems for communication, coordination and integration of activities.
- *Management commitment to knowledge transfer.* The degree to which management promotes and assists knowledge transfer.
- *Knowledge asset management.* The degree to which knowledge, currently existing both inside and outside the enterprise, is valued and used effectively. This includes the effective management of stored knowledge in the organization.
- *Knowledge transfer tools/systems.* Formal and informal systems and tools facilitating knowledge transfer, such as computing infrastructure, video conferencing, travel and group seminars.
- *Human resource management.* Practices in the human resource area that support business performance and, at least indirectly, knowledge management.
- *Innovation.* The degree to which creativity and positive change are valued and supported, the willingness of the organization to adapt and to try new ideas on a continuing basis.
- *External relationships/interfaces.* Boundary-spanning activities encouraging knowledge transfer with customers, suppliers, consultants, government labs, joint venture partners, foreign government labs, foreign governments and academic institutions.

Reviews of these areas indicate that knowledge management is a system. It is made up of very specific processes which can be rearranged, called different things, and be stronger or weaker in various companies, but nonetheless these processes are interdependent and critical to a healthy, sustainable knowledge management system.

INFRASTRUCTURE

I have broken the knowledge management system into three modules. I refer to the modules as infrastructure to emphasize the fact that, just like capital assets, these take time to build and must be maintained. Successful companies don't treat them as projects or programs, but as part of how they do business. There are three types of infrastructure, [6] each with a different primary mission, but interdependent. These missions are:

- To provide for the generation, access, dissemination and application of knowledge throughout the enterprise.
- To provide the business processes to operate the enterprise effectively in the present and position it for the future.
- To provide tools for monitoring, measuring and correcting the enterprise and knowledge transfer systems.

The First Module of Infrastructure

This module deals with those processes that have a primary mission for knowledge generation and transfer. The knowledge system infrastructure can be summarized as follows:

- *Network structure.* Formalized human network, internal and external to the business. The design of the network depends on the needs of the business. It provides known routes and people through which specific types of knowledge flow.
- *Network manager.* Influential person assigned the responsibility of worrying about the knowledge transfer system. This is the individual who is concerned about the processes, efficiencies and future strategy for meeting business needs in this area. He is responsible for removing parts no longer needed and adding parts to meet future needs.
- *Network model.* Formal model of how the network is to work. It is understood by everyone and evolves as the needs change. It may be in several forms, but it serves the purpose of conceptual road map.
- *Networking tools.* Enablers that allow a network to function in a user-friendly manner. These can be educational seminars, electronic information tools, travel budgets, administrative support, team rooms, filing cabinets, etc. These are tailored to the system and culture.
- *Network policies and procedures.* Networks need to be managed and operated in an efficient manner. This can be done only if everyone understands how a network is to be operated and agrees to play by the rules.
- *Formal methodologies for generation and transfer.* Particular types of generation and transfer extend over significant time periods and involve many people from many functions within the business. For these complicated activities, businesses need to use established methodologies or develop their own methodologies.

The first infrastructure is the core of the knowledge management system. However, it will wither away if the enterprise infrastructure is not there to provide a nurturing environment. People are the key to knowledge transfer and people can not sustain this activity in a poor environment. If the personal needs of people are not met, the people will start to focus on them instead of what is needed for the enterprise.

The Second Module of Infrastructure

The enterprise infrastructure tends to feel softer because its descriptors are words we have all used and heard before. It is not soft; it is based on more than forty years of continuing research regarding how businesses can best be operated. Research on how to run a business [7] lives on and builds through consultants' silver bullets and this year's fad. It continues to survive the short-sighted manager who can only rationalize a business, but cannot make it grow. It will be interesting to see if it can survive the institutional investor. The components of the enterprise infrastructure are:

- Culture
- Living vision
- Business strategy
- Plan for implementation
- Organizational design
- Management involvement
- Tools and systems
- People management
- Innovation/creativity
- Understanding corp.-self
- Knowledge asset
- External relationships

There is not room here for further expansion of the enterprise infrastructure; suffice it to say that when further decomposed, the interrelationships can be shown to exist.

The Third Module of Infrastructure

This involves an internalized set of measurements and empowered corrective processes. Monitoring the quality of knowledge management activity requires attention to a set of key indicators—conditions or behaviors enabling and/or supporting knowledge management. The presence of these indicators across several categories suggests the level of knowledge management taking place within an organization.

- *Culture.* Provides a social energy that guides peoples' daily behavior.
- *Vision.* Provides direction and meaning for our strategy, culture, and structure.
- *Strategy.* Provides a road map for everyone to follow.
- *Organizational structure.* Provides the context and rules for daily activities.
- *Management commitment to KT.* Provides the incentive and empowerment for the organization.
- *KT tools/systems.* Provide the basic tool kit to do the job.
- *Human resource management.* Meets the personal needs of the employees and allows focus on the enterprise's needs.
- *Innovation.* Addresses whether it is a learning and thinking organization.
- *Knowledge asset management.* Looks at the amount of the knowledge in the bank, its liquidity and future revenue potential.
- *External relationships/interfaces.* The degree to which the business unit is willing to look outward in acquisition and transfer of knowledge.

Although they are defined individually, key indicators are a tightly interlinked and mutually dependent system. This means that an entity can only achieve excellence in knowledge management through simultaneous movement of all key indicators. Similarly, no individual group of key indicators can move without affecting the others. Ideally, this can build pressure for collateral improvement in all categories. However, the opposite effect is also possible, where motionless or slow-moving indicators check the advance of their more vigorous counterparts. For these reasons, it is crucial that managers accountable for moving key indicators should coordinate their efforts.

The key indicators can be further decomposed to characterize conditions, behaviors, systems and other critical features which have been observed in operations displaying world-class knowledge-management capability.

It is now possible to use these indicators in a survey or audit system to provide information regarding organizational potential or performance in managing knowledge. The following uses can be made of the survey and audit tools:

Survey

- Assess patterns in management across multiple businesses within one corporation (distribution or gap analysis).
- Assess patterns across various levels of businesses (distribution or gap analysis).
- Assess variations in geographical locations domestically or worldwide (distribution or gap analysis).
- Comparisons between corporations,i.e., best in class competitors, suppliers, customers (distribution or gap analysis).

Audit

- Third party evaluation of KTI's in scenarios listed above (provides specific recommendations regarding areas of improvement for management consideration).
- Trained internal audit group, for continuous improvement of KTI's.

There are many tools available and each business must customize tools to provide the appropriate measurement of its knowledge management system.

SUMMARY

Businesses must manage knowledge processes as they do other business processes. They need to understand that people are the key component of the system. Focusing on only one of the infrastructures will discount the value of the desired outcomes. As businesses move toward new paradigms, the transition [8] must be managed. Transitions will still be difficult and complicated, but arduous rebuilding of infrastructure can largely be avoided.

REFERENCES

1. McGowan, W. G. (1991). Revolution in real time. *Harvard Business Review.*
2. *Technology in world civilization* (1990). Cambridge MA: MIT Press.
3. Badaracco, J. L. (1991). *The knowledge link.* Boston: Harvard Business School.
4. *Managing organizational change* (1990). Atlanta: ODR Inc.
5. *Knowledge transfer* (1991). Aluminum Company of America internal report.
6. Kaiser and Associates (1990). Benchmarking the engineering function. Aluminum Company of America internal report.
7. Rubenstein, A. H. (1989). *Managing technology in the decentralized firm.* New York: Wiley-Interscience.
8. Terez, T. (1990). *Managing change in the 1990's.* Charlotte, NC: Arrow Associates.

A MODEL OF TECHNOLOGY TRANSFER

Using Group Decision Support Systems and Electronic Meetings

Suleiman Kassicieh and Raymond Radosevich

R. O. Anderson Schools of Management
University of New Mexico
Albuquerque, NM 87131

INTRODUCTION

Improved utilization of federal technology by private enterprises in the U.S., and by universities and state and local governments as intermediaries in the technology transfer and commercialization process, has become an essential element in plans to improve the U.S. position in international economic competition. American technology represents an underutilized resource which can be harnessed to improve the productivity and innovativeness of industry if effective relationships are developed between technology sources and potential users. A sophisticated system of technology development and transfer entities in not-for-profit organizations and universities has created an infrastructure in most regions and states which provides assistance to individuals and firms seeking technical and business assistance. However, most of the elements in this system have evolved within the last few years and the ultimate impact will not be evident for a while even if proper evaluation is performed. Technology transfer is a contact sport in which success has been elusive and innovative methods need to be explored.

There is a variety of programs under way. Many federal laboratories are creating new incentives for technical innovation by their employees and are establishing technology maturation programs to move proprietary technologies closer to the stage at which receptor adoption is easily performed. In addition to adding value to a proprietary intellectual-property position, commercial packaging of technology through cooperation with business schools, incubators, and other intermediaries is being actively pursued by many government agencies and laboratories. In this rapidly evolving environment of technology transfer, many new and exciting opportunities exist for the addition of new services to the traditional technology-push mechanisms.

As the environment for technology transfer becomes more facilitative, new transfer methods will demand the application of new management technologies. The identification of novel technology applications requires the creative synthesis of technological advancements and industrial problems and opportunities. Expensive travel and face-to-face meetings of experts from technology sources and potential industrial users must be supplemented with cost-effective interaction such as computer-assisted communication and problem solving. The

From Lab to Market: Commercialization of Public Sector Technology,
Edited by S. K. Kassicieh and H. R. Radosevich, Plenum Press, New York, 1994

processes of source/user interaction can be improved by using new information technology approaches. These approaches build on the idea that information systems must support the organizational mission. In the technology transfer and commercialization arena, the availability of information and its dissemination to the right parties within a reasonable time frame is imperative for the success of the efforts. This paper explores some opportunities in the area of information technology.

There has been some progress by federal agencies in the use of information technology to support technology transfer activities. The Environmental Protection Agency, the National Institutes of Health, and the Department of Energy's Office of Scientific and Technical Information have established computer bulletin boards from which interested parties can determine the availability of patents in specific areas. The intent of these systems is to provide an electronic handshake between the technology source and the potential recipient. Our proposed model goes farther by providing the means to support interactions that occur after the handshake stage.

This paper points out the problems of technology transfer, defines and explains the properties of group-decision support systems and electronic meetings, and applies these information-technology techniques to the transfer problems, suggesting a design for such a system.

PROBLEMS IN TECHNOLOGY TRANSFER

Several authors have indicated that the success of technology transfer endeavors has been an elusive goal. Goodman [1990] suggests, as a way to encourage technology transfer, the use of academic-industrial partnerships in which research is funded by the government if it is matched by industrial partners. Gomory [1989] indicates that IBM spent twenty years perfecting an internal technology transfer mechanism, called "joint programs", in which researchers and development people form teams, agree on plans and work together to achieve them.

Cooperative research and development agreements (CRADAs) are touted by Rivers [1989] as a way for the government to provide facilities, equipment and labor and for the industrial partner to provide other resources, such as personnel, to encourage technology transfer from federal labs to industry. Dorf and Worthington [1989] maintain that the sources of technological advancements such as federal laboratories, are not charged with finding the needs of the users. Because of differences in the labs' culture, history and orientation, different facilitation models are needed to improve communication between labs and users. Seven such models are: the information dissemination, licensing, venture capital, large company-joint venture, incubator-science park, ferret and agricultural extension models.

Madu [1990] suggests that technology-transfer failures can be attributed to the structure of the stakeholders' institutions, the effects of culture and motivation, and the appropriateness of the technology. He suggests monitoring the process of technology transfer to ensure its success. Knox and Denison [1990] maintain that technology transfer has three activities: idea initiation, development and commercialization. The interaction between these stages, especially between idea initiation and development, should be participative between the research-and-development organization and the manufacturer.

Baron [1990] suggests that the failure of technology transfer is explained by cultural differences between federal laboratories and industry. Labs do research for its own sake, whereas industry looks at research as a goal-oriented project with some return on investment. Watkins [1989] supports the view that dual-use technologies have not thrived because of differences of culture and structure between civilian and military systems.

It is obvious that, to increase the success of these endeavors, the U.S. needs to bridge the gaps of structure, culture, motivation and communications that exist between participants in technology transfer activities. Information systems can reduce some of the differences between R&D organizations and industry. They also can track the performance of some of the strategies used so that they can be evaluated and modified if necessary to improve the process of technology transfer and commercialization.

GROUP DECISION SUPPORT SYSTEMS

Group decision support systems (GDSS) are designed to support the group decision-making process. DeSanctis and Gallupe [1987] indicate that there are three GDSS levels. The first level removes common communication barriers found in groups by using large screens upon which to display ideas, a program which solicits and tallies group member votes and electronic message exchanges between group members. The second level system tries to improve decision-making, as well as communication. In addition to the communication aids in the first level system, it provides a base for decision-making models and group-structuring techniques such as the Nominal Group Technique discussed by Lewis [1987]. These techniques lead a group through a structured process to a decision and consist of six steps: 1) silent individual generation of lists of items, 2) combining lists, 3) discussion of items, 4) modification of lists, 5) individual ranking of items and 6) merger of rankings. The purpose is to improve the quality of the decision.

The third level GDSS imposes a communication pattern on a group during a group meeting and acts as an expert by helping a group choose appropriate rules for a given decision-making situation. It is more theoretical in nature. Liang [1988] indicates that the main ideas in GDSS are data- and information-sharing capabilities, modeling, negotiation and, to some extent intelligent support.

Watson, DeSanctis and Poole [1988] indicate that groups encounter a "process loss" because of disorganized activity, member dominance, social pressure and inhibition of expression. Jelassi and Beauclair [1987] describe several examples of behavioral issues which affect group decision making: diffusion of responsibility, pressure toward consensus, problems of coordination and "deindividuation". Deindividuation is a situation where decision makers lose a sense of themselves as consequential participants in a process, leading to irrational behavior. Extreme examples of deindividuation are lynch mobs and mass hysteria. A GDSS can reduce diffusion of responsibility and deindividuation by recording and displaying the inputs, voting record and name of each participant. A GDSS can lower pressure toward consensus by allowing anonymous input into the system and providing a forum in which all viewpoints are examined. A GDSS can help minimize coordination problems by recommending an agenda and structuring the process for the group to follow.

Software for GDSS

PLEXSYS is a software package developed for GDSS and used in the decision rooms where these meetings are held. Nunamaker et al. [1988] describe the package as supporting the following activities:

1. *Session director*. This tool is used by the facilitator or group leader to choose software used and to produce an agenda for the session.
2. *Electronic brainstorming*. Used as a support for idea generation this enables members to share comments simultaneously and anonymously.
3. *Issue analyzer*. This helps group members choose the items they want to focus on from a list generated during the idea-generation phase.

4. *Voting.* A voting tool provides a variety of ways in which items can be prioritized. All members cast private ballots and accumulated results are displayed on a public screen.

5. *Policy formation.* To help the group develop a mission or policy statement, this tool uses an iterative process until consensus is reached.

6. *Stakeholder identification and assumption surfacing.* This tool systematically evaluates the implications of a proposed policy or plan. Stakeholders' assumptions are identified, scaled, and graphically analyzed.

PLEXSYS is an integration of models, a database management system, and a user interface supporting decision makers in a decision environment (see Fig. 1). The system is able to handle a large variety of situations. It recommends an approach to follow for a group session, learns about each session, and incorporates this knowledge for future recommendations.

Geographically Dispersed GDSS

Zigurs, Poole, and DeSanctis [1988] discuss additional environments which GDSS can support. A number of small groups may meet simultaneously in separate locations using a network environment to duplicate the decision-room environment. Participants at each meeting site have access to the same technology found in the decision rooms where GDSS are used, with the addition of a network to support electronic, voice, and video communication among the different sites.

GDSS can also be used to support meetings of individuals dispersed at different sites. Individuals working in their own offices can communicate with one another using a local area network. In most cases, this type of group uses electronic communication rather than video or voice channels. The cost of video and voice communication is high, causing electronic communication to be heavily relied on. It is important to note that, in this type of environment, group members have the option of meeting at the same time or working asynchronously.

Both a decision-room environment and a network environment can support varying numbers of participants; the number of participants will affect the type of communication support needed.

GDSS Effects

Since the use of GDSS intervenes in the group decision-making process, researchers hypothesize that group behavior and outcomes will be changed through its use. GDSS seeks to minimize negative group behavior by, for example, mitigating cultural differences and

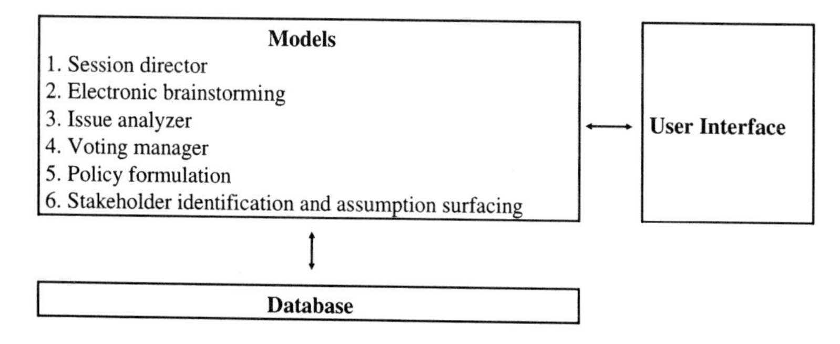

Figure 1. Decision-room components: A decision-room GDSS model.

encouraging positive interaction. Researchers are trying to determine if and how GDSS can accomplish this goal. GDSS research began in the early 1980s, and a major part of the work has been undertaken only during the last five years. For this reason, the impact of the technology is not well understood. The current hypothesis is that the technology has an impact on many aspects of group process and on the outcomes of group meetings. Dennis et al. [1988] and Jarvenpaa, Rao and Huber [1988] listed seven variables which they expect to be affected by the implementation of a GDSS:

1. *Decision quality.* Generally, decision quality is expected to be better when a GDSS is used.
2. *Time to decision.* Groups using a GDSS may spend more or less time deliberating than a group using conventional methods. When more time is used, the benefits accruing from using GDSS might offset the extra time used.
3. *Participation.* In general, researchers believe that a GDSS encourages more equal participation and influence within a group than are found in groups working unsupported.
4. *Consensus.* Since a GDSS is expected to encourage equal participation, it is suggested that it can also help groups reach consensus.
5. *Inhibition.* Comments can be made anonymously, reducing members' inhibitions.
6. *Equity.* GDSS is expected to increase the level of equity within groups where the content of the message, rather than the status of the participant, is the issue.
7. *Satisfaction with process and outcomes.* Increased equality of participation and increased equity should lead to increased group member satisfaction and therefore to repeat interactions.

GDSS has the features needed to address a number of problems faced by technology transfer. In the next section, we will apply those techniques to technology-transfer mechanisms to improve the results.

Table 1. GDSS/TT Activities and Techniques for Stages of Technology Transfer

GDSS levels and activities	Technology transfer stages	Techniques
Level 1 Idea display Solicitation of views and votes Exchange of ideas	Idea initiation	• Bulletin board to introduce parties, list needs and problems, and provide information to interested parties
Level 2 Decision making models and group-structuring	Development	• Development of goals for cooperative scope of work • Exchange of intermediate results • Redefinition of needs or discoveries • Staged revelation of proprietary information • Negotiated scope of work and resource commitment • Project review and revisions
Level 3 Expertise to help group in making decisions	Commercialization	• Manufacturing process design (CAD/CAM) • Competitor technology assessment • Product technology improvements • Product value analysis • Quality assessment

INFORMATION-TECHNOLOGY INNOVATIONS FOR TECHNOLOGY TRANSFER

Technology transfer consists of the basic activities of creating an invention and incorporating the resulting technology into a product, process, service or materials innovation. The large number of inventions that do not result in such innovations suggests that an opportunity to improve the process exists especially when multiple organizations or geographical locations are involved. Part of the problem can be attributed to ineffective information exchange among the innovators, potential users, and intermediaries that may have been engaged in the process.

Several phenomena which significantly influence the process exist in the information exchange of technology transfer:

- A large body of data exists in raw forms, such as abstracts written by the innovators, usually at a level that cannot be easily understood by the majority of readers. These data are not useful unless they have a high information (as opposed to data) content. Information-systems experts suggest transforming the data into information that is useful to the target audience. This requires elaboration, aggregation, dialogue, brainstorming and other techniques.
- Some of the information is perceived as proprietary or classified for security reasons, making it inaccessible to industrial organizations or the general public without substantial bureaucratic delay.
- Researchers and commercialization managers have not, in general, participated in the exchange of information. They frequently consider these activities a waste of time. This perception is caused, in part, by the small number (and percentage) of successes achieved and publicized.
- The exchange of information necessary for innovations to be adopted and adapted by technology receptors and commercialization experts is not a one-time activity. There is strong evidence that extensive discussions are needed between the people who innovate and the people who incorporate innovations into the products. This dialogue also improves communication between the two groups in that they learn to speak the same language more easily. Without extensive discussions, there is a lack of understanding between the two groups which is frustrating and can result in the withdrawal of many participants from all such activities and the eventual characterization of the effort as a failure.

There are certainly other phenomena, but those cited above should provide the reader a sense of the complexity of the communication process in technology transfer. To increase the success rate of technology transfer, improved management of the information exchange is often needed. Since researchers and product managers are often geographically dispersed and place a high value on their time, the application of group decision support systems and electronic meetings represents a significant opportunity to improve the technology-transfer process.

GDSS are designed to support group decision-making in complex business areas where group deliberation leads to more effective decision making, through the use of large idea screens, group voting and an electronic message exchange between group members. The emphasis is on communications, negotiation and intelligent support. Electronic meetings offer all the advantages of attending a meeting while being physically in a different locale.

In the technology transfer context, electronic meetings of technology source representatives and commercialization managers can improve the transfer process through the following proposed procedures:

- Determination of industrial needs through market research and by panels of experts from the technology and potential application areas who can discuss issues in a GDSS/electronic-meeting framework.
- Database searches focused on areas deemed appropriate by technology assessments, market research and the panel of experts. Such searches will identify other experts who could be added or substituted on the panel.
- Listing technology areas that could help an industry or recipient organization. Subpanels of commercialization managers and technologists will follow through electronic meetings, at which product managers might suggest areas where the technologists need to advance or modify the technologies, and technologists can suggest possibilities for improved functioning of products or systems for testing in the marketplace.
- Recording these ideas. If situations are changed by externalities, time can be saved in reviewing these ideas.
- Evaluation of these processes and the use of moderators to determine how future interaction can be improved.

Activities required to set up commercialization and technology-marketing electronic-meetings system are:

- A bulletin-board system providing access to technologists and commercialization managers. The system needs secure communications to protect the privacy and proprietary nature of some exchanges. Several special interest groups could be defined at the beginning to facilitate the process of electronically meeting people with the same types of interests.
- Moderators who can improve communication by helping people meet others with the same interests. Several electronic conferences on some topics could be held to encourage people to be on the network simultaneously.
- Usage tracking and information dissemination about the bulletin board to increase its use by companies in the industrial-needs database and by researchers at technology sources.

HOW THE SYSTEM WORKS

The desired impact of the TT/GDSS system is to increase communication between innovation sources and firms that can use the innovations. The need for cooperation, the motivation to cooperate and the requirements for success are increasingly present at many federal technology sources. This is increasingly true with weapons programs destined for reduction. GDSS will offer a mechanism that reduces the cost of meetings among large numbers of people who can contribute to the generation, modification and implementation of ideas. As communication patterns are improved, some cultural differences will disappear and hurdles will be overcome by making the group's agenda the focus of attention. The first item on the agenda could be support of idea generation to identify potential applications or a technology that can serve a need. Working in both directions might increase the group's ability to solve particular problems in a nonlinear fashion. Idea generation can be anonymous at the start, to encourage equality of participation by accepting and displaying ideas of all group members rather than allowing ideas to be rejected simply because they come from a member of the group who is not particularly influential.

After potential applications or technology-advancement ideas have been generated, an issue analyzer can be used to identify items from the generated list which should become foci. When focus items have been identified, a shared model can gauge factors affecting the need, technology, required modifications and other influences on the technology transfer. All this need not happen during one meeting; the basic idea is that these activities are conducted as part of the process over the necessary time period.

When the effects of potential courses of action are understood, a negotiation tool can help group members reach consensus. Increased commitment will make plan implementation easier and encourage participants to try to make the plan work.

REFERENCES

1. Baron, Seymour, (1990, Jan./Feb.). Overcoming barriers to technology transfer. *Research-Technology Management, 33* (1), 1, 38–43.
2. Dennis, A. R., George, J. F., Jessup, L. M., Nunamaker, J. F. Jr., & Vogel, D. R. (1988, December). Information technology to support electronic meetings. *MIS Quarterly, 12* (4), 591–624.
3. DeSanctis, G., & Gallupe, R. B. (1987, May). A foundation for the study of group decision support systems. *Management Science, 33* (5), 589–609.
4. Dorf, R. C. & Worthington, K. K. F. (1989, February). Technology transfer: Research to commercial product. *Engineering Management International, 5* (3) 185–191.
5. Gallupe, R. B., & McKeen, J. D. (1990, January). Enhancing computer-mediated communication: An experimental investigation into the use of a group decision support system for face-to-face versus remote meetings. *Information and Management, 18* (1), pp. 1–13.
6. Gomory, R. E. (1989, Nov. /Dec.). Moving IBM's technology from research to development. *Research-Technology Management, 32* (6), 27–32.
7. Goodman, D. (1990). A new model for federal-state-industry cooperation: Technology transfer lessons from the New Jersey experience. *Journal of the Society of Research Administrators, 21* (4), 25–29.
8. Jarvenpaa, S. L, Rao, V. S., & Huber, G. P. (1988, December). Computer support for meetings of groups working on unstructured problems: A field experiment. *MIS Quarterly, 12* (4), 645–666.
9. Jelassi, M. T., & Beauclair, R. A. (1987, October). An integrated framework for group decision support systems design. *Information and Management, 13* (3) 143–153.
10. Knox, S. D. & Denison, T. J. (1990, January). R&D centered innovation: Extending the supply side paradigm. *R&D Management, 20* (1), 25–34.
11. Lewis, F. L. (1987, March). A decision support system for face-to-face groups. *Journal of Information Science, 13* (3), 211–219.
12. Liang, T. P. (1988, December). Model management for group decision support. *MIS Quarterly, 12* (4), pp. 667–680.
13. Madu, C. N. (1990, November). Prescriptive framework for the transfer of appropriate technology. *Futures, 22* (9), 932–950.
14. Nunamaker, J. F., Applegate, L. M., & Konsynski, B. R. (1988, Nov. /Dec.). Computer-aided deliberation: Model management and group decision support. *Operations Research, 36* (5), 826–848.
15. Rivers, L. W. (1989, July/Aug.). Why not try Uncle on R&D?. *Across the Board 26* (7), 57–58.
16. Sutherland, D. & Crosslin R. (1989, February). Group decision support systems: Factors in a software implementation. *Information and Management 16* (2), 93–103.
17. Vogel, D. & Nunamaker, J. F. (1990, January). Group decision support system impact: Multi-methodological exploration. *Information and Management, 18* (1), 15–28.
18. Watkins, T. A. (1990, September). Beyond guns and butter: Managing dual-use technologies. *Technovation, 10* (6), 389–405.
19. Watson, R. T., DeSanctis, G., & Poole M. S. (1988, September). Using a GDSS to facilitate group consensus: Some intended and unintended consequences. *MIS Quarterly, 12* (3), 463–476.
20. Zigurs, I., Poole, M. S., & DeSanctis, G. L. (1988, December). A study of influence in computer-mediated group decision making. *MIS Quarterly, 12* (4) 625–644.

A PROPOSAL FOR A FRAMEWORK FOR MEASURING AND EVALUATING TECHNOLOGY TRANSFER FROM THE FEDERAL LABORATORIES TO INDUSTRY

Robert Carr

Technology Consultant
Arlington, VA 22201

This paper, part of a project for the National Technology Transfer Center, briefly describes current measurement and evaluation (M&E) techniques employed in the federal lab system, universities, and industry. It identifies some measurement issues and suggests ways to begin construction of workable systems to measure and evaluate technology transfer from federal laboratories to industry.

MEASUREMENT AND EVALUATION IN FEDERAL LABORATORIES

There is a recognition within the federal system that measuring only the numbers of CRADAs, licenses, patents, and other formal instruments does not provide very useful information, and may ultimately distort the administration of federal technology transfer efforts. Such unqualified counts frequently mask the quality and size of the counted event. For example, a license may transfer revolutionary new technologies or just a modest process improvement. A CRADA agreement may provide for costly ground-breaking R&D in revolutionary new areas or a modest exchange of results in parallel R&D projects. Furthermore, informal interactions between a federal lab and industry may spark a successful R&D project within a firm, even though no "hard" technology was ever formally transferred and therefore never captured in any counting system.

Most internal evaluations of federal technology transfer have been ad hoc counting exercises. Very little good time-series data are available and virtually no economic impact measures have been developed. Efforts to measure and evaluate technology transfer from federal laboratories have traditionally focused on easy-to-count things such as the number of licenses and numbers of CRADAs signed. Several agencies prepare formal reports containing data of varying degrees of utility and accuracy. The best known of these are the Commerce Department's biennial report to Congress on the implementation of the National Technology Transfer Act of 1986, DOE's annual report to Congress, and NASA reports too numerous to list. The Office of Management and Budget collects data on technology transfer from S&T (science and technology) agencies each year, but the data collected are not thought to be complete or reliable.

From Lab to Market: Commercialization of Public Sector Technology,
Edited by S. K. Kassicieh and H. R. Radosevich, Plenum Press, New York, 1994

The General Accounting Office (GAO) has done a number of studies on federal technology transfer. They sometimes suffer from a lack of good data, but do tell us something about developments in the last decade. Patenting and licensing activities of the agencies and their labs were covered in *Federal Patent Licensing Activities* [GAO/RCED-91-80], which provides a comparison of patenting and licensing volumes across agencies from 1981 to 1990. (The fact that the Association of University Technology Managers [AUTM] collected similar data for universities in 1990 provides a gross comparison of federal and university patent and licensing rates as well as royalty income.) The GAO document essentially showed patenting/licensing rates holding steady through the decade, although there was an increase in the percentage of royalty-bearing licenses and license income.

The GAO has a new report on federal lab activities in technology transfer that should be released shortly. It also has a data collection effort under way that attempts to capture a limited amount of data on each CRADA executed by DOE, Army, Navy, EPA, and NIST laboratories. This request seeks basic data on the technology, security sensitivity, and dollar values of the contributions of the government and industry sides.

MEASUREMENT AND EVALUATION IN UNIVERSITIES

The focus at universities is usually on measuring and predicting revenue to the university. However, as evidenced in the AUTM survey many universities track disclosures, patents, licenses, start-up companies, and royalty revenues. Market impact is indirectly measured by jobs in start-up companies and licensing royalties.

The University of California (UC) is planning to use a more direct method to measure economic impact. Based on an average royalty rate and UC's royalty income, the university was able to estimate gross sales from products based on UC technologies. Using well-known macroeconomic relationships, UC analysts were able to approximate (on an industry-by-industry basis) manufacturing costs and therefore jobs and payrolls. Finally, using a (conservative) multiplier of 2.5, they were able to calculate the total impact of the UC technology transfer program on California's economy.

MEASUREMENT AND EVALUATION IN PRIVATE FIRMS

Many assume that the private sector, ever conscious of its costs and productivity, would have devised standard systems of measuring and evaluating its research and development and the transfer of technologies from R&D units to private enterprises. However, this is not so. A study done in the early 1980's indicated that only 20 percent of R&D managers in major firms measured the productivity of their operations, and that of those who did measure productivity, only a few measured any kind of return on investment in R&D.[1]

In 1986, a study of the private sector's R&D evaluation techniques tried to find a system that could be used to measure the economic benefits of federally funded research. The study concluded that industry does not have any generally accepted methodology for the evaluation of R&D and that such methods as may be in use tend to be company-specific. In spite of this, industry has, in fact, done most of the work on evaluation of R&D and transfer of technologies to business units. In general, industrial evaluation systems are grounded in the company's culture and value system and may not be transferrable even to other firms. Nonetheless, industry evaluation systems provide some useful ideas for developing an M&E system for the federal labs.

For example, four techniques used by the General Electric corporate R&D center are particularly relevant, since a corporate R&D facility has technology transfer problems similar to those of a federal lab. The four techniques are:

1. *The "Jimmy Stewart Test,"* based on *It's A Wonderful Life*, where Stewart is asked to think what the world would have been like if he had never lived. In technology transfer, this approach measures the value of R&D by estimating what the commercial position of businesses would be *without* the transfer of R&D Center technologies.
2. *Counting R&D outputs per dollar of R&D.* This is similar to CRADA and license counting, although GE focuses on patents (per megabuck of R&D).
3. *Analyzing technology transactions.* GE performs (using an accounting firm) a rigorous (although somewhat arbitrary) discounted-rate-of-return analysis of technology transactions.
4. *Letting customers vote with dollars by contracting with the R&D center for research relevant to them.*[2]

MEASUREMENT AND EVALUATION ISSUES

Evaluation of technology transfer presents a number of difficult problems. Historically, there has not been a serious constituency supporting technology transfer measurement and evaluation, nor has there been much support for research on the subject. As a result, technology transfer measurement and evaluation is not a well developed science. It may never be. However, developing a functional M&E system for federal technology transfer is not as difficult as finding a cure for cancer. The latter still requires building a considerable body of scientific and technical knowledge through research, whereas we already understand much about the current state of technology transfer. The problem is to pose the measurement problems in the right way and to decide on the compromises we are willing to make over utility and collectibility.

First, there must be agreement on what constitutes success in federal lab technology transfer—what are the goals of federal technology transfer programs? Many laboratory technology transfer officials, after difficult bargaining and negotiating sessions, feel that the completed hand-off of a technology must constitute success. And in a way, it does. After the transfer to a commercial firm, many nontechnology factors, over which the lab has limited, if any influence, affect whether the lab's contribution will have any economic impact. In addition, using a hand-off definition of success simplifies the measurement task and more accurately evaluates the laboratory's skill at making transfers.

However, it is unlikely that such a limited approach will suffice. There is a strong political demand for indications that the federal laboratories are making an impact on the nation's economic well-being and competitiveness. Thus it seems imperative that systems be developed that measure the economic impact of technology transfer from the federal labs. Ideally, such impacts would be measured in dollar amounts of new sales, investment, payrolls, etc. However, in many cases the impact will be felt mostly at the firm level and may take the form of cost reduction or cost avoidance.

Industry itself does not have a coherent definition of successful interaction with federal laboratories. There is anecdotal evidence that some industry managers consider new or improved products and short-term profits the bottom line, and would judge success or failure on that basis. On the other hand, the second survey of Industrial Research Institute members indicates a majority of chief technical officers believe the most important payoffs from interaction with federal labs will come in the form of access to knowledge and expertise, leveraging R&D, sharing risks, and complementing R&D portfolios.[3] These benefits also have economic impact implications but those are visible only in the long term. Some analysts feel that industry views of success may be influenced by company size. Large firms with internal product development and R&D assets tend to seek more generic technologies and expertise from federal labs, while smaller firms more often seek help with products and processes at a point very much closer to commercialization.

Particularly in difficult economic times, political stakeholders in the technology transfer process usually view success in economic impact terms, and often from short-term and parochial perspectives—how many jobs in my state next year? They want to demonstrate that the laboratories produce some economic return from the considerable investment in them, in addition to fulfilling their government-oriented missions. Most observers believe that the labs have (or ultimately will have) a positive economic impact, but these views are almost always based on suppositions and anecdotal evidence.

Second, there must be agreement as to precisely what is being measured and evaluated. The task is to measure and evaluate "Technology Transfer from Federal Laboratories to Industry." In reality, none of these terms describes a unitary whole. Thus, it is necessary to characterize each of the parts precisely. (Technology transfer, as a term, may be losing currency in favor of others such as technology utilization or technology deployment, but the characterization problem remains the same.) Technology transfer, first of all, encompasses many different technology areas, each of which may be transferred in different ways. (The trend in biotechnology is patenting and licensing; in aerospace, it may be use of facilities.)

In addition, technology transfer is accomplished through a number of different techniques. Analysis of federal lab technology transfer activities has tended to focus on licenses and CRADAs. While these are important activities, there are other significant types of interaction between industry and federal labs, including informal professional contact between scientists, workshops, seminars and exchange visits, consulting, employee exchange, the use of specialized lab facilities, information dissemination, contract and sponsored research, and spin-off companies. Many feel that, in the aggregate, informal contacts and information exchange may be the most valuable form of technology transfer. However, this sort of activity will be extremely difficult to capture in a measurement and evaluation system.

The two ends of our technology transfer measurement problem, federal labs and industry, are not unitary groups either. There are 726 (or thereabouts) federal labs, which are very different from one another. They have different missions, different agency ownership, different management types (GOCO/GOGO, or Government Owned, Company Operated/Government Owned, Government Operated), and different technology strengths. Industry consists of large and small firms from very different industrial groups, with different internal competencies and value systems. Furthermore, individual firms sometimes come together in consortia, making for a different sort of laboratory partner in technology transfer.

Third, in this world of Total Quality Management, how is customer satisfaction to be measured? Is the bottom line the only answer, or should we judge whether the industrial partners interacting with federal laboratories are satisfied with the process and results? The latter will be particularly important in cases where cooperative R&D projects or other technology transfer efforts reach a technological dead end (not an uncommon event in R&D) and there are no commercialization results to measure.

A NEW FRAMEWORK FOR MEASUREMENT AND EVALUATION

To meet all these needs, it is clear that a mix of metrics and evaluation techniques will be required in which no single metric prevails, as CRADA counting does at the moment. Some metrics will measure only a specific type of technology transfer; others will be more general. Some measurements can be incorporated into technology transfer processing procedures (e.g., data on CRADAs and licenses, lab and corporate R&D funding and manpower levels). Others will probably require survey techniques (e.g., customer satisfaction, impact of transferred technologies on firms). Some important measures may delay transfer measurement for a considerable time (e.g., royalty income, corporate sales, jobs and other economic impacts).

Steps:

1. Decide What Goals to Measure against. Most observers prefer two goals: economic impacts and actual transfers. The latter are sometimes called "in-process" or "out-the-door" measurements and are either outputs or inputs, depending on one's perspective. "Political measures" can be drawn from both the above categories. The framework being proposed will try to measure success in both actual transfer and economic impact, with the realization that the two, although related cannot be equated. Success measured against "out-the-door" criteria is probably a requirement for, but does not guarantee, positive economic impact.

2. Characterize the Elements of the Phrase "Technology Transfer from Federal Labs to Industry." If we break down each of these collective terms into its component parts (biotechnology, licensing, defense labs, consortia, etc.) and produce a multidimensional matrix of many cells, the task of devising metrics and evaluation methods for individual cells becomes conceptually easier, although the large number of cells adds to the volume of the task. (Actually, many cells will be nulls and large groups of cells will be amenable to use of the same metrics.) Another purpose in characterizing the elements of "Technology Transfer from Federal Labs to Industry" is to ensure the incorporation of data elements that explain other measures. For example, if one lab showed low rates of licensing and royalty income but high rates of facilities use, the fact that the data showed that the lab's technology strength was "aerospace" would provide partial explanation of the former results.

3. Propose a Comprehensive Menu of Potential Metrics and Evaluation Techniques. These can be systematically evaluated for use in one or more cells in the M&E matrix, as well as for ease of collection, utility and impact on the technology transfer system. Single metrics which assume the current importance of CRADA counts may tend to encourage suboptimization and should be balanced by other measurements. In other cases, the imposition of a data collection requirement could alter or inhibit the activity being measured. For example, the requirement that lab scientists record contacts with industry, if not ignored altogether, might well inhibit that activity. Finally, an important part of the M&E menu would be ways to measure customer (i.e., industry) satisfaction.

With the framework proposed in this paper, some elements of technology transfer will only be measured imperfectly, if at all. The goal is not a perfect M&E system, for that is impossible, but rather the best that can be built given the phenomena being measured. The current technology transfer M&E effort is ad hoc and uneven. An improved system of measurement and evaluation is necessary and feasible, and should be developed.

REFERENCES

1. Schainblatt, A. H. (1982, May). How companies measure the productivity of engineers and scientists. *Research Management, 25* (3).

2. Robb, Walter L. (1991, March/April). How good is our research? *Research Technology Management.*

3. Roessner, J. David (1993, March). *Patterns of industry interaction with federal laboratories.* School of Public Policy Georgia Institute of Technology, Atlanta GA, 8.

A FREE MARKET, INDEPENDENT AGENT MODEL FOR TECHNOLOGY TRANSFER

Phillips V. Bradford

Colorado Advanced Technology Institute
Denver, CO 80202

INTRODUCTION

In order to complete a discussion of the role of agents and intermediaries in technology transfer, it may be valuable to examine the analogous roles of agents and intermediaries in other fields where success has been achieved. The author has had an opportunity to observe the role of agents in the literary trades and motion picture production, where governmental subsides are minimal. To the extent that the analogy is relevant, the "Hollywood Agent" model, presented here, offers a free market approach to technology transfer if certain cultural and employment practices among scientists and engineers can be modified.

THE HOLLYWOOD AGENT AND THE ARTIST

The term "Hollywood Agent" is used here as a vernacular term for the person or agency setting up a private business to connect artistic and talented individuals and performing groups to production firms which finance and produce entertainment and educational products for material gain. Its use here includes the profession of the literary agent who connects writers to publishing houses and magazine editors and producers. It may include tour booking agents for everything from rock and roll bands to the Moscow Circus, and agents who provide talent and financing for Broadway and off-Broadway plays. It includes a myriad of sports talent and booking agents, television entertainment, and advertising agents, and, most notably, casting agents who place actors and actresses in roles for motion picture studios. The term "artist" in this article refers to any creative and/or talented individual or group whether in writing, acting, sports, or any of the performing or visual arts.

The practice of Hollywood agents, as defined above, is to garner an abundance of knowledge about the market for certain types of artists for which knowledge they strive to become known within that market. The Hollywood agent then interviews artists with whom contracts may be drafted, to serve as their agent. The terms of such a contract between the Hollywood agent and the artist may be complex, but usually provide for a commission to be paid to the agent based on a percentage of the money paid to the artist for work, whether it is

From Lab to Market: Commercialization of Public Sector Technology,
Edited by S. K. Kassicieh and H. R. Radosevich, Plenum Press, New York, 1994

intellectual property or performance talent. It is important to note that there are two generally observed traditions in the Hollywood agent's practice:

- The agent's fee or commission is contingent.
- The agent's services are not billed to the producers.

There are deep and abiding reasons for these traditions, which deserve some elaboration. The contingent nature of the agent's fee is derived from the freedom of choice held by the talent source. A budding actor or actress is free to apply directly to a film studio for an acting role. An aspiring novelist is free to send his manuscripts directly to any publisher. Almost all film studios and book publishers will gladly accept talented applicants directly and are free to engage them without recourse to agents. For the agent to be successful, it must be plainly visible to the source of talent that the agent's service is potentially worth the cost of a commission charged. The value may be in screening suitable sources of employment for the talented individual or group. It may be in securing a fair or ample remuneration for talent or creative powers in comparison to similar talents available. The agent's value also may be in simply removing the work and responsibility for booking sales and engagements from the creative and talented people who wish to concentrate on improving and nurturing their talents. From the perspective of free market economic theorists, the value of the agent is proven by the fact that the agent competes openly with direct access routes between the artist and the producers.

The fact that the Hollywood agent does not usually bill the producer for his services assures producers that there is no economic disincentive to consider employing the artists whom the agent brings to them. Moreover, it is the producer who is paying the money to hire the artist, and the artist is the one receiving the money in return for services performed. Since the agent desires to be paid in money rather then in services, it seems fair for artists to pay agents with a portion of money that they receive. In the Hollywood agent model the agent is presumed to be intimately familiar with the market for the services provided by the artist, and that knowledge is not as useful or valuable to the producer as it is to the artist, who incidentally, usually has less money then the producer. The tradition of placing the burden of the sales commission on the receiver of money (usually known as "seller") in almost all transactions is common in almost all industries, with the notable exception of stock brokers where both buyers and sellers are charged a commission. Theorists may wish to ponder whether or not the fact that buyers of new initial public issues of stocks are not charged a commission has any significance in this regard.

THE TECH TRANSFER ANALOGY

Readers may question the validity of an analogy between Hollywood agents and technology transfer agents. Let us look at the two areas in more detail. In the analogy, the artist is compared to a scientist or engineer. Both classes of people are trained with various degrees of formal and informal education. On-the-job training plays an important role in both these kinds of profession. An artist's success ultimately may depend on the commercial value of the production in which he is employed. However, it has also been observed that many artists can make a living working for productions that fail as money makers. These artists are often associated with publicly supported cultural performances which are deemed to have socially important values to those of us with the appropriate cultural sensitivities. Performers in Shakespeare festivals and classical symphony orchestras are a case in point. Comparable positions among scientist and engineers are those who work in fields such as astronomy or anthropology, where the potential for direct commercialization is remote and there is a tradition of public support for the benefit of "pure knowledge" which is deemed to have value by the more curious among us. In both cases there are well-established academic programs to support the knowledge base and infrastructure needs of intellectual understanding of the respective fields, whether artistic or scientific.

In the commercial world, on the business side of these two kinds of professions there are also many common elements. A modern motion picture studio might well be compared to a modern manufacturing plant for microwave ovens, for example. The motion picture studio employs artists and the microwave oven factory employs engineers. A major motion picture can cost from $40 to $100 million to produce, and all the money is at risk. Likewise, a new microwave oven design could cost just as much to design and market successfully, and all the money is again at risk. Let us suppose that there is a new technology that can result in an improvement for a microwave oven design. Perhaps it is a sensor that detects movement and prevents food from boiling over or exploding in the oven when the power is too high by reducing the power level. The oven company wants to transfer this technology to a new design of microwave oven with feedback control to prevent such damage to food. Meanwhile, suppose a motion picture producer wants to include a dramatic new idea in a film. Perhaps it is a romance between unusual and unlikely partners that leads to an infanticide when they learn that their child is not accepted. The producer believes that only a powerful actor and actress can properly convey this idea to a mass audience and deliver an acceptable level of tolerance or understanding for such an unacceptable act. In both the microwave oven and the movie, there is a transfer needing to be accomplished and special people are required to perform the task. In the Hollywood case, the producer confers with agents and looks at the various actors and actresses who might fill the bill. Likewise, the microwave oven company looks for an engineer with experience and credentials in sensors and feedback controls.

The difference between the artist and the engineer, relevant to this analogy, is that the artist is employed by the movie studio for a short-term engagement, not a long-term one as is customary for engineering employees. Perhaps this practice needs to be reexamined. Clearly, the Hollywood agent prefers to book his clients for short-term engagements, so that they can be booked again later.

AN HISTORICAL ANALOGY OF DEINSTITUTIONALIZATION

Historically, the Hollywood agent was preceded by the literary agent. In order to identify a role for technology transfer agents in modern times, it may be instructive to look at the structure of the literary agents' profession at the time it developed. If a profession, such as that of an agent, can be said to evolve, perhaps a study of literary-agent evolution will provide insight on how the profession of technology transfer agent may unfold in the future.

Although I have not found clear evidence that any particular historical figure or institution can be credited with the title of " The First Literary Agent," I would suppose that the earliest such agents in Western civilization emerged in Renaissance universities and various church organizations, where they were "gatekeepers" for their institutions. When writers brought manuscripts to these institutions, these gatekeepers may have realized that while many were not suitable for publication by the institution, they might have potential for alternative publication. In the early years of the publishing industry, there were few organizations outside the churches or universities which had the capability for producing books.

The concept of an independent press was made possible by the development of moveable type at Gutenburg in the fifteenth century. In principle, the profession of literary agent should have followed quickly; however, there were formidable barriers. Institutions, such as the churches and universities, sought to preserve their roles as judge and jury over what should be allowed to be published. Literary agents, if they could be named as such, had to be "certified" by these institutions. It was more then 200 years later that a book like *Robinson Crusoe* by Daniel DeFoe, considered to be the first full-length novel without an academic or church agenda, could be produced in quantity for commercial sales to entertain the general public. DeFoe's agent appears to have been a certain theatrical critic named Nathaniel Mist.* Before DeFoe's time in the early 18th century, there were fuzzy boundaries between independent agents. This fuzziness is seen in the continuing debate over whether or not William Shakespeare actually wrote all the works attributed to him. Literature was not firmly established as privately owned intellectual property.

While Nathaniel Mist's contribution to the popularization of literature is obscure, it is evident that he tried earnestly to protect and preserve the original and exact wording of authors and playwrights.** He was often the sole contestant in a losing battle with play producers and magazine publishers who were failing to annotate significant changes and were frankly plagiarizing literary works. Most of Mist's own writing was reserved for scathing attacks on such practices in the *Weekly Journal* a London periodical covering theatrical news and criticism of his time, for which he served as editor. By 1728, the journal was named *Mist's Weekly Journal.* Apparently without conscious awareness, Nathaniel Mist showed the literary world that there was value to the intellectual purity of original works. In short, he seems to have invented the modern idea that literature is intellectual property and deserves the same protection under law and civility as other forms of real property. It should not be altered without due process and annotation, and it should have transferable value with remunerative benefits to its creators and subsequent owners.

The profession of the independent literary agent expanded in the 19th century, when there was a proliferation of independent presses and writers, and a burgeoning demand for literary works for magazines, books, and newspapers. Literary agents were active in promoting works by such luminaries as Charles Dickens, Arthur Conan Doyle, and Mark Twain. Despite the lack of academic programs to train agents, literary agents are remarkably well educated in literature, and often have degrees in relevant subjects from colleges and universities. They are also well informed about the screening practices and creative writing needs of publishers. Literary agents are often former employees of publishing houses hired as editors or "readers" (i.e., gatekeepers) of unsolicited manuscripts submitted for consideration. Some literary agents are also successful writers. It would seem that most successful literary agents have earned considerable experience in the fields of one or both of the parties whom they try to bring together.

All the above would imply, by analogy, that a successful technology transfer agent will not obtain his skill directly from an academic program, but may have formal education in the underlying subjects, will have considerable knowledge of technology relevant to the area in which he is trying to achieve a transfer, will have experience either as a producer of a new technology or as a reviewer of such for institutions that specialize in the relevant technology,

* DeFoe was himself a commission agent in various trades during his early life. His literary agent, presumed to be a man named Mr. Nathaniel Mist, became an unwelcome creditor upon DeFoe's death. DeFoe apparently attempted to avoid payments to Mist by assuming the pen name, Andrew Moreton. The relationship between DeFoe and Mist was complicated by the fact that DeFoe was also serving as an agent for the government, which had some interest in certain manuscripts. The interested reader of history will find DeFoe's life to be more challenging than his fictional characters. If Mr. Mist ranked among the first independent literary agents, he more than earned his fees from the trouble he endured.

** See, for example, "Shakespeare in the Periodicals 1700–1740" by George Winchester Stone, Jr. Published in the *Shakespeare Quarterly* Vol III, No. 4, Oct. 1952, pates 314 and 318. There are examples cited here of Mist's attacks on playwrights who changed (modernized?) certain lines in Shakespeare's plays as performed during that time.

and will be well informed about the practices and creative technological needs of commercializing entities.

The lack of academic programs may be seen in this historical analogy as a part of the deinstitutionalization of intellectual property. Three hundred years ago some literature and music became recognized as noninstitutional forms of entertainment, and developed a commercial value independent of the church and state. It seems reasonable that this would have happened without the blessing of the academic institutions and in fact was discouraged by them, since they were supported by church and state.

In the modern world, universities have reduced (but not lost) their associations with church and government agendas for technology research. A substantial part of the research and engineering establishment is secured by government interest in much the same way literary interests were dominated by church and crown 300 years ago. To the extent that there are independent scientists and engineers, and to the extent that other scientists are made free to capitalize on their own intellectual property whether or not they are employed by universities, government, or corporations, there can be a deinstitutionalized economy of technology transfer handled by independent agents.

Recent trends in government policies designed to encourage technology transfer toward commercial competitiveness are a step in the direction of deinstitutionalization. Recent developments in income taxation policies, which provide far more liberal deductions for corporate taxpayers than for individual taxpayers, may lead to widespread defections in employer-employee relationships, replacing them with professional corporations built upon individual expertise that is contracted to customers (formerly regarded as employers). In this new paradigm engineers and scientists will assign their intellectual property to their own wholly-owned corporations which contract with government university, or corporate customers. If sports figures and Hollywood performers do this, why shouldn't talented scientists and engineers?

A PRIVATE SECTOR MODEL BASED ON THE HOLLYWOOD AGENT

In Hollywood, creative artists—actors and actresses, screenwriters, animal trainers, and the like—often form long-term relationships with their agents. A long-term relationship can reward both the creative artist and his agent. Creative artists enjoy a reliable path to commercialization without taxing their time and energy in marketing their skills. The agents enjoys exclusive representation for their client artists. It is the widespread practice of the publishing industry, and the film and video production industries, to contact agents when they wish to find creative talent of various kinds. There is presently no self-avowed, recognized, private-sector equivalent for technology transfer.

In this comparison, it is also important to note that an agent does not begin to suggest roles for a client, except possibly to characterize general capabilities, until the client has signed what is commonly known as an "agency agreement," which binds the client to pass along all inquires concerning his talent to his agent. This agreement also provides a percentage fee to the agent, which usually requires no advance payment (save perhaps a token "reader's fee"), and which is charged to the client on the contingency that some revenue is obtained. The commercialization entity (publisher, film producer magazine, etc.) does not normally pay the agent.

Let us imagine a new paradigm in technology transfer, where a research scientists or engineer hires an independent agent, who we may assume becomes his agent for a long time by mutual consent. In this brave new world, whenever a corporate executive calls the scientist directly, he would say: "I'll be happy to talk to you about my ideas, but you must first work out the arrangements with my agent." The agent would be skilled at determining how well the company could commercially develop the scientist's idea, and would negotiate the terms for engagement. The agent should be able to auction his client scientist's idea among a group of qualified bidders from competing corporations and advise his client on the best course of action.

The role of the technology transfer agent includes the following:

- Knowledge of the scientist's or engineer's ideas and capabilities.
- An arrangement of trust, confirmed by an agency agreement between the creative client and the agent, qualifying the agent to negotiate on behalf of the client based on contingent income, that helps assure the client that the agent can be effective.
- Ability to negotiate a technology-transfer agreement between the commercialization entity and the scientist or engineer client.
- Ability to auction the technology-transfer opportunity among the greatest number of most interested commercialization entities.

STATE AGENCIES AS TECHNOLOGY TRANSFER AGENTS

State science and technology development agencies provide some of the characteristics of an independent agency for technology transfer. They are close to the economic development communities and consequently have strong ties to industrial parties in their states. They are also familiar with local universities and other technology sources. Furthermore, state agencies are not as susceptible to funding "business as usual"—a charge often leveled against federally subsidized technology transfer efforts. The funding levels of state science and technology development agencies are small in comparison to either industrial or federal budgets. While this may seem a disadvantage, it could be a hidden advantage in that serving as an agent should not require large funding. The pursuit of technology transfer funding as an objective, rather then means to achieve an objective, will not surface as a major problem in most state agencies.

The principal problem for state agencies as technology transfer intermediaries is the limitation of state borders, but there are some interesting regional associations which may mitigate this problem.

INSTITUTIONAL EMPLOYEES AS TECHNOLOGY TRANSFER AGENTS

In the light of the literary and Hollywood talent agent analogy, it is unlikely that employees of corporations, universities, and governments would be good agents for their scientists and engineers to effect technology transfer to others outside their organizations. One might ask, "Why do publishers need literary agents, and why can't they rely on their own employees to perform this function?" The same question can be raised for Hollywood studios. The answer seems to be, in part, that in-house employees are not sufficiently aware of their employer's competitors, whose existence establishes the values for creative materials in the market place. If there were only one company to which one could sell one's technology, then all the screening and evaluating could be done by in-house employees. There are also other reasons. Independent agents are motivated by the prospect of substantial rewards linked to their knowledge of the market, rather than by steady earnings through a salary or wage. The agent's rewards are also linked to the long-term popularity of the writer or talented person, which can continue for many years.

To the extent that the same conditions can be foreseen for scientists and engineers as for creative performers, the analogous characteristics should appear in the comparison of institutional versus independent agents serving on their behalf. In summary, I urge that the "Hollywood Agent" model be viewed as an analog for the technology transfer process in public discussions.

DEVELOPING EFFECTIVE COMMUNICATIONS WITH THE NATIONAL LABORATORIES TO IDENTIFY COMMERCIAL APPLICATIONS

W. O. Anderson, Jr.

Bell Communications Research, Inc.
Denver, CO 80202

INTRODUCTION

Now that the Cold War appears to have ended, there is a golden opportunity for researchers in our national laboratory system to help American companies regain their competitive edge in the world economy. This can be done by aggressively improving the two-way communication between scientists and potential technology-transfer recipients. Communication and technology transfer can occur if there is a proper focus of the research effort which begins with the establishment of an ultimate public-sector goal. When such a goal is established, private business needs can be mapped into the goal, followed by an inventory of underlying technologies and sciences that can contribute to meeting the short and long term objectives supporting the goal.

For example, having a national goal of "Decreasing America's dependance on fossil fuels" will provide the basis for developing meaningful objectives that will help identify the technologies needed to meet that particular goal. When this is done, an inventory of the programs in progress or planned at each national laboratory can provide an excellent starting point. Effective technology transfer from the laboratories requires many policies and procedures to be in place; this paper will address only some of the ways communication networks can be established to aid the technology transfer process.

NATIONAL NEEDS

It should not come as a surprise to anyone that help from the federal government via increased technology transfer from national laboratories would be welcomed by American industry. A recent article in *Business Week* titled "Chipping Away at Japan" related how U.S. semiconductor equipment companies lost market share to the Japanese in the 1980's, but, with our government's help, are slowly regaining the technological edge in some areas: "The turnaround started when U.S. companies and the federal government began—belatedly—to take the Japanese seriously." The article is encouraging, but it cautions "Still, it will be tough for U.S. companies to win back some key areas."(1)

From Lab to Market: Commercialization of Public Sector Technology,
Edited by S. K. Kassicieh and H. R. Radosevich, Plenum Press, New York, 1994

At the same time that many American industries have been attacked by strong foreign competition, most large corporations have focused on cost-cutting measures, with the resulting moves including significant reduction of employees and reorganization of functions. In the process, R & D budgets have been slashed. A article in the June 1992 *Scientific Research* magazine stated: "Although product development and engineering departments have suffered from the R & D cutbacks, in many companies central research has been the most visible target. Projects aimed at developing future technologies—once the crown jewel of corporate R & D—have been whittled down. Many of those that survive are under enormous pressure to justify their keep by producing commercially valuable results." (2) These words succinctly capture the present situation in many American industries and should be the stimulus to energize the national laboratories to work toward aligning some of their goals to support the remaining American research effort.

Arpad Bergh, an executive director in Bellcore's applied research area, provided some applicable insights to perfecting the technology transfer process:

TECHNOLOGY TRANSFER LESSONS LEARNED FROM JAPAN

- Must provide value to an end-user
- Formation of product teams
- Reduced cycle time
- Effective research depends on the strength of development and engineering
- Move researchers steadily into development

There are those who believe that the time is ripe for scientific expertise in the "defense" business to focus more on plowshares than swords. The September 7 1992 *Business Week*, in an article, "The Defense Whizzes Making It In Civvies," stated: "It is about time, say experts such as Jay Stowsky, analyst at the Berkeley Round Table on the International Economy, who argue that too many good minds have been wasted on weapons. At Boeing, engineers are putting materials in Boeing's latest airliner, the 777. And Motorola's Advanced Microcontroller Division has hired dozens of defense engineers—temporarily—to work in software fields such as neural networks and fuzzy logic." (3)

GOALS

No doubt there are already many major goals established at our national laboratories, but perhaps there needs to be an assessment of these goals to make sure that technology transfer to the private sector is *POSSIBLE* and *PLANNED*. In addition to supporting the national defense effort, other major goals could include:

- Reducing America's dependance on fossil fuels
- Addressing environmental concerns
- Rebuilding America's infrastructure
- Increasing productivity in American industry

Goals conducive to technology transfer to the private sector can be developed in many ways. The President could convene a meeting of business and government leaders similar to the economic summit called in early December 1992, or there could be many less formal, high profile methods. Surveys could be sent to hundreds of businesses, followed by regional meetings to discuss the results of the surveys. There are many ways to gather data, but the sense of urgency must prevail, with a maximum time limit of four months to develop a set of major goals and specific objectives. A bedrock idea for each goal is the desire to discover and develop a breakthrough technology that will help American industry.

There needs to be a process for assessing the strategic impact of what is required to meet a "business" goal set by a national laboratory. One model might be:

- Establish a goal
- Determine the business needs
- Identify the near-term technology thrusts required for each need
- Inventory underlying technology and science capabilities at each national laboratory
- Focus effort and fill identified gaps

Establishing laboratory and departmental goals is necessary; providing an organization focus is *each person's responsibility*. Dr. George Heilmeier, president of Bellcore, has provided all employees with his catechism:

- What are you trying to do ?
- How is it done now, and what are the limitations of the current practice?
- What's new in your approach, and why do you think it will be successful?
- If you're successful, what difference will it make?
- How do our customers get paid?
- What are the risks?
- How much will it cost?
- How long will it take?
- What are the mid-term and final exams?

When meaningful goals are established and understood by each individual, technology transfer efforts to the public sector will bear fruit.

APPLICATIONS IDENTIFICATION PROCESS

Successful technology transfer occurs when there has been a thorough identification and mapping of needs followed by effective handoff of useable technology to those who can put it to immediate commercial use. The applied research area (ARA) of Bellcore has worked very hard to develop its technology transfer process to its internal and external clients. For example, a technology called "Superbook " (TM) was developed in four modules to provide a number of technology transfer options. A functional prototype was delivered to Bellcore's Software Technology and Systems (ST&S) division to package for delivery to a specific client. At the same time, separate modules of the prototype were made available for inclusion in other programs at Bellcore or in the Bellcore client companies (BCC).

TECHNOLOGY TRANSFER OPTIONS

The national laboratories could package their "deliverables" in a similar manner, and, when this is done, the critical success factor becomes the communication mechanism with the public sector to identify needs and applications. Next are the methods to distribute information about the technology modules and prototypes that can be transferred to the public sector. This distribution could be accomplished by working on existing linkages and communication networks as well as by establishing new avenues for information dissemination. The author is not familiar with the present method of information distribution, but one model for considera-tion would be a combination of formal and informal networks.

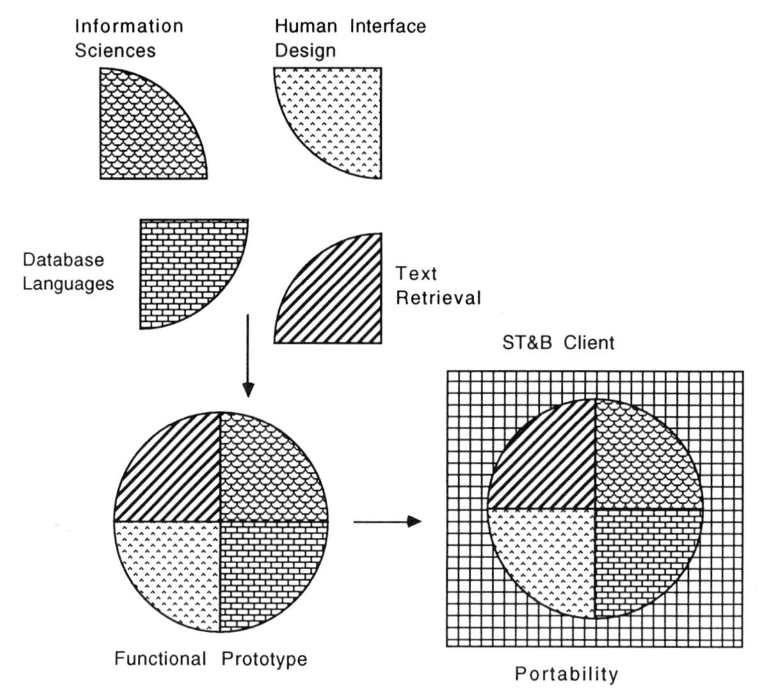

Figure 1. Technology transfer options—Bellcore program.

FORMAL COMMUNICATIONS NETWORK

Any formal advisory council or committee established for the national laboratories would certainly be subject to bureaucratic and political pressures. This could be counter-productive to an effective technology needs identification and transfer process. In spite of these ever-present obstacles, a technology advisory council should be created which would have a preponderance of business leaders as advisors and a minimal amount of government representation outside the laboratory directors. There could also be a representative from the Small Business Administration. Just as the U.S. Senate has equal representation from both small and large states, entrepreneurs and small businesses must have an equal voice along with large corporations in expressing needs and having an opportunity to receive technology transfer.

The committees should be geographic in design, perhaps following the territorial design of the Small Business Administration areas. The exact territory is not a critical factor; however, meeting at locations around the country is essential. For example, a Rocky Mountain area committee could rotate its quarterly meetings between Denver, Phoenix, Salt Lake City and Albuquerque. Easy access for the entrepreneur as well as the corporate executive would be the main goal for periodically changing the meeting location.

Informal Network

Perhaps an even more effective means of gathering business needs and informing a wide array of would-be technology transfer recipients is an *informal* communications network consisting of an electronic bulletin board, e-mail polling, laboratory open houses and technology seminars. The Small Business Administration recently announced its "SBA On-Line" system, which provides information on its services and calendar of events with easy access via an 800 number. Patricia F. Saiki, the SBA administrator, stated that "This is high-tech help for small business owners when and where they need it. It's free, it's fast, and it's easy." (4)

If such a network exists today for communicating commercial application information from our national laboratories, it should be widely advertised. Researchers at the labs should have easy access to the system so they can input "technology scanning " reports in their particular area of study and work. The rapid dissemination of leading-edge, nonclassified technological information to a broader base of American entrepreneurs could expedite the technology transfer process. The possibility of serendipity is increased as researchers nationwide become more closely linked electronically. Corporate downsizing and other recessionary effects have created a number of small, isolated research organizations, and for the national laboratories to take the lead in linking America's research effort would be a significant contribution to our country's well-being.

Technology scanning is becoming more difficult as new developments occur around the world at a rapid pace. Bellcore's applied research area supports its owners' technology scanning effort in a number of ways, one being the issuing of "applied research commentary" documents. A one-to-two page document, discussing a recent announcement of a technological advancement, is written by the researcher(s) best qualified to comment on the subject. For example, a newspaper article on new developments in lasers had a two-page commentary prepared by two SMEs in Bellcore's photonic research laboratory. Similar commentaries could be prepared by researchers in our national laboratories, with a wide distribution of these important documents in both paper and electronic format.

The national laboratories could also issue a "technology transfer catalog" every six months which would describe transfer candidates and the possible impacts on industry. Again, communicating this information to a broad base of American industries would be essential. Actual methods and procedures could be developed by the committees and advisory groups.

SUMMARY

There are a number of options for improving technology transfer from the national laboratories to the private sector, with a critical initial phase being the establishment of laboratory goals that address many of the major challenges facing America. In setting these goals and gaining a better understanding of business needs, the national laboratories will provide a better focus for technology transfer efforts. Establishing both formal and informal two-way communications between the public sector and the labs is a critical success factor in expediting and enhancing the technology transfer. Agreeing to the need and urgency is the first step to be taken. Thirty years ago government and business focused their efforts on putting an American on the moon; it was done in less than ten years. Now is the time for another effort to be led by the national laboratories in addressing the major challenges facing America today and partnering with American business.

REFERENCES

1. Brandt, Richard. *Business Week,* December 7 1992, p 120.
2. Corcoran Elizabeth. *Scientific Research,* June 1992, p 102.
3. Schine, Eric. *Business Week,* September 7 1992, p 88.
4. Holzinger, Albert G. *Nations Business* December 1992, p 8.

TECH TRANSFER AND THE ENTREPRENEURIAL TEAM

The Need for Balance

Gary M. Lundquist

Market Engineering International
Parker, CO 80134

Technology transfer is a time-honored practice in business, whether from internal R&D labs or through alliances. Recently, technology transfer has been promoted as a partial solution to the problem of increasing national competitiveness. In particular, the federal laboratory system, which employs over 15% of all scientists and engineers in this country, is increasingly seen as a primary source of new technologies for new products.

Industry labs have downsized and are looking more to alliances to meet their technology needs. Government has put technology transfer into law and has begun the jobs of changing policies, establishing infrastructures and evolving cultures to enable technology transfer. Many universities now encourage professors with tangible benefits for patenting and licensing.

In all of these cases, there is a fundamental reliance upon the entrepreneurial spirit. Somewhere there is a vision that a particular technology can be developed and that it will either be *directly valuable* to a specific market or will *enable products* to open new markets or take competitive advantage in existing markets.

At least two types of entrepreneur can be identified on the commercial side of technology transfer. Very simply stated, the *technical entrepreneur* perceives that, "It can be done," and the *business entrepreneur* perceives that, "It is worth doing." All successful businesses grow out of a balance between the two.

A basic premise of this essay is that successful technology transfer requires this type of balanced entrepreneurial team on both sides of the transfer.

To simplify the discussion, we will refer to the "company team" and the "lab team". The company team is focused on a product or product line, and may be an entire small company or a group within a larger company. The goal of the company team is to produce commercial-quality products for use by end customers.

The lab team is often focused on a technology, not a product. This team is any viable source of that technology, including federal labs, universities, R&D within the same company, private R&D labs, and companies who use technology transfer as a strategy for commercializing R&D funded by SBIRs (Small Business Innovation Research grants) or other sources.

In its purest abstraction as technology team vs. product team, this perspective can deliver valuable insight into all product development, even down to the individual inventor-entrepreneur.

From Lab to Market: Commercialization of Public Sector Technology,
Edited by S. K. Kassicieh and H. R. Radosevich, Plenum Press, New York, 1994

For the purpose of discussion, it will help to have a generalized picture of product development, and to settle on a particular type of technology transfer. In Figure 1, the development and proof of technical concepts are identified as tasks for R&D. Note that a proof of concept is not a complete product, even if it is as complete as a fully usable prototype. To get that technology into use by customers, the lab must do technology transfer.

The company team also has conceptual and proof tasks. Though the company may do much of its own product development, we will assume in this case that the team's job is to transfer technology in and then to complete and deliver a commercial-quality product.

To justify investment in a product, this team must demonstrate that the company can make money on the product. A well done proof begins by proving the value to a single, typical customer, and then shows that there are enough viable customers in target markets to generate required revenues. Both value and economic proofs must consider competition, time to market and much more as discussed below.

This model shows places for technical entrepreneurs on both sides of a technology transfer, but indicates the need for the business entrepreneur only on the company side. As a practical matter, the business side of lab teams is often virtually nonexistent, and the technical side is often not very entrepreneurial in nature.

Since the company's entrepreneurial team is much better understood, we will seek to abstract lessons for the lab by examining entrepreneurial roles in the company team. In effect, we will show that this starting model for value delivery is incomplete. The premise, again, is that both teams need both technical and business entrepreneurs to succeed. This essay will consider what role greater lab entrepreneurship might play in improving technology transfer.

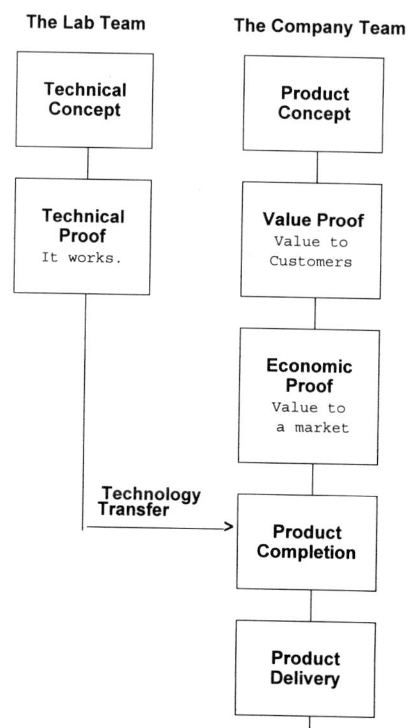

Figure 1. The value-delivery process: Unbalanced-team model.

THE COMPANY TEAM

The company team has a clear profit motive. No matter where its members get their technology, they must engineer it, manufacture it, package it, market it, distribute it, sell it, support its customers, warranty it and make continuous improvements in order to be competitive. And, they must do all of this at a profit.

Technical entrepreneurs must have both the vision of what can be done and the drive to marshal the resources to make it happen. These people are typically product-driven. That is, they focus on what the product is or does—on what the product can be or can do when finished. Technical entrepreneurs are features-oriented and concerned with functionality. They rely strongly on R&D for the research and engineering that implement their visions. Technical entrepreneurs are often those who seek opportunities, and always those who manage the mechanics of technology transfer.

Business entrepreneurs work from an entirely different paradigm. Their vision is of products used by paying customers, and their drive pushes for rapid delivery of unique and desired products to customers at a profit. The best of the business entrepreneurs are value-driven. They focus on what a customer can do with the product, on what benefits the product can deliver, and on how the company can profit.

Peter Drucker notes that, "The entrepreneur always searches for change, responds to it, and exploits it as an opportunity." From this perspective the technical entrepreneur implements change, and the business entrepreneur makes change commercially viable.

When considering a new technology or a new product, a balanced entrepreneurial team will consider a range of questions, as shown below boiling down to, "Can we do it?" and "Should we do it?" It is clear that these two questions are distinct and separate. The company can successfully transfer a technology that does not produce or enable anything of value to customers. That it can be done is no reason to do it.

Balance of the two entrepreneurial activities is essential. A significant percentage of high-tech companies and products fail, but the problem is more often inadequate business skills than lack of technical skills. Table 2 gives two parallel examples of product failures due to lack of a balanced company team.

The final characteristic of successful companies is that the entrepreneurial vision is shared. There is a common vision of something worth working for and, perhaps, even worth sacrificing

Table 1. Technology Transfer Concerns of the Company Team

The Technical Entrepreneur	The Business Entrepreneur
Does our product need it? Why?	Do our customers need it? Why?
Does our product need it now? Why?	Do they need it now?
Will it add uniqueness to our product?	Will it add value to our product?
How can we be sure it will work?	Is it within our focus?
How can we be sure they can deliver?	What corporate goals does it meet?
Do we know how to do it?	Is there a market for it? Who will buy it?
Can we afford to do it?	How big is the market?
Do we have the resources to do it?	Who else serves the market?
	Can we serve the market?
	Can we serve it better than others?
	Can we serve it soon enough?
	Can we get significant market share?

Table 2. The Need for a Balanced Company Team

WILDCAT: In 1988, British Petroleum licensed Wildcat, a major software system to process seismic data, to Sierra Geophysics, Inc., a software company serving the oil industry. At BP's encouragement, Sierra hired BP's technical entrepreneur, thus assuring that Sierra had the highest possible expertise and experience on the technical side. The technology was successfully transferred; the product was competitive; resources were available for both marketing and further development; and a demonstration/service facility was established in Houston. Sierra did not, however, assign a business entrepreneur to this product. Commercialization efforts were minimal, and the product failed.

STARPAK: About the same time, Texaco licensed its seismic processing package to Geco, an oil-industry service company. Geco assigned a commercialization team who developed brochures and sales presentations, showed the system at trade shows, and marketed StarPak worldwide. On the other hand, Geco did not assign a technical team and did not incorporate StarPak in their own processing centers. Technical efforts were minimal, and the product failed.

for. Both business and technical elements have the same vision, and that vision is known and understood by all team members.

THE LAB TEAM

The lab team also works from a profit motive, but that motive is often less clear. Ultimately, survival of the lab, and of the jobs within the lab, depends on its ability to demonstrate that it delivers value to some significant customer base. Technology transfer is a primary vehicle for delivering value, so the lab has a profit motive for doing technology transfer.

Lab technical entrepreneurs will have characteristics similar to those of the company counterpart: vision, drive, technology focus, and the ability to work within or around the system to make contact with potential licensees. If the system is too difficult to work with, they may resign to pursue their vision in their own entrepreneurial companies.

The true business entrepreneur in the lab is a rare person. Most people put into business or marketing positions have come out of R&D with neither the training nor the experience required to attract and sign licensees. Lack of commercial experience means that labs consis-

Table 3. The Benefits of a Balanced Lab Team

STARCUT: In 1991, Texaco and Market Engineering established a program for transfer of oil-production technology from Texaco's Houston R&D labs. A push/pull strategy was implemented. That is, a value proof was done and used to develop brochures for communication to end customers to create demand (or *pull*) for the technology from vendors. The economic and value proofs were used in a prospectus for communication to potential commercial allies to *push* the technology toward the vendors. A tradeshow presence was used for both push and pull.

The pull strategy creates vendor interest, enhances the reputation of the lab with customers, and gives the lab control over how its technologies are presented to its industry. The push strategy focuses on desired alliances, presents information not given to customers, and establishes a basis for futher contacts. A combined *push/pull* strategy is a powerful way to bridge both gaps and get potential alliances to the negotiating table ready to work on win-win license agreements.

This program balanced the technical energies inside the lab with a value-driven marketing program. All technologies in the program were licensed, and the up-front money for one exclusive license paid for the entire project.

tently overvalue their technologies and undervalue the effort and costs of completing and commercializing the product.

Typically, the lab team is unbalanced and the interaction with potential licensees is therefore compromised. Those in the lab are not accustomed to thinking in terms of the questions listed in Table 1, and they are even less well equipped to answer those questions.

Should the two sides find each other and decide to make a transfer, the natural bond of language and situation between technical entrepreneurs may effect a successful transfer. In most cases, however, technology transfer is inhibited by the problems of getting the two sides together, meeting the credibility and profitability tests of the company team, and resolving the high-royalty expectations of the lab.

BUSINESS ENTREPRENEURING IN THE LAB

Drucker sees the entrepreneur as an agent of change. In federal labs, the potential business entrepreneurs might be the ORTAs (Officers of Research and Technology Applications); in private labs, they might be the new business managers; in technology companies using technology transfer as a strategy, they might be the president or business manager. In universities people from patenting/licensing offices might take this role, although they won't be as close to the technical entrepreneur as desired.

In Drucker's focus on change, the questions then become: what changes might the lab business entrepreneur attempt? What roles might these people assume? Three classes of change are obvious:

1. Change in processes. Complete all three proofs—technical, value and economic. The profit and customer orientations of company teams mean that they, in effect, speak a different language than do lab teams. They ask different questions and need different types of answers than the labs ordinarily give. These communication problems amount to barriers to technology transfer.

One way to develop both a language and answers to questions is for the labs to do their own value and economic proofs, and to use elements of those proofs in outreach programs and marketing efforts.

Value proofs cause technology developers to think in terms of end-customer needs and how the technology might meet those needs if implemented as a complete product. Simply looking at the technology from a customer perspective can provide strategic direction to further developments that will result in technology better suited for both licensing and for end-customer use.

Development of economic proofs puts the lab and company on the same wave length. Technology transfer becomes a business activity with specific benefits to both sides. Each side can respect the situation and realities of the other, and negotiations can operate in an environment of mutual understanding.

Performing the proofs will enable better communication, more licenses, better licenses, more niche licensing, greater revenues, and a better reputation for the lab. When lab performance is measured, the lab will be able to achieve better metrics.

2. Change in modes of communication. Present the proofs more clearly. Knowing answers is not enough. Information is only as valuable as the quality of communication. Labs are notoriously feature-oriented and technology-driven. To get the greatest benefit from the three proofs, the labs must improve their abilities to communicate.

One might argue that the company team must take the initiative in seeking the technology. Hasn't the lab has done its share by having done the three proofs? This is the classic "mousetrap syndrome; " the world will *not* beat a path to your door. If the gap between company and lab is to be closed, it is necessary for the lab to do the work.

The need for high-quality communication is a need for presentation of the right messages and for clarity in the presentation. Labs often produce very expensive brochures and catalogs that do not communicate the *value* of what the lab does. The mission of the business entrepreneur must include clarifying that value and presenting it in language clear to potential licensees.

Better communication will lead to greater awareness in the community by potential licensees, greater appreciation of what the lab offers, fewer false leads, more efficient license negotiations, and more technology transfer.

3. Change in culture. Provide leadership in making technology transfer a valued part of the lab mission. The processes of proof development will result in some cultural change as individuals learn how to bridge the gaps between themselves and the commercial world. The business entrepreneur can move the lab culture toward a more value-driven style simply by involving a wide range of lab technical professionals in the development of the proofs.

It must be understood, however, that usually it is not in the best interests of scientists and engineers to become business entrepreneurs themselves. Their expertise and focus must remain in the technical domain. The business entrepreneur should relieve technical staff members from pressure, not add pressure to them. The business entrepreneur must implement the business activities required to facilitate technology transfer.

On the other hand, as agents of cultural change, the business entrepreneurs can and should communicate the value of lab technologies to upper management and to sources of funding. The three proofs can be used for *internal marketing* as well as for outreach to potential licensees. That is, business entrepreneurs can prove the value of the labs themselves to the institutions in

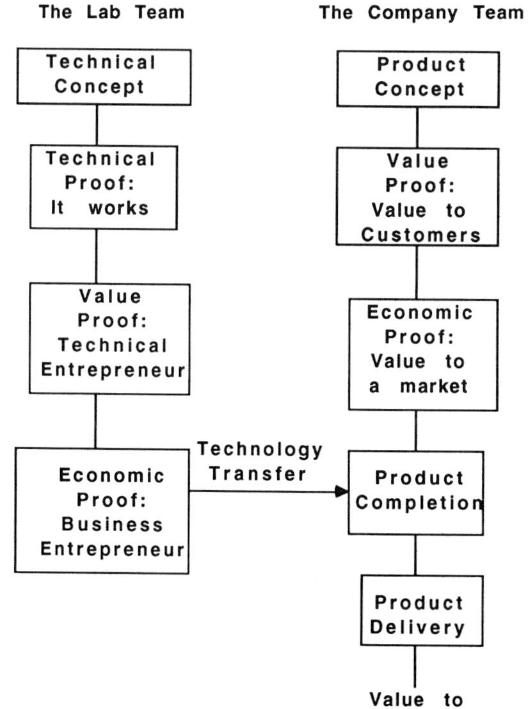

Figure 2. The value-delivery process: Balanced-team model.

which the labs operate and to external sources of lab funding. This will lead to greater support for lab R&D, greater autonomy of the lab scientists and engineers, greater confidence in longer term funding of programs, and greater management acceptance and support of technology transfer as an essential lab activity.

THE NEED FOR BALANCED ENTREPRENEURIAL TEAMS

This essay has presented a model that predicts greater overall success for technology transfer when commercial high-tech companies and R&D labs each have balanced teams of entrepreneurs. The balance requires teamwork between both technical and business entrepreneurs.

In companies, this balance is well understood to be fundamental to commercial success. In labs, however, the need for a balanced team is neither well appreciated nor well implemented. Consider an analogy: the lab might be viewed as a technology engine. The mission of lab management, then, would be to keep the engine running, to keep the production of technology high, and to keep the fuel of funding and resources flowing. The issue arises at the output end. What happens to proven technology in an R&D lab? Where does it go? Can it, in effect, plug up the engine by not having a natural outlet? Do scientists and engineers hold on to completed projects because there is no implementation strategy for their technologies?

Technology transfer is a mechanism for moving technology out of the lab and into organizations equipped to turn prototypes into completed, delivered products with value to end customers. It is also a mechanism to keep the lab technology engine running smoothly by creating an outlet for lab creativity and productivity.

To improve the efficiency of the lab engine, lab management is advised to create business entrepreneurs within the lab, to train them and provide them tools to work with, and to support their activities. The lab business entrepreneur will facilitate the creation of more technologies, technologies better suited for the purposes of potential licensees, greater lab competitiveness, greater company competitiveness, and, in the long run, greater national competitiveness.

REFERENCE:

Drucker, Peter F., (1985). *Innovation and Entrepreneurship: Practice and principles.* New York: Harper & Row.

DEFINING VALUE

Translating from the Technical-Ese

Gary M. Lundquist

Marketing Engineering International
Parker, CO 80134

INTRODUCTION

Clear communication of value is one of the most fundamental strategies of commerce. Listen to Andy Grove, CEO of Intel: "First, you have to start with a better product. Second, it has to be better in a way your customers can appreciate. And third, your customers have to know it's a better product in a way they can appreciate."

This is true whether you are a commercial vendor selling to customers, an R&D lab transferring technology to licensees, a group or team of people seeking funding, or even an individual looking for a job or a raise. On the other hand, clear communication of value is perhaps the greatest stumbling block faced by technical professionals. We are taught how to do science and engineering, we learn how to manage both ourselves and others, and we even learn how to deal with budgets and other financial realities. But we are not taught, and typically do not know how, to define and communicate the value we offer.

The problem is more complex than learning better writing and presentation skills. We have significantly different perspectives on our value than do our customers, so effective communication of value requires a translation from our technical language into their user language or investor language.

However, translation isn't that simple. We are too close to what we do to be effective translators. We are too accustomed to our historic ways of thinking about our technologies and our team. Almost never are we capable of truly seeing ourselves from a customer's perspective. Thus, we need a qualified outside translator to help us define our value.

This essay will investigate the paradigm gaps that interfere with good communication, and will then place our problem in the context of a model for communication. Finally, a process for the definition of value will be introduced that will, in the hands of a good facilitator, help us translate our value into customer language.

The situation and requirements established in the essay may seem daunting, but the task of defining value is neither difficult nor time consuming. Furthermore, it can be done best by using the strengths of scientists and engineers. The key is that the bridge between technical professionals and their customers is a clear understanding by customers of the value that has been or can be received.

From Lab to Market: Commercialization of Public Sector Technology,
Edited by S. K. Kassicieh and H. R. Radosevich, Plenum Press, New York, 1994

PARADIGM GAPS

If a paradigm is a mental model for how we interpret what we see and hear, then a paradigm gap is the difference between models, the difference in how information is interpreted. As technical professionals, we see the world differently than do the users of our technologies. We therefore have paradigm gaps between ourselves and those with whom we must interact. To communicate effectively, we must bridge those gaps.

For simplicity, and to get the greatest polarization, let us consider the paradigm gaps between R&D lab scientists and engineers and the "customers" of the lab. As used here, customers may be end users, potential licensees for technology transfer, or internal customers in other departments of the company or government. By this generalization, customer jobs may include applying transferred lab technologies to create new products for end users. That is, customers may even be other R&D professionals.

The following mismatches contribute to paradigm gaps between labs and their customers:

Mismatch of job function. Scientists and engineers focus on developing technologies. Customers focus on doing their jobs. End-user jobs usually have nothing to do with technology development. In cases of technology transfer, the job functions are closer, but recipients of lab technology are much more focused on meeting the needs of their commercial customers than are the labs.

Mismatch of motivations. The lab is motivated toward science and engineering and technology is the end product. Customers are motivated to do their own jobs, and technology is just a tool.

Mismatch of language. The lab defines value in terms of quality of science and innovativeness of engineering. Customers define value in terms of solutions to problems. Customers don't care about technical merit, only about how well and how fast and how cost-effectively their problems can be solved.

Mismatch of expectations. The lab sees its technology as a finished product. Customers require that technology be easy to acquire, easy to use, focused on their particular needs, priced appropriately and guaranteed to work. The lab usually does not know or care what it takes to make a commercial-quality product.

Mismatch of environment. The paradigm gaps listed above can be difficult to bridge, but if the lab is a government agency, the perspectives become even more distant. Government labs often perceive their "customer" as the government, not the public. They work for whoever

Table 1. Elements of Paradigm Gaps

Mismatches in:	Mismatches between:
• Job function	• Industry and government
• Motivations	• Industry and university
• Language	• Military and non-military
• Expectations	• Small business and large
• Needs	• Entepreneur and mature
• Education	• Producer and consumer
• Culture	• Technical and non-technical
	• Local focus and global focus

funds them and they may not see that the lab and its funding agency could operate as a team to serve the nation better.

Other elements of lab-customer paradigm gaps are listed above and can be developed, but these examples demonstrate the need for translation from one paradigm to another. Furthermore, these examples indicate the skills required of the translator.

It will help to have a mental model (a paradigm) by which to view paradigms and their impact on communications.

THE REACTION SEQUENCE

The simple model in Table 2 indicates the steps between observation and action. When it is applied to both sides of an interaction, it clarifies the need for both translations and translators.

Observation includes all input to the mind: what we see, hear, smell, feel or sense in any way. Decoding is the process by which we convert observation into a mental image. That image is always and unavoidably filtered and shaped by our paradigms.

Perceptions are what we believe, but they are not necessarily correct, because they are a mixture of actual truth and our paradigms. Right or wrong, perceptions are very personal and are guarded closely by each individual. It is not easy to change perceptions. Because perceptions are influenced so strongly by observation, we can best affect what customers think by managing what they can observe.

Decisions are made in the mind, not on paper, in the computer or in conversations. Decisions may be objective and logical or subjective and dependent upon feelings, but decisions are always based upon perceptions and therefore upon paradigms.

Customer *actions*, according to the model, are influenced by effective communication. But, because every step in the sequence is affected by paradigms, our communication will not be effective without translation into language compatible with the customers' paradigms.

A second value of the reaction-sequence model is the picture of communication as a decoding *and* encoding process. Every interaction with customers involves at least two paradigms. We might be able to encode our communications into language that customers can understand and appreciate, but it is almost impossible to decode our concepts without filtering them through our paradigms.

Perhaps worse, technical professionals may try to work in their perception of a marketing paradigm. They may move away from scientific integrity toward the hype that is seen in much of advertising. The value core must be clearly defined to provide a focus for quality communication with customers.

It is thus essential to have the help of an unbiased translator whose own paradigm is broad enough to encompass both sides.

Table 2. The Reaction Sequence

Observation	Input to our minds
Decoding	Filtering by paradigms
Perceptions	What we believe
Decisions	What we expect to do
Actions	What we do

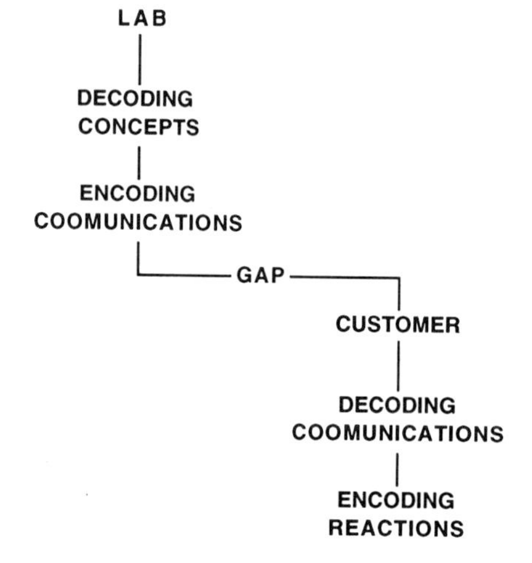

Figure 1. Two paradigms for every communication.

DEFINING VALUE

We have established the need for both translation and a translator, but we still need to state what the process might entail and how the translator might facilitate the process.

The Market Engineering approach is to define value by asking the development team a specific set of questions. This cooperative brainstorming approach takes advantage of the natural strengths of technical professionals by making value definition a form of intellectual inquiry. Better definitions result from team synergy and cross-functional experience, and team commitment results from participation.

The job of the facilitator is to know what questions to ask, to know when sufficiently complete answers have been given, and to keep the team focused. The facilitator literally picks the brains of the team, records their data on flip charts, helps wordsmith statements developed by the team, and then produces a report based on the brainstorming. Translation is accomplished by the nature of the questions asked and by the skills of the facilitator.

The process described below can be used to define the value of products, services and teams. It can be used on completed products ready for market or as milestone review to justify further work on technology under development. Further, it can be used to define the value to end-user customers, to potential licensees and to sources of funding.

Focus
Customers
Needs
Benefits
Differentiation
Purpose
Identity
Name

Figure 2. Marketing product analysis: Definition of value.

Several different views on the product are investigated during an analysis. The multiple-view approach is necessary to balance the tendency of individual developers to believe that they know everything about their product.

Each topic listed below contains several questions and may take as much as two hours to complete. A full marketing product analysis takes about one and a half days of team time.

1. *Functional description.* Marketing-oriented product description—what the product is and does, the class in which this product or technology competes.

2. *Customer description.* Market segmentation and prioritization, followed by a description of those in each market who are involved in purchase or licensing decisions. If distinct markets are to be addressed, customer descriptions and steps 3–5 below must be done for each market.

3. *Needs analysis.* Need: requirement for something essential that is lacking. In surveys, customers rank an understanding of their needs as highly important to marketing success, yet few feel that vendors take the time to understand their needs or to relate needs to product values. R&D labs always rank well below commercial vendors in terms of understanding end-user needs.

4. *Benefits analysis.* Benefit: result of satisfying a need. It can be argued that customers don't buy products, they buy "need satisfaction". (Not drill bits, but round holes.) During this analysis, benefits are derived from both customer needs and product features.

5. *Differentiation analysis.* Competitive analysis: that which makes this product unique and preferable to alternatives. Need and urgency apply to the problem, not to the solution. There may be many credible solutions, so we must find, develop and communicate differences to establish preference in customers' minds. This step can be expanded to include analysis of competitive strategies.

6. *Mission statement.* The purpose to be fulfilled by a product or team. A team mission is a commitment to produce and deliver value as eventually recognized by customers. A product mission is a commitment by the team to deliver specific benefits to specific customers through specific technologies.

7. *Desired product position.* Definition of how a team wants itself and its product to be perceived by customers. The position statement is a key tool for achieving mental impact, an important theme for marketing communication, and a focus for product development. Experience in consumer and high-tech marketing (e.g., personal computers and software) shows that use of a positioning strategy to improve product acceptance creates a strong competitive advantage.

8. *Product name.* Key link between the product and the mind. The best names are those which preempt competition by becoming a symbol of the product's competitive class.

Several of these steps may require confirmation by market research. If so, product analysis clarifies objectives and makes the subsequent research more efficient and more effective. If these steps are completed early, the results can guide product development to make sure that the finished product has the desired impact on customers.

It is useful to think of marketing product analysis as homework to be done before communicating with customers. Technical professionals would not start manufacturing a widget or writing a software program without extensive design, so the concept of a design step for marketing is intuitively appealing once it is understood.

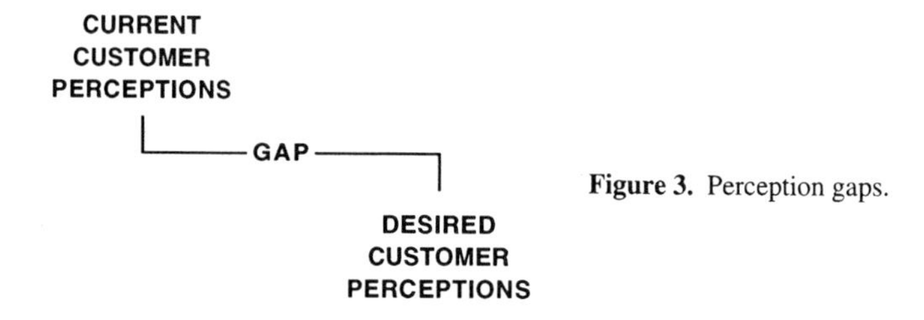

Figure 3. Perception gaps.

TRANSLATOR SKILLS.

Translation as described here involves a range of skills beyond listening and rewording. The following characteristics and skills are worth considering when looking for a qualified translator:

- Ability to
  - perceive the *technical vision* and expand it, using fundamental marketing principles.
  - perceive the *value in a product*, and create and communicate clear definitions of that value.
  - *translate concepts* from the development team into language compatible with the motivations of customers.
  - grasp *technologies* quickly.

Leadership. Translations take place in dynamic, interpersonal environments. The translator must demonstrate initiative, be willing to listen, maintain conviction in fundamental principles, and demonstrate the ability to cooperate to achieve defined goals.

- *Marketing orientation.* An instinctive understanding that the purpose of every job is to make and keep satisfied customers by delivering value to them at a profit.
- *Technical and/or product development experience in the target markets.* The ideal translator will be able to relate to product developers on their own terms.
- *Ability to build teams* and facilitate decisions in a brainstorming environment.
- *Excellent communication skills.* Abilities to organize and analyze information and to communicate well in both written and verbal presentations.

COMMUNICATING VALUE

Application of value definition to communication is the topic for a separate essay, but it is worthwhile stating the goals of communication in terms of bridging gaps.

As an exercise, assemble your team and brainstorm a list of current perceptions, what you think customers now believe about you or your product. Then make another list, this time of desired perceptions, what you want customers to think. The differences between matched pairs on the two lists are perception gaps.

Common perception gaps are: ignorance (customers simply don't know anything about the topic), skepticism (customers know something but don't believe it) functionality (customers don't know what it does or what it is good for) quality (customers don't believe it is as good as they have been told), value (customers don't feel it meets their needs), and ivory-tower (customers believe the lab is out of touch with the real world). Many other situation-specific gaps will exist.

A primary goal of communication is to reduce or eliminate perception gaps. The question is, how? If there is skepticism, how can it be reduced? If there are misunderstandings, how can they be remedied?

Approaching the problem logically ignores the truth that many perception gaps are not logical. Approaching the problem by delivering information ignores the fact that changing perceptions is much harder than making them, and changing paradigms is even more difficult. Approaching the issue with standard tools ignores the concept that those communications must bridge paradigm gaps at the same time.

That is, it is easy to focus on changing perceptions without doing the homework of translation. We wouldn't go to Germany or Japan without an awareness of the language problem, but we often go into marketing communications without paying any attention to the paradigm problem.

Our communications and therefore our effectiveness at closing perception gaps are significantly improved by definitions of value. Further, the definition process naturally leads to a more value-driven style in both product development and interaction with customers.

CONCLUSIONS

Translating value from the language of the developer to a language compatible with the motivations of the customer is necessary because:

- Clear communication of value is essential to both sales and licensing.
- Paradigm gaps interfere with good communication and must be bridged.
- Perceptions, decisions and actions all depend upon what can be observed, so clear communication is the most effective way to influence actions.
- The translation process helps us see the technology or product in new ways.
- The translation process helps technical professionals understand marketing and achieve a more value-driven style.
- The translation makes possible communication that can close perception gaps.

Translators are essential to the translation process because:

- Technical professionals have no training in value definition and little training in communication skills.
- The technical team is too close to its technology to have an objective view.
- It is difficult for technical professionals to create communications that are not filtered by R&D paradigms.
- Translation may not be the job of technical professionals, and changing paradigms may not be in their best interests. It may be much more important for them to focus on science and engineering and to assist the translator as necessary.

The bridge across paradigm gaps and the power to close perception gaps is a clear understanding by customers of the value that has or can be received.

REFERENCE

1. Hadjian, Ani, (1993, February 22). Andy Grove: How Intel makes spending pay off, *Fortune*.

INDEX